GIS RESEARCH METHODS

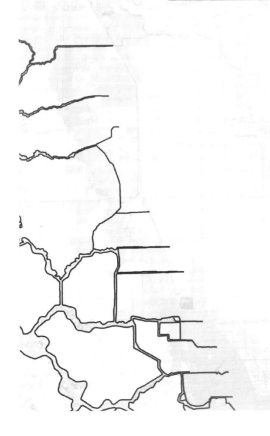

Incorporating
Spatial
Perspectives

Sheila Lakshmi Steinberg
Steven J. Steinberg

Esri Press
REDLANDS | CALIFORNIA

Cover image Esri Map Book, Volume 25 *(Redlands, CA: Esri Press, 2010), 38; courtesy of U.S. Government; Department of Interior; Bureau of Reclamation.*

Esri Press, 380 New York Street, Redlands, California 92373-8100
Copyright © 2015 Esri
All rights reserved. First edition 2015

Printed in the United States of America
19 18 17 16 15   1 2 3 4 5 6 7 8 9 10

Library of Congress Cataloging-in-Publication Data

Steinberg, Sheila L.
 GIS research methods : incorporating spatial perspectives / Sheila Lakshmi Steinberg, Steven J. Steinberg.
    pages cm
 Includes bibliographical references and index.
 ISBN 978-1-58948-378-1 (pbk. : alk. paper) 1. Geographic information systems. 2. Geographic information systems--Research. 3. Spatial analysis (Statistics) I. Steinberg, Steven J. II. Title.
 G70.212.S74 2015
 001.4'20285--dc23
                                                                                  2014046261

The information contained in this document is the exclusive property of Esri unless otherwise noted. This work is protected under United States copyright law and the copyright laws of the given countries of origin and applicable international laws, treaties, and/or conventions. No part of this work may be reproduced or transmitted in any form or by any means, electronic or mechanical, including photocopying or recording, or by any information storage or retrieval system, except as expressly permitted in writing by Esri. All requests should be sent to Attention: Contracts and Legal Services Manager, Esri, 380 New York Street, Redlands, California 92373-8100, USA.
The information contained in this document is subject to change without notice.

US Government Restricted/Limited Rights: Any software, documentation, and/or data delivered hereunder is subject to the terms of the License Agreement. The commercial license rights in the License Agreement strictly govern Licensee's use, reproduction, or disclosure of the software, data, and documentation. In no event shall the US Government acquire greater than RESTRICTED/LIMITED RIGHTS. At a minimum, use, duplication, or disclosure by the US Government is subject to restrictions as set forth in FAR §52.227-14 Alternates I, II, and III (DEC 2007); FAR §52.227-19(b) (DEC 2007) and/or FAR §12.211/12.212 (Commercial Technical Data/Computer Software); and DFARS §252.227-7015 (DEC 2011) (Technical Data – Commercial Items) and/or DFARS §227.7202 (Commercial Computer Software and Commercial Computer Software Documentation), as applicable. Contractor/Manufacturer is Esri, 380 New York Street, Redlands, CA 92373-8100, USA.

@esri.com, 3D Analyst, ACORN, Address Coder, ADF, AML, ArcAtlas, ArcCAD, ArcCatalog, ArcCOGO, ArcData, ArcDoc, ArcEdit, ArcEditor, ArcEurope, ArcExplorer, ArcExpress, ArcGIS, arcgis.com, ArcGlobe, ArcGrid, ArcIMS, ARC/INFO, ArcInfo, ArcInfo Librarian, ArcLessons, ArcLocation, ArcLogistics, ArcMap, ArcNetwork, *ArcNews*, ArcObjects, ArcOpen, ArcPad, ArcPlot, ArcPress, ArcPy, ArcReader, ArcScan, ArcScene, ArcSchool, ArcScripts, ArcSDE, ArcSdl, ArcSketch, ArcStorm, ArcSurvey, ArcTIN, ArcToolbox, ArcTools, ArcUSA, *ArcUser*, ArcView, ArcVoyager, *ArcWatch*, ArcWeb, ArcWorld, ArcXML, Atlas GIS, AtlasWare, Avenue, BAO, Business Analyst, Business Analyst Online, BusinessMAP, CityEngine, CommunityInfo, Database Integrator, DBI Kit, EDN, Esri, esri.com, Esri—Team GIS, Esri—*The GIS Company*, Esri—The GIS People, Esri—The GIS Software Leader, FormEdit, GeoCollector, Geographic Design System, Geography Matters, Geography Network, geographynetwork.com, Geoloqi, Geotrigger, GIS by Esri, gis.com, GISData Server, GIS Day, gisday.com, GIS for Everyone, JTX, MapIt, Maplex, MapObjects, MapStudio, ModelBuilder, MOLE, MPS—Atlas, PLTS, Rent-a-Tech, SDE, SML, Sourcebook•America, SpatiaLABS, Spatial Database Engine, StreetMap, Tapestry, the ARC/INFO logo, the ArcGIS Explorer logo, the ArcGIS logo, the ArcPad logo, the Esri globe logo, the Esri Press logo, The Geographic Advantage, The Geographic Approach, the GIS Day logo, the MapIt logo, The World's Leading Desktop GIS, *Water Writes*, and Your Personal Geographic Information System are trademarks, service marks, or registered marks of Esri in the United States, the European Community, or certain other jurisdictions. CityEngine is a registered trademark of Procedural AG and is distributed under license by Esri. Other companies and products or services mentioned herein may be trademarks, service marks, or registered marks of their respective mark owners.

Ask for Esri Press titles at your local bookstore or order by calling 800-447-9778, or shop online at esri.com/esripress. Outside the United States, contact your local Esri distributor or shop online at eurospanbookstore.com/esri.

Esri Press titles are distributed to the trade by the following:

In North America:
Ingram Publisher Services
Toll-free telephone: 800-648-3104
Toll-free fax: 800-838-1149
E-mail: customerservice@ingrampublisherservices.com

In the United Kingdom, Europe, Middle East and Africa, Asia, and Australia:
Eurospan Group
3 Henrietta Street
London WC2E 8LU
United Kingdom
Telephone: 44(0) 1767 604972
Fax: 44(0) 1767 601640
E-mail: eurospan@turpin-distribution.com

Portions of this work were originally published in *Geographic Information Systems for the Social Sciences: Investigating Space and Place* by Steven J. and Sheila L. Steinberg, SAGE Publications, Inc., 2005.

To our son, Joshua

Contents

Preface xi

Acknowledgments xix

About the authors xxi

1. **Why think spatially?** 1
   Using spatial knowledge 2
   What is GIS? 4
   A new approach to research methods 6
   The spatial advantage for research 10
   Spatial analysis 13
   Spatial thinking in research 16
   Multiple research methods approach 19
   Sociospatial thinking 20
   GIS as a useful tool 22

2. **Spatial conceptualization and implementation** 25
   The *G* in GIS 26
   The *I* in GIS 33
   The *S* in GIS 35
   Conceptual data model: Incorporating GIS 37
   Analytical approach: Phases of abstraction 40
   Determining project goals 47
   Guiding questions 48
   Steps in the research process 53
   Moving forward 57

3. **Research design** 59
   What is the purpose of your research? 60
   Deductive versus inductive approach to research 65
   Stages of sociospatial research for deductive research 66
   Grounded theory: GIS using an inductive approach 80
   Sociospatial grounded theory using GIS 82

4. Research ethics and spatial inquiry 91
Research ethics and GIS 92
Errors caused by analysis 98
Errors in human inquiry 102
Ecological fallacy 104
Ethics and data collection 105
Ethics and data sharing 112
Ethics and data storage 112

5. Measurement, sampling, and boundaries 117
Moving beyond your personal experience 118
Choosing a sampling method for your spatial analysis 121
Concepts, variables, and attributes 123
Different data types: Matching geographic and social variables 126
Data sampling and GIS 130
Study area and sample unit boundaries 134

6. Using secondary digital and nondigital data sources in research 141
Evaluating data sources 142
Searching for secondary data 142
Evaluating data suitability 143
Obtaining GIS data from the Internet 144
Choosing GIS variables 146
Validity and reliability 152
Obtaining data from offline sources 158
Using news as a source of data 160

7. Survey and interview spatial data collection and databases 165
Developing your own data 166
Spatializing your survey or interview questions 172
Using GIS in the field, with and without a computer 172
Data collection considerations 173
Unit of analysis 179
Database concepts and GIS 180
Rules for GIS database development 181
Creating GIS-friendly data tables 182

8. Public participation GIS 191
   - Public participation GIS and participatory GIS 192
   - Using public participation GIS as part of mixed methods 194
   - Does using GIS mean I have to be "high-tech" in the field? 198
   - Volunteered geographic information 200
   - Maps of your research area 200
   - Qualitative data and GIS files 203
   - Conducting a PPGIS data collection 203
   - Preparing for your own PPGIS session 211

9. Qualitative spatial ethnographic field research 219
   - Sociospatial documentation 220
   - Integrating GIS into field research 222
   - Ethnography 227
   - Case study research 229
   - Oral history interviews 230
   - Participant observation 232
   - Data cataloging 233

10. Evaluation research from a spatial perspective 237
   - What is evaluation research? 238
   - Why do evaluation research? 238
   - Sociospatial evaluation research 246
   - Presenting the spatial evaluation 259
   - The challenges and benefits of evaluation research 263

11. Conducting analysis with ArcGIS software 265
   - Approaching the analysis 266
   - Analysis techniques 267
   - Cartographic classification 268
   - Buffer and overlay 271
   - Spatial interpolation and simulation 290
   - Modeling 295
   - When to use GIS as a problem-solving tool 298
   - Potential pitfalls 299
   - Spatial statistics 302
   - ArcGIS Spatial Analyst 304

12.	Spatial analysis of qualitative data	309
	Qualitative data and GIS	310
	What are qualitative data?	310
	Spatial qualitative analysis	311
	Steps for spatial qualitative analysis	316
13.	Communicating results and visualizing spatial information	329
	Keys to effective communication	330
	GIS output	339
	Selecting the mode of communication	347
	Preparing the final product	349
	Conclusion	351
14.	Linking results to policy and action	355
	GIS and visualizing policy	356
	What is policy?	356
	Challenges to creating good policy	358
	A fire example	361
	Coordinating data	366
	Decision support systems	366
	From maps to action	368
	How to create good place-based policy	372
	Final thoughts	375
15.	Future directions for geospatial use	379
	Imagine the future	380
	Geospatial agility	383
	Image versus data	384
	A rebirth of spatial awareness	386
	GIS is an art form	387
	GIS as change technology	388
	The role of geospatial crowdsourcing	390
	New directions for GIS-based research	391
	Parting thoughts	395
	Suggestions for student research projects	396
Index		401

Preface

In the more than half-century since geographic information systems (GIS) came into existence, GIS has grown from a backroom computer analysis tool used by large government agencies and specialists in a few fields to a widely used tool across almost every discipline today. Applications of GIS can be seen in diverse fields of inquiry, including business and economics, health care, emergency management, criminology, and social services, and in more traditional applications in natural resource management, demographics, and planning. Since the turn of the millennium, and particularly with the widespread availability of mapping applications on the Internet, GIS (or, to the lay public, simply computer-based mapping) has gained broad recognition as a valuable tool for practitioners and researchers in these, and many other, fields of inquiry.

As GIS software has become more affordable and easier to use, we have witnessed wider interest in and acceptance of this technology beyond the traditional areas of the natural sciences. The value of GIS and spatial analysis techniques is expansive and limited only by the creativity of the people who use it in their own work. Of course, regardless of one's field of study, almost all of the data we collect and analyze can be connected to location. Considering spatial relationships is a very natural and intuitive process. We consider the best route to drive to our destination; a preferred set of criteria when considering where we want to live; or why it is that every time we go to certain parts of town, we feel a bit uneasy.

As we wrote this book, GIS training and course work continued to become more widely available at a variety of levels. Although GIS has long been taught on university campuses, in the last decade, we have witnessed an expansion of course work and interest in disciplines that previously may not have considered GIS approaches relevant. Numerous community colleges, high schools, and even elementary and middle schools now integrate spatial thinking and GIS into their curricula. Professional organizations and local GIS user groups, and hack-a-thons, now provide opportunities for active professionals to become familiar with spatial analysis and related tools relevant to their work.

Although this book introduces the underlying theory and applications of GIS, it is not intended as a manual for GIS software. If you are already an experienced GIS user, we hope this book will increase your understanding of the capabilities of GIS and its approaches in your own research applications. For those just beginning to use GIS, we designed this book to help you understand how spatial research approaches may strengthen and enhance the work you are already doing. We address key considerations in planning and carrying out your own GIS analysis. However, because GIS is an ever-changing technology, it is not unusual for many of the specific commands, menus, and tools in the software to change and improve as new versions of the software are released. This typically results in multiple possible approaches to accomplishing any given task.

With the explosive growth of GIS, numerous books now introduce the technology to practitioners in specific disciplines, joining countless introductory texts for GIS and specific software applications. Incorporating GIS into qualitative research is somewhat less well-charted territory; methods for incorporating GIS have only recently begun to emerge. What has eluded us is a text specifically addressing the fundamental topics of GIS research methods. A unique aspect of this book is that we focus specifically on how to integrate GIS into both qualitative and quantitative research. Our objective in writing this book is to provide a foundation for GIS research methods and, more specifically, to integrate spatial thinking and spatial analysis into a research tool with clear methodological techniques. The book is useful to anyone, from the student, researcher, or practitioner to the consultant, environmental scientist, city planner, or community leader who wants to establish such a skill set.

GIS is a continually evolving technology; a wide variety of companies and groups produce GIS software and tools across a variety of platforms. Clearly a text of this nature could never begin to cover all possible operations, commands, and capabilities of the technology. Instead, our goal is to provide readers an introduction to some of the core concepts and steps necessary to perform GIS-based research, using Esri's ArcGIS software as an example. However, it would be incorrect to presume that this book comprehensively covers everything that is possible with GIS. A high-powered GIS platform, such as ArcGIS, makes available more concepts and commands than any one person could possibly hope to master, let alone cover in a single text. We encourage you to explore this text alongside additional titles and resources, many of which we mention in this book.

Over the years, we have worked and collaborated with people and communities interested in spatially based research and problem solving. However, we have consistently found that people who work in fields unfamiliar with GIS grapple with understanding how spatial analysis methods can be applied to their own work. Integrating GIS into your own research projects can truly enhance the value of the work in many ways. Namely, GIS enhances your ability to collect, analyze, and, perhaps most importantly, communicate and convey the findings of research in a visually accessible manner. The visual outputs of GIS can effectively cross traditional barriers of culture, language, and literacy.

We encourage you to use this text as a springboard into your exploration of conducting your research with GIS. Because first conceptualizing a research question is critical to subsequent data collection, analysis, and output, it is essential for researchers to consider all aspects of the research process from beginning to end. This book is designed to assist researchers in the process of conceptualizing space as a part of the research process. We hope that you find this book a valuable resource in your exploration of GIS research methods across various disciplinary boundaries.

Organization of the book

This text is organized into fifteen chapters, each beginning with a brief description of the chapter contents, followed by a list of primary chapter objectives and outcomes. Each chapter highlights and defines important terms, which are **bolded** in the text. Chapters also include components designed to guide the reader to specific topics of importance, including chapter objectives, key concepts, examples with ArcGIS software, review questions, additional readings and references, and relevant websites. To provide the reader flexibility in exploring specific topics, each chapter is written as a stand-alone unit; although, references to related chapters are provided when appropriate.

Chapter 1: Why think spatially?

Chapter 1 presents a new approach to thinking spatially about research questions and methods. Why think spatially? We address this question by illustrating the versatility and wide-ranging applicability of GIS. This chapter sets the stage for using GIS with a brief, contextualized review (with examples) of GIS and geospatial techniques as a component of research. It explores the added value spatially based research methods bring to enhanced scientific investigation. We address how GIS research methods fit into an overall research framework to provide a more complete picture of the topic under study. The chapter serves as a foundation for later sections of the text.

Chapter 2: Spatial conceptualization and implementation

In chapter 2, we first examine the strengths and challenges associated with GIS. This is done through a step-by-step process that closely discusses each aspect of the GIS terminology: *geography*, *information*, and *system*. This chapter focuses on the connection between conceptualizing and understanding the broader picture and the values of the logical data model and physical data model. The chapter explores determining project goals and defining concepts and parameters important to a study question. Once we define key variables, we provide guidelines on how to implement them, with an emphasis on spatial analysis in ArcGIS.

Chapter 3: Research design

Chapter 3 explores the different purposes of research. It also presents two different approaches (deductive and inductive) to research design. The chapter includes a step-by-step process for designing spatial research using either a deductive or inductive approach and discusses key research concepts such as baseline data. The traditional scientific method underlies the deductive research approach, and the notion of grounded theory drives the inductive approach to social research methods.

Chapter 4: Research ethics and spatial inquiry

Chapter 4 addresses research ethics and the unique opportunities and challenges that incorporating spatial research methods into a research design brings. We begin the chapter with a discussion of ethical considerations when doing research in a spatial context. Researchers must keep in mind a number of important social, cultural, and political considerations, particularly when data, which may be sensitive, are linked to a map. We then discuss the use of existing or secondary data sources and potential errors that can arise when using these, owing to the data, processing operations within the software environment, or other human errors in the research process. The chapter also presents approaches for maintaining confidentiality and anonymity of data, masking data, and managing research data.

Chapter 5: Measurement, sampling, and boundaries

In chapter 5, we address issues of measurement, sampling, and boundaries. Topics include how to choose a sampling method for spatial analysis; the difference between primary and secondary data; and a discussion of concepts, variables, and attributes. From a sampling standpoint, we focus on using probability and nonprobability sampling and on spatial sampling considerations for stratification, data interpolation, and modeling. The crux of the chapter is how to incorporate spatial elements into a sampling design.

Chapter 6: Using secondary digital and nondigital data sources in research

In GIS-based research, a significant amount of time and effort goes into data acquisition and preparation. Data may come from a variety of existing sources or could be newly collected.

With the widespread use of GIS, a growing number of digital data sources are becoming available. Therefore, it is common for many research projects to incorporate a substantial amount of existing, secondary data. Examples may include GIS layers or tabular data representing demographic, health, economic, or environmental information. Data may be acquired from a variety of sources, including local, regional, statewide, national, and international governments; nonprofits; and private organizations. Chapter 6 outlines how to locate, assess, and gain access to existing or secondary data relevant to the researcher's project.

Chapter 7: Survey and interview spatial data collection and databases

Chapter 7 covers data collection via development of survey and interview instruments for use in a spatial analysis framework. It leads the reader through a series of steps and questions in the interview creation process that will allow researchers from a variety of backgrounds and disciplines to develop useful and spatially based interviews and surveys and the resulting databases. Additionally, this chapter leads the reader through how to approach data collection using ArcGIS in the field with and without a computer. Furthermore, it covers various data collection considerations, units of analysis, and factors to consider in creating a spatial database.

Chapter 8: Public participation GIS

Chapter 8 explores various aspects of spatially based public participation GIS (PPGIS) methods and volunteered geographic information. We address methods for organizing and collecting PPGIS data in the field that make use of simple, low-technology, and computer-based approaches. We highlight the collection of spatially based and spatially linked data through community engagement. We also discuss methods for integrating data from qualitative social contexts into measurable spatial variables.

Chapter 9: Qualitative spatial ethnographic field research

Chapter 9 explores approaches to conducting ethnographic research that has a spatial component. Specifically, we focus on how to accomplish spatial qualitative data collection in the field. Although GIS is a spatial computer program for ethnographic field research, we advise using a simple, low-technology process. Adopting a low-technology approach reduces the risks to data collection security and provides more data collection flexibility. In this chapter, we highlight the collection of primary, spatially based data via on-site ethnographic data collection methods, including case studies, oral histories, and participant observations. In essence, these are all forms of sociospatial documentation.

Chapter 10: Evaluation research from a spatial perspective

Chapter 10 discusses integrating spatial thinking and analysis with evaluation research. Over recent years, evaluation research has become increasingly common in many disciplines. Studies taking on an evaluation research approach seek to assess how well staff, projects, programs, and organizations accomplish and meet their goals.

Chapter 11: Conducting analysis with ArcGIS software

In chapter 11, we discuss using analytical tools in ArcGIS to analyze data prepared from both quantitative and qualitative sources. We provide examples to show how you might find the valuable geographic element in your data. We introduce various forms of analyses, including those you will likely want to apply as you begin to use GIS technology. Topics include buffers, overlays, networks, map algebra, raster analysis, interpolation, simulation, and modeling. We also provide an overview of analytical methods, extensions, and spatial statistics. The chapter addresses the means for linking external and discipline-specific quantitative and qualitative analyses with ArcGIS spatial outcomes. Finally, the chapter discusses common pitfalls to avoid when analyzing and interpreting results.

Chapter 12: Spatial analysis of qualitative data

In chapter 12, we discuss the use of qualitative analysis techniques and their link to spatial data in ArcGIS. Specifically, this chapter focuses on aspects of spatial content analysis such as coding, content, data type, manifest and latent spatial data collection, inductive approach and the deductive approach. This chapter explores how to handle qualitative data and how to analyze such data using spatial concepts. The chapter presents a series of steps used in the spatial qualitative data analysis process, including the process for data theming and coding. Different forms of qualitative data are examined such as hard-copy and digital data. The ideas of variable definition tables and how they can be used in the analysis process are also explored.

Chapter 13: Communicating results and visualizing spatial information

Chapter 13 introduces key considerations in presenting your research findings. Communicating your message in an effective and appropriate manner and considering your audience and their needs can make all the difference to a project's success. Of course, when using ArcGIS, one of your main means of communication will be a map. In this chapter, we present examples of excellent visualization using ArcGIS and strategies for effective communication using spatial technology. We offer guidelines for putting together a final presentation of your data that effectively incorporates cartographic visualization tools in ArcGIS. We also explore some of the other ways GIS can be used to communicate and share important research findings, including outputs beyond the map, with a variety of audiences.

Chapter 14: Linking results to policy and action

Chapter 14 explores how to translate your GIS analysis and findings into action. Throughout this book, we explore various ways to use spatially based research methods. Carrying out an analysis in GIS is just the first step in answering spatial questions and informing decision making with your results. As a GIS analyst implementing a study, the effort you make to ask appropriate questions, identify relevant data, employ analytical methods, and present results effectively may provide essential information to policy makers. Conversely, if you find yourself in the role of decision maker, you will benefit from understanding the underlying methods used in the project. In this chapter, we lay out steps to follow to achieve solid, spatially based policy.

Chapter 15: Future directions for geospatial use

In chapter 15, we explore directions, trends, and emerging new applications in spatial technology. Although it can be difficult to accurately predict the future, particularly as the future may evolve around computing technology, a number of trends are already incorporating spatial information in new ways. Not so many years ago, Global Positioning System–enabled smartphones were uncommon, and in just a span of a few years, they have become prevalent. We are now seeing the emergence of wearable computers, in the forms of glasses and "smart clothing," that not only have location intelligence but may also include additional sensors to monitor environmental or individual conditions such as the heart rate of the wearer. The Internet of Things has also begun to emerge, forming a world in which almost everything is in some way "connected."

Acknowledgments

First, we would like to thank everyone at Esri Press for assisting us throughout the development and writing of this book. We would also like to acknowledge our friends and families—and especially our son Joshua—for their patience on those many days when the book took priority. Additionally, we want to acknowledge the many communities, research teams, and students with whom we have worked throughout our careers. All of those people, places, and experiences have directly informed our writing. Without the opportunity to develop our thinking about GIS in research with the valuable living laboratory of the classroom, we would never have had the opportunity to implement, improve, and bring these concepts together in a coherent fashion. Finally, GIS has played a major role in our personal lives. It was GIS that initially led to us meeting as new faculty members early in our careers and, ultimately, to marrying. We regularly refer to this phenomenon as "GIS Love," and perhaps in sharing our own passion for GIS here, a little bit of "GIS Love" will come to our readers, too.

About the authors

Dr. Sheila Lakshmi Steinberg
Sheila is a full professor of social sciences at Brandman University. She completed her bachelor's degree at the University of California, Santa Barbara; her master of science degree at the University of California, Berkeley; and her doctorate at the Pennsylvania State University, State College, Pennsylvania. She enjoys guiding students in research and teaching them about the important role research plays in effective policy creation. Sheila's research interests include interdisciplinary research methods, environmental sociology, applied sociology, community, geospatial research, culture, and policy. She has conducted field research in Nepal, Guatemala, New Mexico, Pennsylvania, and California. She is also a former US Peace Corps volunteer, having served in Guatemala, where she was involved in community development and taught classes at the Universidad de San Carlos de Guatemala, in Huehuetenango. Throughout her career, Sheila's primary research focus has been the examination of people and their relationships to space and place. Her research examines the intersection of community, people, place, and the environment through a policy lens. During her career, Sheila has taught at Western New Mexico University, Silver City, New Mexico; Humboldt State University, Arcata, California; and Chapman University, Orange, California. In 2013, she joined Brandman University, Irvine, California, and part of the Chapman University system. She currently teaches courses on research methods, senior capstone, diversity, and environmental science. Sheila enjoys Brandman University for its innovative approach to teaching and learning.

Dr. Steven J. Steinberg
In 2011, Steve joined the Southern California Coastal Water Research Project, a public agency for environmental research in Costa Mesa, California, where he is a Principle Scientist leading the research and development of geospatial and data collection, management, analysis, and visualization systems. Steve received his bachelor of science degree from Kent State University, Kent, Ohio; his master of science degree from the University of Michigan, Ann Arbor, Michigan; and his doctorate from the University of Minnesota, Minneapolis, Minnesota. From 1998 to 2011, Steve was a professor of geospatial science at Humboldt State University, Arcata, California and he introduced the first geographic information system (GIS) course at Chapman University, Orange California in 2011. Steve was honored as a Fulbright Distinguished Chair (Simon Fraser University, Burnaby, British Columbia, Canada) in 2004 and a Fulbright Senior Scholar (University of Helsinki, Finland) in 2008. He continues to remain involved with the Fulbright program

as a disciplinary reviewer in geography. Steve is active as a member and leader in multiple professional geospatial organizations, including the American Society for Photogrammetry and Remote Sensing, the Urban and Regional Information Systems Association, the GIS Certification Institute, and the California Geographic Information Association. Steve has been a Certified GIS Professional (GISP) since 2008. He also serves on a number of statewide and regional workgroups that address data management and visualization of spatial data. Through each of these opportunities, he remains actively engaged in the practice and application of GIS in both environmental and human contexts and in actively exploring opportunities to incorporate the integrative power of GIS across a variety of disciplines involving space and place.

Chapter 1

Why think spatially?

In this chapter, you will learn a new approach to thinking spatially about research questions and methods. You will explore the following questions: Why think spatially? What does thinking spatially really mean? Why should I incorporate spatial analysis into my research methods? You will learn about geographic information systems (GIS) and how they are used as a component of research. You will also explore the added value that spatially based research methods bring to enhance scientific investigation and how GIS research methods fit into an overall research framework to provide a more complete picture of the topic under study. This chapter serves as a foundation for later sections of the text.

Learning objectives

- Learn about spatial thinking
- Learn how GIS is useful to various forms of research
- Learn the definitions and relationship between space and place
- Learn about sociospatial, informal, and formal spatial analysis
- Learn the value of a multiple methods approach
- Learn the historic context for spatial thinking

Key concepts

formal spatial analysis	place	spatial advantage
home range	policy	spatial analysis
informal spatial analysis	sociospatial	spatial thinking
multiple methods	space	

Using spatial knowledge

A headline on a local news website reads, "Westside Mugger Caught!" Given that you work on the Westside, you feel a great sense of relief as you begin to read the article. The past few weeks have seen a rash of muggings; every couple of days, another victim was attacked, and it seemed as though the assailant was a step ahead of the police. You have always wondered how the police catch up with criminals, and as you read the story, you come across a sentence that piques your interest: "We never would have caught the person behind these attacks without our new CompStat system," stated the chief of police. The article goes on to explain that CompStat is a computer-based analysis system built around crime statistics mapped in GIS.

Interesting. You begin to wonder exactly what the journalist means by mapping crime statistics. How would that help catch a criminal? You have always found maps to be interesting, and they certainly help you find your way when traveling. You have even heard about those maps to movie stars' homes you can buy in Hollywood, but you don't recall ever seeing a map to criminals' homes (figure 1.1).

It turns out that the GIS behind CompStat wasn't exactly used to find the home of the criminal, but almost. The police took advantage of a variety of basic information, or data, about the area in which the crimes were occurring (figure 1.2), along with information about the locations of each of the muggings as they were reported. As the locations of the crimes were mapped, some interesting patterns began to develop.

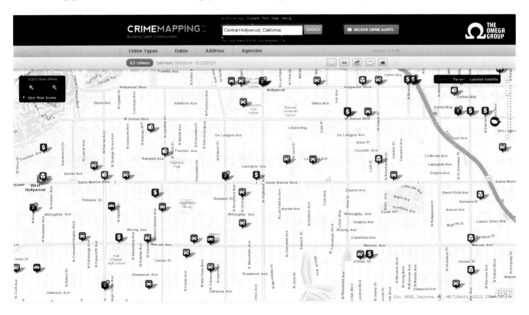

Figure 1.1 An example of a publicly available crime map from the Hollywood area of Los Angeles, California. This web-mapping site integrates crime data from police departments around the country and is powered by the Esri ArcGIS for Server. Courtesy of the Omega Group, San Diego, CA. Basemap data from Esri, HERE, DeLorme, IPC, METI/NASA, USGS, EPA.

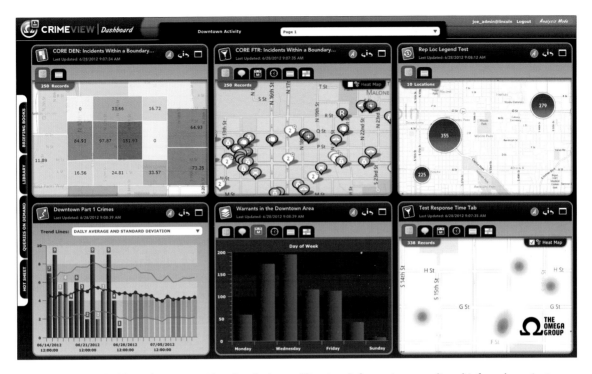

Figure 1.2 The dashboard view provides detailed, specific crime information to police chiefs and precinct commanders. Courtesy of the Omega Group, San Diego, CA.

For example, all of the muggings occurred within two blocks of an ATM machine. That seems sensible to you; the mugger might well have been targeting people who were getting cash. The attacks were always late in the evening, after ten o'clock, and the victims were always confronted on streets that had little traffic. What streets don't have lots of traffic at that time of night? Perhaps residential areas, where folks are in bed? Maybe. But wouldn't someone hear the commotion? A more likely area is around the financial district, where everything closes at five in the evening and there's not a lot going on at night.

You start to realize that by looking at some basic map information, you might be able to narrow down areas that meet a particular profile that seems to be developing. But as you think about it, you wonder, "Aren't there lots of areas on the Westside where there's little activity in business districts at that time of night? And ATMs? It seems there's one on almost every corner. What else could have helped the police get the bad guy?" It turns out that the crimes were clustered in a ten-block area. Perhaps the mugger lives near that area, or, better still, he probably lives near the middle of that area, so he didn't have to walk too far to find his victims.

Of course, the police knew other things that helped them narrow down the suspect. They knew who in that area had a record for mugging, robbery, burglary, or other similar crimes. They knew if any recent parolees lived nearby. They may have had other clues that matched the modi operandi of known offenders. Odds are this wasn't someone who woke up one day and decided to become a mugger—someone like this probably has a history.

As you ponder all this, you begin to understand how a system like CompStat would be so helpful. Of course, if you could somehow put all of these data together on a map, defining areas that meet the given criteria, you might be able to narrow the search area down to something manageable. Sure, you might not come up with the criminal's home address, but you would certainly know where to put extra police on the beat to catch him. But one thing still bothers you: the complexity of getting all of this information onto a map and doing the analysis to get to this point. Wouldn't that be a major task? It was hard enough for you to draw a readable map for your friends to find their way to your new apartment for your last Super Bowl party! It's much easier to direct your friends to one of the many online mapping tools to find your house.

It must be that computer thing the police chief mentioned in the article, that geographic information system, that performs such a complex task. It all sounds very complicated. But you are intrigued and want to find out more about these geographic information systems. Maybe they could be useful in other ways. After all, if you can use them to narrow down locations of criminals, what other kinds of analysis might they be useful for?

What is GIS?

Although you may already have some familiarity with what GIS is, it is useful to start with a definition. A **geographic information system (GIS)** is a specialized computer database program designed for the collection, storage, manipulation, retrieval, and analysis of spatial data. GIS provides far more than the ability to create maps; although maps are a common output of GIS, they are not the only outcome of analysis, and sometimes not even an essential end product. If this surprises you, consider that GIS technology was originally developed in the 1960s, when computer graphics were virtually nonexistent and output was more often printed on hard copy than displayed on monitors. GIS serves as a powerful data collection, organization, exploration, and analysis tool that can assist the researcher in multiple ways. Perhaps its greatest value lies in its ability to help us understand, draw parallels, and see connections between factors and/or variables with an eye for spatial relationships: understanding any situation, problem, or issue necessitates gaining information, and the best way to gain information is through a variety of channels, not by relying on a single source of information that could be error filled. The next section explores what is meant by the term *GIS*, which facilitates spatial thinking.

In its simplest form, **GIS** is designed to store, manipulate, analyze, and output map-based, or **spatial**, information. In practice, the functions of GIS can be carried out by hand, using only paper, pencil, and a ruler (as a surprising number of people still do). Of course, this is not practical or efficient for many research applications.

When we refer to spatial information or data, we mean that the information is linked to a specific location, such as a street address. Figure 1.3 provides an example of a real-life view of the world, as represented in an aerial photograph. This photograph is tied to associated data about the world similar to what you might collect or analyze in a study. These tabular data are related to the world via their location.

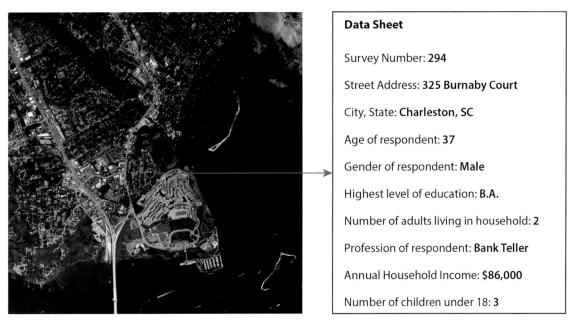

Figure 1.3   On the left is an example of a US Geological Survey (USGS) aerial photograph of a suburban location. This shows the world in much the same manner as if you were looking out the window of an airplane. When collecting data in this area, you could record the area's street address, census block, or neighborhood. These are examples of spatial information. On the right are tabular data associated with one surveyed household, as recorded on a survey form. Used together, spatial and tabular information would be useful in doing GIS-based analysis. Figure by Steven Steinberg, color infrared imagery, USGS National Aerial Photography Program (NAPP), Charleston, SC, acquired February 6, 2007.

Although no single definition of GIS exists, GIS professionals do agree on some general principles. First, GIS requires a combination of computer hardware and software tools. Second, GIS requires **data**, and these data must possess a spatial or location component. Third, GIS requires knowledgeable individuals to develop the **database** and carry out the data processing. Although GIS software has become much easier to use since the introduction of **graphical user interfaces**, GIS programs, and much of the underlying geographic theory, require people to have a basic understanding of maps and map analysis. Anyone with a little basic computer knowledge, which we discuss in this book, can accomplish most GIS tasks. However, for more complex data and analysis, it is often helpful to work with a GIS analyst with in-depth knowledge of GIS and data.

Last, and perhaps most important, GIS is a system for analysis; that is, GIS is useful for examining, displaying, and outputting information gleaned from the data that are stored and maintained in the system. This book explains the necessary mapping concepts and **spatial analysis** you need to do GIS-based research.

Understanding geographic information systems

To best understand a GIS, you need to understand GIS terminology and how GIS apply to various analysis situations. In particular, how can your area of interest and the associated data be placed into a GIS context? How can GIS technology enhance your analysis and understanding of data? You can use GIS to study issues with real data as well as conceptual data. The concept of space exists in different dimensions: the actual and the perceived. *Space* is defined as distance and time between locations and is often used to determine position. For example, an interview script asking individuals about their homes, communities, relationships, or other interactions inevitably will include phrases such as "in our neighborhood," "around the corner," or "over in the next valley." While investigating social relationships, you might come across examples of conceptual geography. For example, the strengths of social ties between individuals might be represented in statements such as "I'm very close to my younger brother" or, conversely, "We found ourselves drifting further apart with each passing year." These statements, although not tied to physical locations, nonetheless may be mapped and analyzed using many of the same techniques that one might apply in traditional GIS analysis.

A new approach to research methods

The value of spatial relationships, patterns, and connections represented with maps has a long history in many disciplines across the natural and social sciences. However, it is only more recently that we can take this information and put it all into a computer analysis environment using GIS, which can account for and analyze space in meaningful ways.

Everyone thinks spatially on a daily basis. At the beginning of the day, when we navigate our way to work, school, or the grocery store, we think about our destination and how best to reach it. This spatial thinking occurs in our minds, based on our knowledge of the surrounding environment (figures 1.4 and 1.5). We may choose a route to a particular destination based on what is most familiar or what we have found to be the most effective path in the past.

Figure 1.4 A map of essential landmarks from the perspective of our son at age seven. This map features locations he found important at the time, including our home, his school, parks, and other significant locations. Geography is not accurately represented for either distance or direction from home. However, features closer to home (to which he has personal experience walking or riding his bike) are more accurately represented than those that must be reached as a passenger in a car. Map courtesy of Joshua Steinberg, 2011.

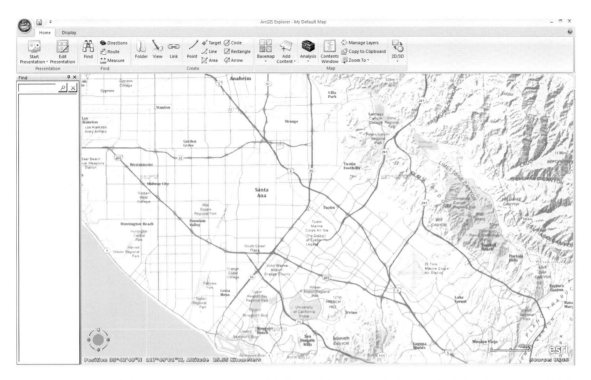

Figure 1.5 A view of the same region represented in figure 1.4 as a map built with ArcGIS software with USGS basemap data. The landmarks highlighted in our son's map actually stretch across a region in excess of 300 square kilometers. Map courtesy of Steven Steinberg. Data from USGS.

When you are in a new place or city and do not know where you are, how do you find your way to where you want to go? You could rely on digital technologies, such as Global Positioning System (GPS) (figure 1.6) or web-based mapping tools, to assist you in finding your way to a specific location. You could get there by following a paper map (figure 1.7), although not many people do that anymore. You could find your way by asking local people for directions. Asking for directions will most likely produce a variety of answers, depending on whom you ask and that person's own experience with travel and mobility in the city.

Figure 1.6　An in-vehicle GPS receiver can help you navigate to your desired destination with customized GIS data, including locations of streets and various landmarks of interest to travelers. Rafal Olechowski/Shutterstock.com.

　　This book will help you integrate the spatial thinking you do every day into your research process. This approach incorporates GIS analysis tools, which make spatial thinking much easier than it has been in the past. Spatial thinking used to require hand-drawn maps to accomplish what can be achieved today quite easily with the use of GIS technology. In the chapters that follow, you will learn how to use GIS to conceptualize, create, and apply a unique, space-based approach to research methods. Examining and analyzing information about the environment of your study provides you with a wealth of contextual knowledge. Using GIS, you can integrate, overlay, view, and apply different information layers simultaneously, leading to an accurate and realistic understanding of situations. Applying a spatial perspective using GIS can facilitate more efficient and effective decision making.

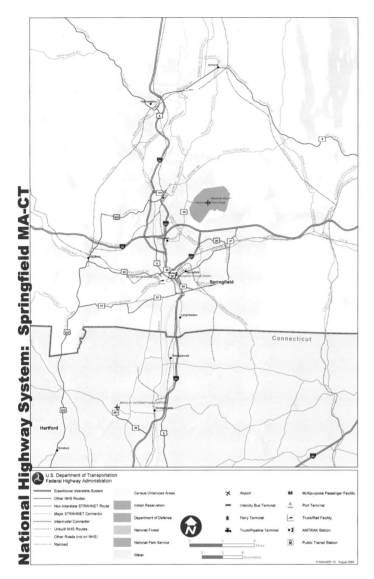

Figure 1.7 Paper maps, such as this urbanized area map of the National Highway System (NHS) near Springfield, Massachusetts, can help you find your way around an unfamiliar place. Courtesy of US Department of Transportation, Federal Highway Administration.

The spatial advantage for research

When you use GIS to locate, analyze, and assess information and, ultimately, make better decisions, you gain a **spatial advantage**. Why would you want to incorporate

GIS into your research methods? Because using GIS facilitates better, more comprehensive, and effective research. GIS integrates different types of data and presents a more realistic picture of the situation. It is perfect for multiple methods research. Locations can be explored at multiple scales ranging from the most micro level, such as the inner workings of a biologic cell, to the local-level relationships of artifacts within a homesite or at an archeological dig. Or they may be explored at larger, macro-level spatial scales, such as customer demographics for a business in a city or emergency responders at a regional level. The study area of natural scientists may scale from an individual habitat or ecosystem all the way to global effects of climate change, whereas social scientists may focus on communities or places defined by political boundaries, such as states, provinces, or sometimes even countries. Almost everything has a spatial component—it's just a matter of identifying what that is and determining how it relates to your research question.

GIS is a powerful technology and has much to offer. Like any new tool you encounter, GIS must be handled with care so as not to misuse it. But with a little patience and practice, GIS can serve you well and provide a new set of capabilities that would otherwise be missing from your analytical repertoire.

Spatial thinking

Spatial thinking is the application of geographic principles, such as place, time, and distance. People travel to different places every day. Accomplishing daily tasks requires interacting with spaces, people, and places over a certain period of time. Some of these places are a part of our daily patterns of mobility (such as getting coffee or going to the gym before work), which we could say are a part of our "home range"; other paths we choose to travel may involve going to a new location or place. **Home range** is a term commonly used to describe animal mobility patterns in the wild, but we humans all have a home range, too. Think about your own daily patterns of mobility and the different places that you visit throughout the course of the day; these would compose your home range. If you were to chart these data points using GIS, you would actually be able to see the spatial patterns of your own home range. Thinking spatially for research methods means being able to conceptualize a research question in light of place, distance, and time.

Spatial thinking is an approach that often has been used by people in environmental resource management, forestry, business marketing, the health field, and social services to accomplish their jobs in a more strategic, efficient, and targeted geographic manner. GIS provides the ability to use a variety of data in a holistic and spatial manner. Any organization, agency, or business can be more efficient and save resources by using a spatial perspective to locate their user needs and existing resources. From a business perspective, this might include locating a business's markets and supply chains.

Everything exists in time and space and is a certain distance from somewhere. In the research world, the science of collecting, analyzing, and reporting data has often involved a spatial component, even if it was not formally recognized and incorporated into the research process, such as in collecting survey responses. Information regarding location may have been collected as a part of research, but most likely was not used in the most efficient manner. The difference now is that, with GIS software, it is becoming increasingly easy, for example, to map out a variety of responses to survey questions and identify patterns.

Space and place

A researcher can begin to understand a particular issue or problem by asking questions, reading the local newspaper, observing the community, or employing all of these methods while tying in a spatial perspective. **Space** can be thought of as the distance between places. What is place? **Place** can be thought of as "meaningful location" (Creswell 2004). In other words, place is a location that has some importance. This importance can vary by individual or by group. When you think about the way that different geographies are named by different groups, this makes sense. For example, many Native Americans and the US government refer to the famous battle fought between General Custer's Seventh Cavalry and the Lakota, Northern Cheyenne, and Arapaho in different ways: many Native Americans refer to it as the Battle of the Greasy Grass, whereas those of nonnative descent refer to it more often as the Battle of the Little Bighorn or Custer's Last Stand. For the Native Americans who participated in the battle, it was a clear victory; for the losing side, the US Army, the battle was conceptualized as a great loss. Perhaps this explains why, for many years, many referred to the site as the "Custer Battlefield." Place-names clearly depend on the group who is doing the naming. To be named and/or defined as a place, a location has to have some meaning.

Places can either be conceptual or real. An example of conceptual space is social status or hierarchy. You could be neighbors with someone in a different class or social circle than you, which may mean you have little to no interaction with the person. Real, physical space is easier to measure than conceptual space because it is more easily defined and demarcated as distances and directions between locations.

Spatial analysis

Spatial analysis explores the relationships within and between data in space and provides the ability to define the common geographies and their characteristics as they relate to other information that has been collected. It is common for spatial thinking to be informed through our direct experience and knowledge of a local environment or place. When you find your way around town to various locations, you are using **informal spatial analysis**. When you thoughtfully and methodically assess spatial data to determine patterns, you are using **formal spatial analysis**. GIS facilitates spatial analysis.

Although GIS is a relatively new tool, the concept of conducting a geographically based spatial analysis of social issues is fairly old. Past researchers from a variety of disciplines, including anthropology, forestry, public health, history, biology, political science, urban planning, geography, economics, and sociology, have incorporated spatial analysis into their research projects.

Historic examples

The geographically based spatial approach was used prior to the existence of computers and GIS software. In the following sections, we highlight just a few of these historic studies as examples.

Spatial thinking and social inequality in Chicago slums

Florence Kelly was an applied researcher, a women's rights activist, a child labor law advocate, and one of the original founders of the National Association for the Advancement of Colored People (NAACP). Kelly believed the power of research could help improve conditions in society. To this end, in 1893, Kelly used laborious geographic analysis, without a computer, to map the "Slums of the Great Cities Survey Maps" (figures 1.8 and 1.9). Kelly, a resident of Chicago's famous Hull House, led a federally funded study to identify poverty in urban areas (Brown 2004). Hull House, run by Jane Addams, provided aid and social organizational skills to Chicago's poor. Kelly's study involved interviewing all of the residents who lived near Hull House in Chicago to find out whether they lived in tenements, rented rooms, or owned homes (Addams 1895).

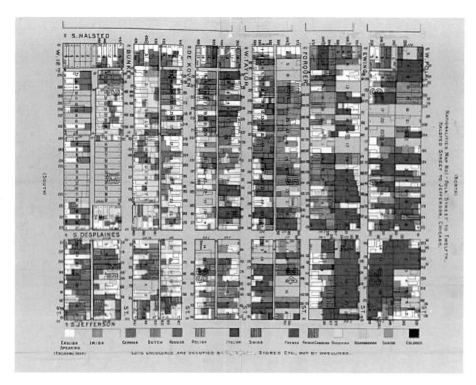

Figure 1.8   Among the Hull House Settlement research projects were a series of detailed studies of its neighborhoods. These settlements were home to many recent immigrants to Chicago. The majority of these individuals emigrated from central, southern, and eastern Europe. This map originally appeared in "Hull-House Maps and Papers" and shows the nationalities of residents of the neighborhoods bordered by Polk, Twelfth, Halsted, and Jefferson Streets. Hull House Map (Nationalities), 1895, Northwestern University.

Figure 1.9   Rear houses near the West Side area, off an alley near Hull House, circa 1910. Courtesy of University of Illinois at Chicago Jane Addams Memorial Collection, JAMC neg. 1054.

A very important aspect of Kelly's study was that she recorded the geographic locations of the respondents she interviewed. Kelly was unique in that she transferred much of the sociodemographic, employment, and housing data collected about individual households onto maps of the city to illustrate geographic patterns of poverty. She was a pioneer in integrating geographic variables into the study of social inequality. This was one very early application of spatial analysis techniques, albeit without the benefit of GIS, to study social inequality. Kelly believed that the scientific documentation of social inequalities could help lead to improving the working and housing conditions of the poor. Thus, geographic location can be a very important factor to consider when studying social problems.

Today, researchers can employ the same approach, but the work will be much easier given the advances of GIS technology and software.

Railroads as indicators of civilized society

Mark Jefferson is another early pioneer in using geographic variables and spatial analysis to study social issues. In 1928, Jefferson, a geographer, used the geographic concept of buffers to analyze the influence of the railroad on settlement patterns in different societies (figure 1.10). His study was unique because he operationalized the modernization or "civilization" of different societies based on the networks of rails (Jefferson 1928). He attempted to analyze societies' levels of civilization based on their proximity to railroad lines (Corbett 2004). His assumption was that the presence of railroad networks was indicative of "civilized" society.

Figure 1.10 Railroad tracks served as an important means of transportation and thus a spatially relevant feature as towns and resources developed across the country. Yana Gayvoronskaya/Shutterstock.com.

Jefferson drew buffer zones (mapping a ten-mile distance) around the railroads in different countries throughout the world (Jefferson 1928). The notion of buffers was used to extend the examination of the geographic lines the railroads caused. Jefferson's use of buffers in the study helped people understand where railroads both existed and where they did not exist (Arlinghaus, Goodman, and Jacobs 1997). In other words, those areas outside of the buffer zones represented areas that would not directly benefit from their construction. Buffering around the railroad tracks spread the width of the line, which assisted in mapping and comprehending the significant impact that railroads had on communal development and the "civilization" of societies.

Jefferson found that places that had more extensive railway systems were more "civilized" than places with fewer railway systems. Today, such a study would be viewed as ethnocentric and biased because it favors traits of development that are partial to Western societies (the railroad). Nevertheless, Jefferson was an early applied researcher who adopted a creative approach to incorporating geographic spatial variables into a research project.

Spatial thinking in research

Developing the ability to integrate different forms of data is essential to conducting good research. Developing this ability is about seeing the connections, themes, and patterns in the data and tying the various parts together. A good researcher will seek to develop an understanding of the complete picture of the issue under study, while also being keenly aware of the parts that compose the whole. A good researcher has creativity, vision, and patience, which describes in essence the ability to think spatially.

Spatial thinking is especially valuable to the research process because it enables you to visualize, communicate, and implement based on a unique and integrated research process. The spatial perspective allows you to contextualize the problem you are studying, to integrate different forms of data, and to develop a holistic understanding of the particular issue you are considering. Integrating GIS into research methods is natural in our increasingly interconnected world.

Visualize and communicate

Engaging in spatial thinking enables you to better envision your situation. Why? It allows you to picture, in a clear, visual manner, an issue, problem, or situation that needs to be confronted. You can do this using a computer screen or a printed map. More important, the strength of spatial thinking, particularly using GIS, lies in bringing the different types of information or data layers together. Doing this can create a clearer picture of what is going on.

Being able to visualize data or different types of information means that you can convey the information more easily. The visualization GIS enables allows you to represent complex data in a pictorial format, which may be more quickly and easily understood than tables of numbers and textual descriptions. The spatial portrayal of data is effective in conveying information to people regardless of their cultural background or spoken language. As researchers engaged in research working with communities and policy makers, we have found GIS to be an extremely effective tool for communication with a variety of audiences. Various community groups with which we have worked have used maps to communicate with policy makers, to "tell a story" or, rather, "document" their situation, which is all part of the research process. Time is often of the essence when people are trying to communicate or share information, and a well-constructed spatial output can quickly convey a story. GIS can greatly enhance the capacity to share complex information in an effective and visual manner.

Contextualize

Every research project has a context, which is the setting for your issue or project. To understand a situation, you need to understand the place where it occurs. Where is your issue located? Is it in a far-away place or an urban area? Who lives there? What do they have access to? What is the surrounding environment like?

Let's assume for a moment that you are an anthropologist studying health for a rural indigenous society in Brazil. Part of understanding the group and its culture involves understanding the environment or context in which the group functions. For instance, is the group you are studying living in a rain forest? Is there a health clinic where the group is located? If not, what is their proximity to a city or town that has a health clinic? How long does it take to get to the nearest city or town? Do the people travel by river or road? Do the people collect herbs and other products from their environment to stay healthy? Where are these products located in proximity to the village?

Capturing the context of a people is a very important factor when considering almost any social phenomenon. As an example, if you were to map the original settlements of Californian indigenous tribes, you would see that their settlement patterns often followed surrounding watershed boundaries. In other words, the physical landscape features influenced the clustering patterns and social communities, including the topography of the land, water, trees, and wildlife. All of these factors together effectively determined the physical locations of the tribal communities and their access to resources.

Integrate

Probably the most important advantage of incorporating GIS into research methods is the ability to integrate different types of knowledge and information. Because GIS is built on a database platform (hence "information system" is a core part of its name), what we are really considering is an analytical software tool that allows us to work with any type of data, including quantitative and qualitative data, that can be coded and organized in database format. Data are not limited to coded text or numeric values but may also include digital files, such as sound recordings, digital images or video, or any other digital data, that can be stored in a computer for further analysis. Because of this, GIS programs can tell a story. Such features in the software greatly expand the usability of the software. Most important, when you have the ability to share information in a number of ways, visually and verbally, the potential impact can be much greater. Ultimately, many of the qualitative data collected in such studies can be, and usually are, coded in some fashion, either by hand or with the help of a data analysis software package. Much of the information that might be collected in the course of an interview, observation, or survey embeds some sort of geographic component.

Holistic understanding

We research because we want to understand a situation or issue in order to act on it or make a decision about it. **Policy** describes methods, laws, or courses of actions used to make decisions about something. To understand a problem or issue enough to make a policy about it, it is useful to think holistically. GIS allows you to do this by giving you the ability to integrate multiple data forms. GIS facilitates data analysis that moves beyond the silo approach, wherein researchers consider a topic within its own context and fail to consider multiple contexts. It is common for people to develop a research project or study based on one method. GIS is useful in developing a larger, more overarching perspective because it facilitates examining multiple layers of information. GIS gives you the ability to adopt macro- and micro-level perspectives simultaneously. Many aspects of research involve moving toward or identifying a holistic or synergistic (meaning the whole is greater than the sum of the parts) understanding of a particular research question.

Implement

Visualizing your data using GIS can help you put your ideas into action and implement policy. The first step in developing and enacting any sort of policy or solution is to assess the situation. The next step is to develop viable solutions or policies that can be used to improve the situation. The final step is to take some action to implement a potential solution for the situation. Incorporating GIS into research can provide you, policy makers, and the general public with the tools you need to engage in better decision making.

Multiple research methods approach

Research methods are used to identify and gather information and patterns. Why? Although research occurs for different purposes, such as to test a hypothesis, to better understand a phenomenon, or to evaluate the effectiveness of a program, all research shares the common link of investigation. Different types of research methods can be thought of as tools in a toolbox. As a researcher, the greater the diversity of the tools to which you have access, the more adept you will be at investigating your research question.

Multiple methods research involves using both quantitative and qualitative research practices together (Creswell and Clark 2007). For the researcher, using multiple methods is a major strength for a number of reasons:

1. Providing multiple views of a problem or issue allows a broader picture of the issue under study to be developed.
2. Using multiple methods presents the opportunity to include a greater variety of variables in the study.
3. Multiple methods can provide multiple measures of the same concept, which increases the validity of the project.

Multiple methods research helps you delve into various aspects of a topic to better comprehend it. Involving more variables means that you can capture a more true-to-life picture of the situation you are studying through the diverse data sources. It is common for researchers to focus on only a few key variables. Doing this can often limit the conceptualization of the issue at hand. It can also limit the real-world applicability of findings. However, adopting a multiple method approach to research requires the interest and desire to learn different research methods.

GIS is a natural fit with multiple methods research because of its unique ability to integrate different types of data. Incorporating GIS into your research allows you to examine multiple data sources at different points in time, space, and place. Later, we will show you how to use multiple research methods to collect data and better understand your research question.

Sociospatial thinking

The term *sociospatial* accurately captures the importance of the topic here. For the purposes of this text, **sociospatial** is defined as an integrated examination of space, place, and social indicators in a holistic fashion (Steinberg and Steinberg 2009). This means that one can simultaneously examine multiple indicators that relate to a particular problem or issue under study, and in a holistic way. Why would someone want to employ a sociospatial approach? In any project, resources are tied to the success of the project. The resources may be your time as a researcher or the cost of gathering, analyzing, and interpreting the data.

Historic poverty example

Of course, examples of sociospatial research occurred well before GIS and computer technology existed. In this section, we examine a historic example of sociospatial thinking and multiple methods in urban England. Between 1886 and 1903, Charles Booth conducted an in-depth study of poverty in London, England. His approach involved conducting field research by visiting every street in London. Through this approach, he was able to document the social, economic, and environmental conditions for residents (Fearon 2008). He was interested in examining the relationship between space, time, and social class. His grounded-theory approach to research eventually *disproved* original statistics that had underestimated the poverty rate in London at 25 percent. Booth's detailed, on-the-ground examination of poverty illustrated that poverty levels were really closer to 35 percent (Fearon 2008).

His research project took a number of years to complete and is a good example of sociospatial research that considers space, place, and social indicators in a holistic fashion. Booth methodically visited all homes within a designated geographic area to assess urban poverty (figure 1.11). Today, that same work could be undertaken more quickly using computers rather than manual analysis. As a researcher, Booth employed both quantitative and qualitative methods (he took detailed field notes on the survey-gathering process of poverty data). Booth was definitely ahead of his time as a researcher because he incorporated interviews with survey data and field observations. Additionally, he simultaneously achieved a holistic and spatially specific review of poverty using this process. Had Booth had access to GIS in 1903, he could have provided interesting analyses to policy makers, and it would not have taken twelve years to do so.

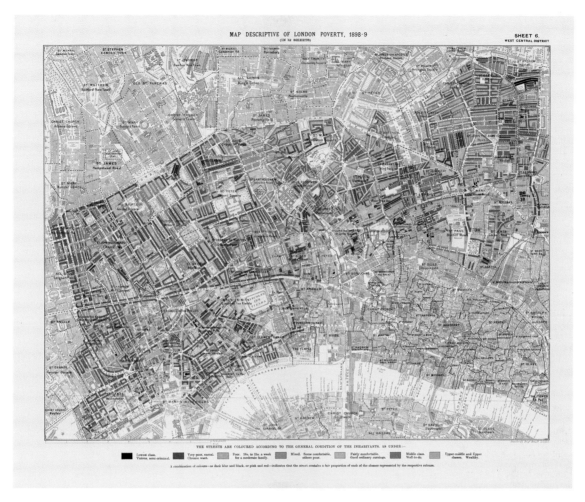

Figure 1.11 **A map created by Charles Booth that indicates levels of poverty in London, 1898–99.** Courtesy of the Library of the London School of Economics and Political Science, reference LSE/BOOTH/E/1/6.

The interesting thing is that social inequalities always exist within a particular environmental context. In other words, every community has its haves and have-nots, just like in the days of Charles Booth and his study of poverty. People from different income levels will vary in terms of their ability to access different resources and in their general life chances. Incorporating GIS today into research projects on inequality can open up an entirely new way of looking at a topic. Using GIS allows a researcher to consider and actively model the spatial context and the inequalities that exist within that context. Not only can GIS model real environmental and spatial inequalities, it can also model "perceived" inequalities. This becomes extremely important to the social scientist interested in perception and how perception influences action. In Booth's case, there were real manifested inequalities that he did not perceive. Being able to physically map out and "document" these conditions made the issue of urban poverty more real for the larger population.

GIS as a useful tool

As a research tool, GIS provides a wide range of opportunities for examining relationships in space, often by incorporating additional information not traditionally considered in research methods. As GIS continues to become more accessible to nonexpert users, there are a wide range of new and creative opportunities in research. As with any new tool, GIS can be used incorrectly and provide misleading results. Any researcher familiar with standard statistical analysis software can relate to this concept. Use of the tool is not simply a matter of collecting a dataset and pushing buttons in the software. Good research necessitates collection of appropriate data in an appropriate fashion and then conducting an intelligent, well-thought-out analysis that makes sense for the hypothesis or question being addressed.

The following chapters discuss several of the essential concepts all users should keep in mind when developing a dataset for analysis in a GIS. They also address the major GIS data formats, along with their strengths, weaknesses, and applications. With these few but important pieces of information, you will be better prepared to use GIS as a tool in your own research applications, while avoiding many of the common pitfalls new users of these tools encounter. Chapter 2 examines spatial conceptualization and implementation, which are central to any research methods approach.

Review questions

1. What does it mean to think spatially? Provide an example.
2. What is the difference between informal and formal spatial analysis?
3. Name four of the advantages to spatial analysis as discussed in this chapter.
4. What is the definition of *space*? What is the definition of *place*? Explain how these two concepts differ.
5. What does the term *sociospatial* mean? What is a topic that you could study using this perspective?
6. Is spatial thinking a new or an old concept? Explain by drawing on examples of spatial thinking used in this chapter.

Additional readings and references

Addams, J. 1895. *Hull-House Maps and Papers. By residents of Hull-House, a social settlement, a presentation of nationalities and wages in a congested district of Chicago, together with comments and essays on problems growing out of the social conditions*. New York: Crowell.

Arlinghaus, S., F. Goodman, and D. Jacobs. 1997. "Buffers and Duality." *Solstice: An Electronic Journal of Geography and Mathematics* 8:2. http://hdl.handle.net/2027.42/60252.

Babbie, E. 2013. *The Practice of Social Research*. 13th ed. Belmont, CA: Cengage Learning.

Brown, N. 2004. "Florence Kelly: Slums of the Great Cities Survey Map, 1893." Center for Spatially Integrated Social Science. http://www.csiss.org./classics/content/35.

Corbett, J. 2004. "Mark Jefferson: Civilizing Rails, 1928." Center for Spatially Integrated Social Science. http://csiss.org/classics/content/12.

Creswell, T. 2004. *Place: A Short Introduction*. Malden, MA: Blackwell.

Creswell, J. W., and V. L. Clark. 2007. *Designing and Conducting Mixed Methods Research*. Thousand Oaks, CA: Sage.

Fearon, D. 2011. "Charles Booth: Mapping London's Poverty, 1885–1903." Center for Spatially Integrated Social Science. http://www.csiss.org/classics/content/45.

Goodchild, M. F. 2011. "Spatial Thinking and the GIS User Interface." *Procedia Social and Behavioral Sciences* 21: 3–9.

Jefferson, M. 1928. "The Civilizing Rails." *Economic Geography* 4: 217–31.

Steinberg, S. L., and S. J. Steinberg. 2009. "A Sociospatial Approach to Globalization: Mapping Ecologies of Inequality." In *Understanding Global Environment,* ed. Samir Dasgupta, 99–117. Mahwah, NJ: Pearson Education.

Relevant websites

The first two websites listed are portals to all things related to GIS. In addition to basic information about GIS, they link to information about software, training, data, and a variety of other useful resources.

- **Esri website (http://www.esri.com):** Esri is the creator of ArcGIS, a popular GIS platform. Many resources are available on its website, including examples of how to apply GIS and details regarding the different types of software.
- **GIS.com (http://www.gis.com):** This is a GIS portal managed by Esri. It offers a variety of general information and resources relevant to getting started with GIS technology.

- **The GIS Lounge (http://gislounge.com):** This is a GIS portal offering a variety of general information and links related to GIS technology, software, data, and other resources.
- **Cartographic Communication (http://www.colorado.edu/geography/gcraft/notes/cartocom/cartocom_f.html):** This website about cartographic communication is part of a larger Geography Educational website developed by Kenneth E. Foote and Shannon Crum, the Geographer's Craft Project, Department of Geography, University of Colorado at Boulder.
- **Center for Spatially Integrated Social Sciences (CSISS) (http://www.csiss.org/):** CSISS recognizes the growing significance of space, spatiality, location, and place in social science research. It seeks to develop unrestricted access to tools and perspectives that will advance the spatial analytic capabilities of researchers throughout the social sciences. CSISS is funded by the National Science Foundation under its program of support for infrastructure in the social and behavioral sciences.
- **PhoneBooth (http://phone.booth.lse.ac.uk/):** This website provides mobile access to the "Charles Booth's *Maps Descriptive of London Poverty*" 1898–9, and selected police notebooks that record eyewitness descriptions of London street by street.

Chapter 2

Spatial conceptualization and implementation

In this chapter, you will discover the strengths and challenges of GIS by examining each aspect of the GIS terminology: *geography*, *information*, and *system*. You will learn about the connection between conceptualizing and understanding the broader picture and the value of the logical and physical data models. You will learn how to determine project goals and define concepts and parameters important to a study question. You will also learn how to define research objectives. Once you identify the key variables for a research project, you can follow the guidelines in this chapter for implementing them focusing on spatial analysis in GIS.

Learning objectives

- Comprehend the challenges facing geography, information, and system aspects of spatial research
- Discover trends in the data
- Learn the phases of abstraction and how they relate to research
- Learn the differences between a logical data model and a conceptual data model
- Understand the importance of data aggregation to spatial research
- Learn what questions to ask about geographic data, location, and analysis

Key concepts

basemap
cartograms
code data
computer animation
conceptual model
coordinates
data aggregation
data dictionary
datums
digital
fuzzy GIS
index
key informant
latitude
logical data model
longitude
oral history
phases of abstraction
scale
social networks
trends in the data
variability
visualizing data

GIS are best understood by breaking down the terminology and learning how to apply GIS to various analyses. In particular, how can your area of interest and the associated data be placed into a GIS context, and how can GIS technology enhance your analysis and understanding of data? Here we review GIS in detail, letter by letter, to establish this understanding, before discussing data conceptualization.

The G in GIS

The geographic component of GIS is simultaneously obvious, confusing, and difficult to master. From an early age, we all develop an understanding that the locations of people and places can be marked on a map, and furthermore, that connections can be made between these locations. What we may not have is a good understanding about the scientific basis for mapping—that is, the numerous issues of **scale, coordinates**, control **datums**, and so forth. Other than mapping professionals, very few people have a deep conceptual understanding of the mathematical algorithms behind these concepts and the potential errors that result from various combinations and interactions of such data.

Fortunately, most of these underlying issues are addressed for us through the GIS software, so it is not essential to have a deep understanding of them. It is important, though, that you pay attention to a few essential concepts, even if you do not know exactly how they work. You can think of this as analogous to knowing the difference between a CD player and an MP3 player—you know these are different tools with different strengths and weaknesses, but selecting the right one does not require you to understand their inner workings. What is important is that you know which format to ask for so that the medium selected fits the player you own.

GIS has been used to locate and manage natural resources and is a well-embraced technology among many in business and marketing. Although GIS is very valuable to social science research, it has not been incorporated as frequently into this field. Many social science studies focus on social, economic, cultural, and survey data, asking questions such as, Do pregnant women who are better educated or wealthier receive higher-quality prenatal care? Perhaps more probing questions would ask about the locations of prenatal clinics relative to available public transportation, child care, and so on. You might use census data to conduct a statistical analysis of census block groups and levels of prenatal care but may miss an important locational component. Often, when explored in conjunction with other map-based location information (e.g., where are the blocks located relative to other important components?), a more complete understanding of the causative relationships can be obtained. Furthermore, from an applied standpoint, the geographic component can help in determining where to best locate and spend limited resources to help improve the situation.

In reality, almost all information researchers collect about people, their communities, and their environments can be tied to some geographic location. For example, you may survey people at their home address or by some geographic unit such as a census block or city of residence. All of these locations can be easily mapped. Furthermore, if privacy is a concern, you can engage in **data aggregation**, which means using a larger geographic unit to mask specific, personal information. In short, if you can answer the question, Where were the data collected?, then GIS is an appropriate means for storing and analyzing the data.

Difficulties with the G

The geographic context may be difficult to collect because determining the exact location of a piece of data on the ground is not always easy to accomplish or, for reasons of privacy, may not be permissible. When mapping people, we face an additional challenge: people may move around, may be without a home, or otherwise may be difficult to tie to a particular location. However, because geographic data are the heart of GIS, knowing a location of some kind is an essential part of the GIS process (even if it must be spatially degraded or detached from the exact, true location).

For example, if you were doing a study of homeless individuals, it might be better to define their location at the level of a particular neighborhood they call home than at a particular street address. Furthermore, even in studies where mappable locations are available, privacy issues may necessitate degrading that information. In other words, even if you have specific addresses of your respondents, you might choose to degrade the data to census blocks, to neighborhoods, or even to the city level to maintain the privacy required for ethical research. Choosing the level of spatial detail is an important part of the GIS process.

Conceptually mapped features, such as data about perceptions, ideas, or interactions (figure 2.1) are perhaps more difficult to map; although, they are equally important as physical locations in research. For example, social networks or interactions between individuals may be mapped in such a way that people who are emotionally close would be located conceptually close together, whereas individuals who are casual acquaintances might be mapped at a greater distance. Lines connecting people on the map could represent social distance rather than true geographic distance. On such maps, referred to as **cartograms**, the distance between mapped data is scaled to a variable or index value other than distance. In the case of social ties, this might be an index representing the strength of a particular relationship.

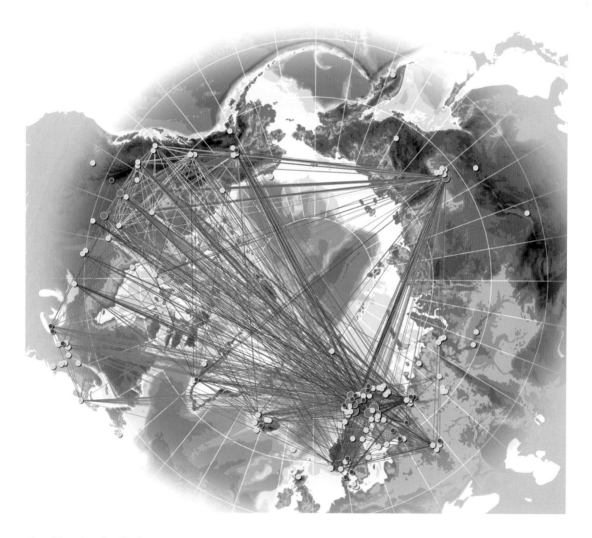

- Member institutions
- Non-member participants
- Network hosts (size indicating number of networks hosted)

Figure 2.1 University of the Arctic thematic networks map. This map presents collaboration networks between higher education and research institutions in the northern hemisphere. The map's goal is to illustrate the networks of activity and the geography that these networks cross. Courtesy of Hugo Ahlenius and Veli-Pekka Laitinen. From *Esri Map Book*, vol. 28 (Redlands, CA: Esri Press, 2013), 29. Data from University of the Arctic.

A second difficulty in GIS mapping relates to the **variability** that occurs in time and space. Most data are collected as a snapshot in time. We have a more difficult task obtaining data over a span of time to reliably map changes or **trends in the data**. Furthermore, because many of the things that we may map—especially individual people—will move over time, there is an added dimension of analysis to consider. Do we locate a survey respondent based on her home address or her place of employment, or perhaps based on where the individual is most likely to be at a particular time of the day or week? This decision would be most significantly influenced by the question under study; there are no set answers.

Using **computer animation**, you can change map data from static to dynamic. However, this type of mapping is still limited by the difficulty and expense of collecting data at a high frequency (temporal scale) as well as by software limitations for incorporating data instantly as it is collected. Fortunately, only a few social science applications necessitate true real-time analysis. Your primary goal as a researcher considering GIS as an analysis tool is to make such decisions before collecting the data.

You also need to consider the spatial representation of your data. Often, in research, privacy is of the utmost concern. Researchers typically lump data to mask individual data points representing individual respondents. Lumping, or degrading, data in this fashion results in a serious trade-off: the true, raw data may be permanently lost and no longer available for future research. As a result, researchers may collect an enormous amount of redundant data when the simple **recategorization** of existing data in different but equally valuable combinations would have allowed them to explore different questions.

For example, say you are looking at the populations of 426 incorporated cities in California. The cities range in size significantly, from the city of Vernon (population 80) to Los Angeles (population 3.4 million). In examining these cities for research purposes, you should consider numerous methods of categorization. As an illustration, consider an example using five categories for city size, as in figure 2.2.

Figure 2.2   An example of categorical classifications for the size of cities. Size classes can be defined a variety of ways, depending on the objectives and preferences of the map author.

How you choose to organize your data into these categories will have a direct effect on the outcome of analysis. Optimally, you will have access to the actual numbers so that you have a choice in the matter. If not, the metadata should define how cities were assigned to each of these categories.

Typically, your GIS software will have default settings for categorizing and representing these data, as in figure 2.3, which shows portions of Los Angeles and Orange Counties in southern California. The data categorization is based on the defaults used in ArcGIS (a popular GIS software package produced by Esri)—five categories based on the natural breaks within the dataset.

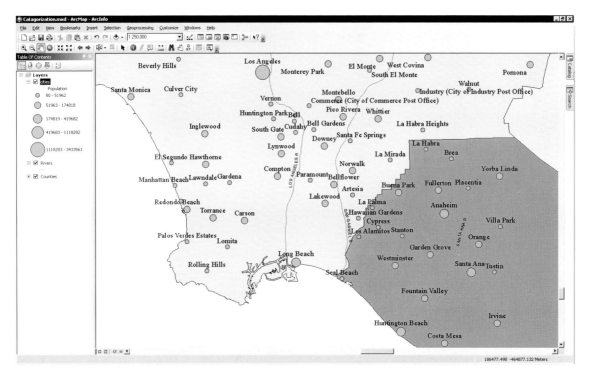

Figure 2.3 A map of city populations symbolized using default settings in ArcGIS. This map illustrates the various population sizes of some southern California cities. Map by Steven Steinberg. Data from US Census and State of California.

Although using the defaults in your software may produce a nice map, they may not be appropriate to your study data and objectives. Therefore, it is important to understand and define data categories that make sense for your needs. Perhaps there are legal or regulatory definitions for the sizes of cities you should consider. Or there may be statistical justification for how you examine your data. Changing the categories, of course, changes the map and the analysis results.

The map in figure 2.4 retains the five categories from very small to very large but uses a geometric interval as the basis for the categorization. Notice how the distribution of city sizes appears differently on the map.

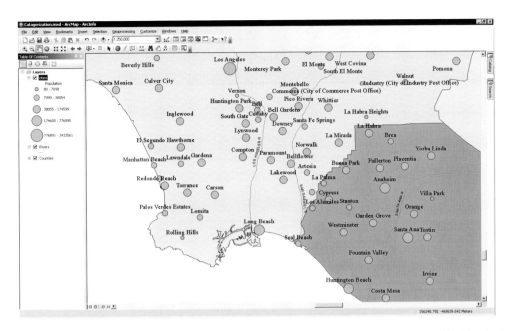

Figure 2.4 **A map with city sizes symbolized using geometric categorization in ArcGIS.** Map by Steven Steinberg. Data from US Census and State of California.

And finally, the map in figure 2.5 uses five quantiles, again changing the appearance and categorization of city sizes. Quantiles are a method of classification by which the data are divided into a specified number of equal-interval categories.

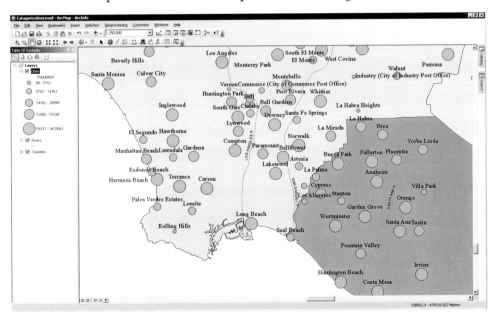

Figure 2.5 **A map of city sizes symbolized using five quantiles of geometric categorization in ArcGIS.** Map by Steven Steinberg. Data from US Census and State of California.

2 Spatial conceptualization and implementation   **31**

Although all of these examples are drawn from exactly the same dataset, they each represent the data differently. If you receive data that have been categorized in advance, you may find that the data are difficult or impossible to use in a study with a different set of questions. For example, what may be a medium-sized town to the person creating the original dataset may be a small town in your study. Another simple example of data degradation is the grouping of income levels into categories, which is a common practice in survey research. Categorical information, such as <$15,000 and $15,001–35,000, provide no means for a later study to distinguish individuals with incomes between $20,001 and $30,000. In a mapping context, it could be useful to link people or ideas to specific locations, but more commonly, data are collected by larger geographic regions, such as census blocks or other political boundaries; however, a census block doesn't show the internal distribution of data in the census block (e.g., are the households equally distributed across the area, or is clustering of the households hidden in the simplified data?).

Where data are provided in categorical form using category definitions that do not meet your requirements, you may need to locate an alternative data source or even collect your own primary data. Data that are degraded can no longer be recategorized to explore new or different questions. Of course, these are not simple issues to address because anonymity is an essential component of many social science questions; however, to the extent possible, when data are maintained in near-original, detailed form, the possibilities for analysis both within and outside the GIS are much greater.

Expanding the G

Mapping attitudes, ideas, **social networks** (connections that exist between people), and countless other human constructs should be viewed as equally valid as mapping the **latitude** and **longitude** of a data point on the ground. Numerous opportunities, limited only by the creativity of the researcher, will allow GIS systems to extend into realms not envisioned by the traditional geographies originally programmed into the software. The question that remains to be addressed is how one can develop an appropriate mapping context to represent concepts such as social interaction, desirability of a community, or social ties. Developing an index value or relationship between data points that can be used in place of physical distance as traditionally mapped is one means for visualizing data in a mapping context. **Visualizing data** means being able to portray the data in a visual format.

The *I* in GIS

The **information** component of GIS relates to the database aspect of the software. Databases are specialized software programs designed for storing, organizing, and retrieving information. GIS software packages can read or directly interact with data from almost any data management and analysis software. Some data translation may be necessary to facilitate the movement of an existing research database into a GIS software package. Many of the fundamental baseline datasets you might need to answer a question are already available in GIS-ready formats.

In particular, data from the US Census as well as many state- and local-level datasets are available through online sources or via a visit to the appropriate government office. Numerous university sources also provide GIS-ready data, as do a variety of private firms. Many of the data are freely accessible, including a wide range of spatial data from government agencies and other organizations. Additionally, a number of commercial databases are available for purchase. Such databases typically compile specialty data that are in some cases not available through other sources. Commercial data may be updated more frequently and may already be compatible with your GIS software.

The decision whether to use free or commercial data most often comes down to your needs and experience. Esri provides users access to a wide array of both free and commercial datasets designed to operate seamlessly with ArcGIS. Many other third-party providers also sell a wide range of economic, infrastructure, business, and social data types. Commercial data are reviewed for quality and presented in formats that are compatible with popular GIS software applications. Many free sources are similarly prepared, whereas others may require more effort and manipulation to make them compatible with a particular GIS package. Choosing your data source may simply be dependent on your budget. If you do not have the resources to purchase data or to collect your own, new data, you may need to explore free options. Of course, if you use free data, there could be a trade-off in the time it takes to prepare the data for use in your analysis, or the necessary data may simply not be available.

Of course, you can also use your own data, whether collected via survey, interview, observation, or almost any other means. In fact, so long as you intend the data to end up in a computer in **digital** form, regardless of the particular software involved, it will be accessible to a GIS. In fact, with a little foresight, perhaps through one or two additional questions or notations on the data sheet, you can collect data in a way that facilitates easy incorporation into GIS. These additional data can take the form of either real or conceptual locational information tied to a **basemap** that you choose (figure 2.6). Thus, the informational aspect of GIS is the easy part because almost everyone working in social sciences is already familiar with the process of collecting and coding data. Most are also familiar with entering data into a computer for analysis.

Basemaps

Topographic

The Topographic map includes boundaries, cities, water features, physiographic features, parks, landmarks, transportation, and buildings. This basemap features the World Topographic Map (aka the community basemap) which features the best available data from thousands of users and partners around the world. The World Topographic Map is designed to be used as a general purpose background map to support a wide variety of applications.

Learn more about the Topographic Map

Streets

The Streets basemap presents a multiscale street map for the world. This basemap features the World Street Map, which includes highways, major roads, minor roads, one-way arrow indicators, railways, water features, administrative boundaries, cities, parks, building footprints, and landmarks, overlaid on shaded relief for added context. The World Street Map is designed to be used as a general reference map and is ideal to support geocoding and routing applications.

Learn more about the Street Map

Light Gray Canvas

The Light Gray Canvas basemap is designed to be used as a neutral background map for overlaying and emphasizing other map layers. This basemap draws attention to your thematic content by providing a neutral background with minimal colors, labels, and features. Only key information is represented to provide geographic context, allowing your data to come to the foreground. The basemap includes two map layers, a base layer and reference layer, so that you can display your map layers above the base layer but beneath reference labels.

Figure 2.6 Esri provides a wide range of basemaps through its online map services. Some basemaps may require special licensing. Similar basemaps can be obtained from government agencies and other commercial data providers. Esri.

Extending the *I*

Making information more accessible to GIS requires upfront organization and structure of data storage as well as data coding and formatting. These tasks are not unique to GIS; rather, they are essential considerations for all data collection and analysis. Computers and GIS provide additional storage opportunities. Multimedia capabilities of the computer allow you to link photographs, sound bites, movie files, and scanned information, providing significant opportunity for raw data preservation and thus maintaining complete, detailed information.

For example, you could record a key informant interview or oral history on tape in its entirety, or you could record a traditional dance on video. (A **key informant** is an individual who is knowledgeable about the particular issue or topic under study. An **oral history** is a qualitative research method in which a researcher gathers individuals' stories and histories associated with a particular place, event, or issue.) You can convert these records into sound and video computer data files and link them to a GIS map, as discussed in chapter 6. When a user clicks on the location associated with a video or sound file on the GIS map, the complete recording becomes available.

Data coded or summarized from open-ended surveys or interviews for purposes of analysis are simultaneously available in their entirety to a researcher who may opt for a different coding scheme or analysis at a later time. To **code data** means to assign a meaningful symbolic numbering system to represent the answers in the data. For example, researchers collecting answers using a rating scale of 1 to 5 may reserve an additional code to represent questions that the respondent skipped, 0 or 99 being the most commonly used codes for this purpose.

Thus, beyond the many analytical and data presentational benefits we explore in this text, the GIS can store a variety of digital datasets collected during a particular study. If these data can be maintained in raw form, they can be used in new and different analyses at a later time. If you consider how your data can be used in the future when you build your library of GIS datasets, you can greatly extend the life and utility of any individual dataset.

The *S* in GIS

The **system** necessary to carry out development and analysis in GIS includes a variety of hardware and software components, in addition to people who can make these components work. Most academic institutions and government agencies, and many private consultants, may have these capabilities. The cost and training required to build a GIS from the ground up varies greatly. What is important for you to consider are the trade-offs between doing the GIS work on your own versus turning over aspects to a more highly trained GIS analyst. Regardless of what you decide to do, you need to ensure that everyone working on the GIS understands the data structure and format needed to achieve compatibility with minimum difficulty.

Difficulties with the *S*

GIS were largely developed with traditional geography in mind. Since the initial development of GIS, the use and capabilities of GIS have expanded across a variety of disciplines and applications in both the public and private sectors. However, these systems are limited by the map **data model**, consisting of **points, lines**, and **polygons**. This model assumes that all data can be linked to a specific, discrete location and that lines can be drawn to explicitly delineate the boundaries between data categories. Of course, many datasets are not so clearly defined, especially those collecting social science data, which tend to be less geographically specific, either because the data must be degraded to protect privacy or because social science research typically involves analyzing **conceptual maps** as opposed to geographic maps.

Although there have been efforts to develop **fuzzy GIS** systems that allow locations and boundaries to be less definite, these systems are not yet available in the mainstream. Therefore, the existing GIS data models do not necessarily fit the analysis being conducted, and you may be forced to come up with a creative solution to make nondiscrete data fit into a discrete data model. One caution is that it is easy to allow the capabilities of the software to dictate the analysis you carry out, or conversely, the analysis you never attempt. This is one reason why researchers in some areas of social science have taken longer to integrate GIS technology into their research than researchers in the natural sciences (who have been using GIS for about fifty years). Nonetheless, GIS began to find its niche in social science applications soon after its early development, most notably as a tool for collecting, storing, and analyzing US Census data, beginning with the 1970 US Census.

Those early years of GIS focused on natural resources for two simple reasons. First, natural resource managers in Canada, with vast amounts of information to map, originated GIS technology in the 1960s. It is relatively easy to map the location of a forest, which doesn't move and doesn't change quickly, and the computer is an excellent tool to accomplish that task. Working out methods to map nondiscrete data, such as mobile human populations, concepts, and attitudes, more typical of social science research would come later, thus explaining why GIS has taken substantially longer to make its way into daily use by certain researchers.

Furthermore, unlike today, early GIS systems were expensive and complicated to use. This made GIS a difficult technology to bring to small community groups or individuals, and a researcher often had to act as an intermediary between the data provider and the GIS analysis system. Even today, a researcher who wants to make use of the additional capabilities of a spatial analysis needs to understand spatial data characteristics, a multitude of data types, spatial and topological analysis, and spatial modeling. Social scientists must be able to discuss how GIS technology is applied to their data, even if they employ GIS experts to develop the databases, perform the analysis, or output the results from the GIS. Fortunately, powerful personal computers are now relatively inexpensive (especially compared to computers of the 1960s and 1970s, when GIS technology was in its early development). In combination with easier-to-use GIS software, the technology is much more accessible. You can now answer any spatial analysis questions that may previously have required the expertise of a GIS specialist. Learning the ins and outs of any GIS program, as well as the essential underlying concepts, is still a relatively time-consuming process. However, with some time and effort learning the fundamental concepts and organization of GIS and spatial analysis, you can go a long way toward answering interesting and important questions in research methods.

As GIS has grown in popularity, the number of trained GIS professionals has also expanded tremendously. Thus, when your questions or analysis go beyond your own experience, you can turn to a number of different people for additional assistance. In particular, you can find GIS experts at local universities, community or technical colleges, or government agencies. In many areas, GIS user groups meet regularly to share ideas, assist with questions, or show off new projects or the capabilities of their favorite GIS software. The creator of your GIS software can often provide information on these local groups and access to online communities of users who can assist you.

Conceptual data model: Incorporating GIS

The **conceptual model** takes analysis goals from reality to a spatial conceptualization. In accomplishing this task, the key components of the analysis are determined and the appropriate data identified. To illustrate the process of conceptualization, consider as an example the accessibility of prenatal care for low-income families at public health clinics.

A number of datasets would be useful in an analysis of this sort. The conceptualization phase is the point at which you begin to identify appropriate data by working through the "thought problem" of your analysis. Identifying data layers involves more than simply determining "I need a layer 'X' for my study"; it also includes specifications regarding what must be in the database to complete your analysis.

In this example, what data do you need? You need some demographic information, perhaps from the most recent US Census, which should give you some sense of socioeconomic characteristics as well as the location of these characteristics (to the census block level). A map of clinic locations is also essential. If no such map exists in a GIS format, you will need to explore other options for creating this layer (e.g., geocoding an existing map by street address). If detailed information about clinics exists, it may be possible to more specifically identify those in the database that provide prenatal care services. If the data on clinics do not exist, you will need to do some field research to create your own database.

Don't forget a basemap; you will need to determine an appropriate scale for the analysis. Are you studying clinics in one city, an entire region, or an entire nation? For this example, assume a local-level study. Perhaps a city street map from a local city or county planning department would be most appropriate and up to date.

As a last, and likely most conceptually difficult, step, you will need to develop a working definition of accessibility that you can use throughout your study. We propose a few possibilities here, but in all likelihood, you could include additional conceptualizations in your analysis. One aspect of accessibility is related to transportation. You might expect a significant number of individuals in this study to rely on public transportation to get to the clinic, so route maps and schedules could be important data. Additionally, there may be time-related accessibility issues, such as work schedules, the ability to take time off from work, school start and release times, and so on. How do these time and transportation issues relate to where the clinics are located and open? All of these issues have a social-psychological component related to an individual's willingness to go to the trouble of getting to the clinic. How far are prospective clients willing to travel, and how much time are they willing or able to spend in actually going to the clinic?

The list could go on to consider accessibility related to knowledge. Do the prospective clients of the clinics know the clinics are available? Are clinic staff members fluent in the appropriate languages for the populations they serve? Are printed materials at an accessible reading level for the clients? What about funding for the clinic? Can the clinic handle the demand? How long are people kept in the waiting room? Do clients know what health coverage is available to them, and can they navigate the paperwork to get it? Suddenly, what at first may seem like a simple concept rapidly balloons into something very complicated; of course, this is the nature of social science research.

In this laundry list of concepts, many are spatial, particularly those related to where the client base is located and how people will get to the clinic. Are people likely to be coming from their homes or from their workplaces? When looking at the willingness to travel, it might be possible to develop some values that make sense. For example, if you determined that a person would be willing to travel for one hour on public transportation, you could convert that time into a map of bus routes that are within one hour of clinics. In this way, the concept of willingness (with a series of variables that inform that concept) is operationalized into a map with a distinct spatial component. In other words, the GIS is used to integrate a set of concepts into something we can more easily analyze.

When going through thought exercises such as this one, it is sometimes helpful to sketch a figure or write an outline about your concept. Having a picture of your concept provides a framework of other important information, such as the type and source of the data you will need to create or find to complete your study. If the conceptualizations of your analysis result in datasets that cannot be located from an existing source, you will have a better sense of how much new data collection will be required.

This is also a good time to assess the feasibility of the study. Taking time to reevaluate each component of your developing model early can save a significant amount of time, money, and energy later in the analysis. After all, there is no sense in incorporating data that contribute only minimally to the overall model or that will present significant uncertainty in the final results. As in any study, relationships between variables should make sense, not just conceptually, but also practically. Some possible questions you might consider at this point are as follows:

- Will the results of the model be useful?
- Can actual, on-the-ground decisions be implemented based on results of the model?
- Are the relationships in the data statistically significant?

Once you have considered each component, data availability or your ability to create it, and the overall use of the model, you are ready to move beyond thought problems. But remember, until you complete the thought problem, it is probably not a good idea to go rushing forward with a GIS analysis. As mentioned earlier, many people gloss over this stage of the GIS process and end up going down dead-end analysis pathways that result in wasted time, money, and effort. Once you are content with your conceptual model (figure 2.7), then you are ready to move on to the third phase of abstraction: the logical data model.

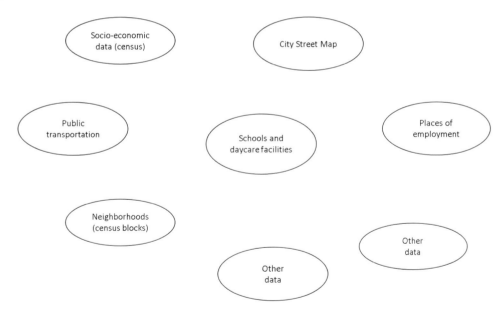

Figure 2.7 Sketching out the components of the conceptual model can greatly assist in identifying and organizing the information that will be necessary to complete a GIS analysis.

Defining terms or concepts is essential to developing a sound research project. This is especially true for names used to refer to phenomena. We advise you to conduct a complete literature review before beginning any type of research project to ensure you are aware of how different words are defined by others. It is likely that any concept you might want to name or investigate has a given name, and many concepts have multiple and, sometimes, competing definitions. The first step is to identify your concept, name it, and then define it. A literature review helps one to better understand the relationships that exist between different variables. Once you have identified your dependent variable, a literature review will further assist in explaining the relationships between the different variables. This is especially useful when you explore relationships between variables that have not been examined in the past. Once you have accomplished all of this, you are ready to plan your analytical approach.

Analytical approach: Phases of abstraction

When preparing an analysis in GIS, you must work through four distinct **phases of abstraction**, or modeling, in sequence:

1. Evaluate the real-world situation you intend to analyze.
2. Conceptualize in terms appropriate to a computer-based analytical approach.
3. Organize the logical approach to the analysis.
4. Implement the specific software steps.

Although many people conducting a GIS analysis tend to address the first three of these informally, taking the time to expand in some detail can greatly reduce missteps and dead ends as you carry out your analysis. In other words, taking a little extra time up front to plan the analysis process will yield benefits in the long run. We address each of these phases in the following sections.

Before you perform these four phases, keep in mind that, at this point, you are only considering the GIS aspects of the analysis. We assume that you have already determined the general nature of your study and that incorporating a spatial component through GIS will benefit your overall process.

Reality

Before initiating an analysis with GIS, it is essential that a GIS represents a model of reality—the computer does not understand reality. Although this may seem obvious, all too often, as people start to consider how they will work with their projects inside a GIS environment, they allow the GIS environment and their data to dictate their approach. Of course, knowledge of the data, the project, the disciplinary expertise, and local knowledge should drive the approach to the study.

Although it can be quite enticing, you should not allow the capabilities of a particular GIS software program or available data to dictate your analytic process. This is especially true when time, money, and personnel available to conduct an analysis are limited. Unfortunately, we cannot offer you a simple solution. Every project requires finding a balance between the real world that you are analyzing and the abstraction that is required to make that reality fit into a GIS. It is important to recognize that there is a distinction between reality as it truly is and reality as represented in the model. When assessing a GIS model or its result, keep yourself firmly planted in the true reality, where your results and decisions affect real people.

Beyond the issues of how the reality fits into the GIS, you will need to consider (as a precursor to this) what reality is for your study. Our discussion of reality isn't philosophical but rather practical. Defining reality is not always quite as simple as it may seem. For example, if you are doing a survey of household income, where is the breakpoint between economic classes? How might this vary in different geographic regions? Are only dollars and cents relevant? What about bartered services? What about work performed in the home by a member of the household for no pay? As you can see, there are numerous possible components to how income might be assessed and analyzed.

As the analyst who is going to locate, collect, or use these data, how do you define your variables and classes? When looking for archival data, referring to the data dictionary and the metadata may help, if they exist. A **data dictionary** is a collection of information and details about the data that includes variable names. When collecting your own data, these are the kinds of issues you need to ponder and, in many cases, debate with colleagues, community members, or stakeholders. In short, defining reality is no small task, but it is by far one of the most important.

Logical data model

The **logical data model** adds specific processing steps to the conceptual model developed in the previous step, thus specifying the analytical procedures necessary to complete the analysis. That is, we want to add a workflow to the conceptual figure. This step can be represented graphically by creating a flowchart that specifies each step in specific detail. One unique characteristic about the logical model is that it does not include software-specific processing steps, only the logic behind them.

For example, in our prenatal clinic model, we know we need to extract specified socioeconomic data related to the prospective client population within our particular study area (city). The data source we have identified for this example is the US Census; however, because the census data contain much more information than we really need, we want to pare that down. We want to narrow the geographic extent and, via a database query, extract only those attributes we require. Graphically, this portion of the flowchart might look similar to figure 2.8.

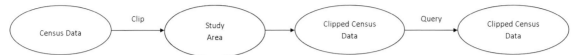

Figure 2.8   A portion of a logical data model showing the steps necessary to reduce the full US Census dataset to the appropriate spatial and attribute components necessary for our study. The logical model first indicates a clip, which is a spatial operation to cut out only the required geographic area, much like a cookie cutter. The resulting dataset is then queried to obtain only those records or attributes needed in the research analysis.

In this example, we specify a logical sequence of steps but do not concern ourselves with how these steps are achieved in the particular software. It is important to have a sense of the possible processing options. Once we have developed a logical flow for the entire analysis, we reach our next evaluation point: determining the feasibility of accomplishing the necessary processing steps using available software tools (figure 2.9). Tools may vary based on the particular software licenses available for data collection, aggregation, and analysis as well as those that may be imposed by outside forces (e.g., field equipment used, versions of software).

Just as data may be brought into the GIS analysis from a variety of software environments, they can also be analyzed and processed in a variety of environments. You may determine that certain steps are better accomplished in a specific database program, a statistical software package, or a variety of other software options. In fact, it is not at all uncommon to incorporate multiple software programs and discipline-specific models into an analysis. For example, aspects of a project in ArcGIS may include data tables produced in Microsoft Excel or imported data from field equipment such as Global Positioning System data from Trimble Pathfinder Office. Many software packages commonly used in a GIS project are easily integrated with ArcGIS, either directly or using ArcGIS tools or other software.

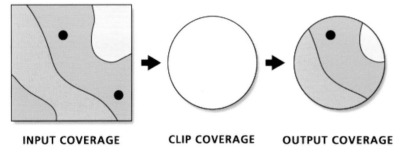

Figure 2.9   An example of a data clip as illustrated in ArcGIS documentation. The dataset on the left covers an area larger than needed. The study boundary (center) is used as a "cookie cutter" to clip the data down to the study region. Esri Resources: Clip (Analysis).

The idea behind thinking through your logic, independent of the software to be used, is to avoid one of the most common pitfalls in GIS analysis: allowing the software, or your knowledge of particular tools, to dictate how you analyze your data. After developing your logical data model, you might have a good basis for selecting the most appropriate GIS software and add-ons (potentially including some new extensions or add-ons) for your analysis. This is akin to every problem looking like a nail when you are only familiar with a hammer. If, in fact, logic dictates using a screw, then obtain and learn to use a screwdriver.

Of course, you can't completely change software tools for each new analysis. Once your organization has committed to a particular program, you need to take advantage of what it offers, perhaps adding individual extensions or additional tools as needed. Most software, such as ArcGIS, includes thousands of tools and options, more than the typical user can keep track of. So, even if you are not sure how to implement your logical model when you develop it, there is usually room to explore options you may not have considered to determine the best approach to analyzing your data following the logic developed at this step of the abstraction process.

AN ASIDE ON CHOOSING THE RIGHT TOOLS FOR YOUR ANALYSIS

Selecting the right tools for your analysis is essential to conducting research in GIS. It is important to avoid tunnel vision when it comes to the software you incorporate into your analysis. The value of the logical data model is that you think through what you want to accomplish without being distracted by the question of how you will accomplish this from a software perspective. Once you have worked out the logic model, or *what* you wish to accomplish in your analysis, you can turn to *how* you can best accomplish it with the physical model.

Selecting the physical tools for the job may require you to go beyond the software tools you already know. Although GIS software packages, such as ArcGIS, provide a relatively complete, out-of-the-box solution, it is not unusual for additional software tools to be required to carry out your analysis.

ArcGIS is compatible with optional software components, called extensions and add-ons, that enhance its capabilities for a particular type of analysis (table 2.1). Extensions and add-ons come in three primary varieties: Esri-developed add-ons, third-party add-ons, and user-contributed add-ons. Although these products can sometimes be expensive, they can pay for themselves in time and effort saved on routine or repetitive analysis tasks. A substantial number of freely available extensions and add-ons are also available. Finally, a wide array of stand-alone software tools (both commercial and free) may be used in conjunction with ArcGIS.

Table 2.1 Extensions and add-ons available from Esri

Extension/add-on	Key benefits
Analysis	
ArcGIS 3D Analyst	Analyze your data in a realistic perspective.
ArcGIS Geostatistical Analyst	Use advanced statistical tools to investigate your data.
ArcGIS Network Analyst	Perform sophisticated routing, closest facility, and service area analysis.
ArcGIS Schematics	Represent and understand your networks to shorten decision cycles.
ArcGIS Spatial Analyst	Derive answers from your data using advanced spatial analysis.
ArcGIS Tracking Analyst	Reveal and analyze time-based patterns and trends in your data.
Business Analyst Online Reports	Directly access demographic reports and data from Business Analyst Online (BAO) for trade areas and sites created in the desktop.
Productivity	
ArcGIS Data Interoperability	Eliminate barriers to data use and distribution.
ArcGIS Data Reviewer	Automate, simplify, and improve data quality control management.
ArcGIS Publisher	Freely share your maps and data with a wide range of users.
ArcGIS Workflow Manager	Better manage GIS tasks and resources.
ArcScan for ArcGIS (included with ArcInfo and ArcEditor 9.1 and higher)	Increase efficiency and speed up raster-to-vector data conversion time.
Maplex for ArcGIS (included with ArcInfo 9.1 and higher)	Create maps that communicate more clearly with automatically positioned text and labels.
Solution-based	
ArcGIS Defense Solutions (includes ArcGIS Military Analyst, Grid Manager, and MOLE)	Create workflows, processes, and symbology to support defense and intelligence planning.
Esri Aeronautical Solution	Use the full power of GIS to efficiently manage aeronautical information.
Esri Defense Mapping	Efficiently manage defense specification-compliant products.
Esri Nautical Solution	A GIS-based platform for nautical data and chart production.
Esri Production Mapping	Standardize and optimize your GIS production.

Table 2.1 (continued)

Extension/add-on	Key benefits
No-cost add-ons	
ArcGIS Editor for OpenStreetMap (download)	Contribute to the OpenStreetMap project by adding, editing, and deleting data within the familiar ArcGIS 10 editing environment.
ArcSketch (download)	Quickly create features in ArcGIS with easy-to-use sketch tools.
Districting for ArcGIS (download)	Create defined groupings of geographic data, such as census tracts, ZIP Codes, and precincts, by creating a districting plan.
Free Geoportal add-ons (download)	Catalog geospatial resources within an enterprise and provide quick access to those resources regardless of location or type.
Geodatabase Toolset (GDBT) (download)	Manage your scalable geodatabases with diagnostic performance tools.
OLAP for ArcGIS (download)	Create, view, use, and manage connections to OLAP databases in ArcGIS for Desktop.
US National Grid Tools for ArcGIS (download)	Support disaster relief and search and rescue with a coordinate system that can be standardized across agencies.
WMC Client (download)	Open Web Map Context (WMC) files directly in ArcMap.

Courtesy of Esri.

You can also develop or add custom scripts, analytical models, or other tools to supplement the built-in capabilities of ArcGIS. These custom components may be available through online user forums for the specific GIS package or by searching on the Internet. Many of these components are just as good as commercial products, and they are inexpensive or free; however, they may lack formal technical support.

If you are working with others, using a common software platform and version and data standards can save headaches when it comes to data compatibility and analysis. In recent years, geospatial data formats have become much more consistent under the leadership of the Open Geospatial Consortium, which boasts more than four hundred members, including geospatial software companies, government agencies, universities, and other organizations from around the world (http://www.opengeospatial.org).

Open-source GIS
Several open-source GIS software tools are available. Open-source software is free to download from a variety of sites on the Internet, and many of the software programs have supportive user communities happy to help out via e-mail lists. In most cases, you need to compile the program to work with your computer, a process that may require additional open-source components that you download independently. However, open-source software can be an excellent option if you are comfortable working with software with its configurations, and sometimes these tools provide analytical capabilities not readily available with other software.

Physical data model

The final phase of model abstraction is the physical data model. At this point we lay out the software-specific steps—the specific commands or menu options to accomplish the processing necessary for the analysis. Although many GIS programs use similar terminology for particular processes, there are variations.

Here, again, it can be helpful to annotate the flowchart that has been developed to this point. Where lines between data bubbles are located, you can indicate the specific command for your particular software package (figure 2.10). You should note specific names of the data layers used and created at each step. Although this may sound a bit tedious, it can pay off in helping you to keep track of your data at each step along the way. The alternative, trying to keep track of everything in your head when sitting at the computer, is asking for trouble.

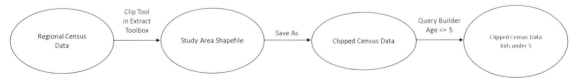

Figure 2.10   A portion of the logical data model showing the steps necessary to reduce the full US Census dataset to the appropriate spatial and attribute components necessary for a study of children under age five in the study region. The logical model first indicates a clip, which is a spatial operation to cut out only the required geographic area, much like a cookie cutter. The resulting dataset is then queried to obtain only those records (or attributes) needed in the study.

Even a simple GIS analysis may result in dozens of individual data layers. Developing a flowchart to help you to know what each one is and the software options used to create them can be essential to keeping track of the steps in your analysis. There is nothing worse than accidentally deleting a file you needed. Equally frustrating is when you finally get the solution to your question only to realize you have no idea which steps and software options you used to get there.

Determining project goals

Any person interested in conducting research ultimately begins by determining the main idea, or purpose, of his or her research. For any discipline, the possibilities are endless, especially when it comes to incorporating GIS into the research plan. Many different factors can affect the purpose of your research, such as your personal interests and the interests of the sponsoring organization or agency. Applied research projects may have more than a single purpose, setting out to answer a series of questions or to work toward a solution for specific social problems.

Any time you are going to engage in a research project, you need a clear picture of your goals. A surprising number of people carry out an analysis without asking, What is the purpose of this study? Having a clear project goal is very important because, beyond being the reason for engaging in research in the first place, the goal directly informs the data to locate, collect, and prepare for eventual analysis. A project may have more than one goal—different participants in and funders of the project process may have different goals. If this is the case, it is useful to develop a set of overarching project objectives and a list of needed data prior to beginning the data collection process. One of the most troublesome things that can occur in carrying out an analysis is realizing after the fact that an essential and easy-to-acquire piece of data was not collected as part of the original design. Going back to fill in missing data after the fact, assuming it is possible, can be time consuming, expensive, and, of course, embarrassing. In the GIS realm, you will also need to determine how you will encode the data in a GIS-compatible format that simultaneously works within the GIS and effectively meets the requirements of your study. Chapter 5 addresses these issues in more detail.

Some questions you might ask yourself in developing goals for your study include the following:

- Why are you engaging in this study?
- What sorts of relationships do you want to examine as a part of your study?
- What causes or organizations might your study benefit?
- Who is funding your study?
- Are you studying patterns of human behavior?

If you are able to answer these questions clearly and up front, you will have a much easier time defining the specific goals of the project and thus the data and variables you will require in achieving your objectives.

Guiding questions

The following sections describe four sets of questions you can ask yourself to determine how a GIS could be integrated into your study. The sections include questions about concept, data, location, and analysis.

Questions about concept

The questions in this section focus on the ideological structure of your study. These questions are designed to be asked early on in the research process, usually before you begin implementing your research methodology. When answering these questions, some background research may be necessary; this research should be done before beginning the actual project and going out into the field. You may be able to conduct much of your background research through a literature review and an Internet search or through talks with local experts and others who are familiar with the study site to gain a good understanding of the existing relationships. However, your contact with people in your study site will depend on whether you have established social ties there, how accessible or remote your study site is, and the availability of and access to technology (phones, Internet, etc.) at the site. Sometimes a small pilot study is useful to ensure your approach is acquiring the necessary data and is not confusing to those doing the collection or to those responding to questions. Pilot testing the analysis can also help ensure that the data can be analyzed as expected.

Of course, it is important to collect the proper information from each location, respondent, or unit of analysis in your study. This information is commonly referred to as a **variable**. What is a variable? A variable can be defined as a domain of attributes that relate to a particular concept. For example, the variable gender might be defined as the three attributes male, female, or other. So what is an **attribute**? An attribute is a characteristic that is associated with a particular object or person. The following questions might help you define the concepts to be used in your study:

- What is your research question?
- What is your main dependent variable?

- What are your independent variables?
- What are the main hypotheses of your study?
- Are there any geographic features that you have already identified as variables in your study? If so, what are they? Are they manufactured or natural features?
- Could any of the geographic features contained in your study site potentially affect the issue that you are studying?

Questions about data

The questions in this section relate to the data collection portion of your project. Once again, it would benefit you to answer these questions prior to going into the field. This information should be available from government agencies (local, state, county, federal) that have responsibility for the area that includes your study site (figure 2.11). In addition, contacting private businesses or local interest groups may turn up even more of the information you require. As mentioned earlier, one of the keys to successfully using GIS is to think innovatively and seek creative avenues to find the data that you might need for your study. Although you may want to simply search the Internet for agency and organizational websites to get all the data you need, this process will rarely provide anything close to a complete list of the data that are actually available. More likely, you will need to make direct contact with the appropriate people at each agency or organization to answer these questions. Because collecting and developing quality datasets is time consuming and expensive, some people will be reluctant to give their data to you. Therefore, it is often worthwhile to have something to offer in exchange for data. Data trades, access to related data to be developed during the study, or access to the final report results are all valuable offers you can make to those who assist you. When contemplating the use of GIS for any particular study, some items to consider include the following:

- Do GIS data exist in any form for your study site?
- If not, what sort of data exist that might be GIS compatible?
- Do you have access to these data, and if so, under what conditions?
- What was the original purpose for which the data were collected?
- What sorts of variable attributes are included in the dataset that you are interested in examining?
- How old are the data? Were they collected at only one time, or are they longitudinal?

- What metadata exist for these data?
- Has any type of accuracy assessment been completed, or have these data been ground truthed?

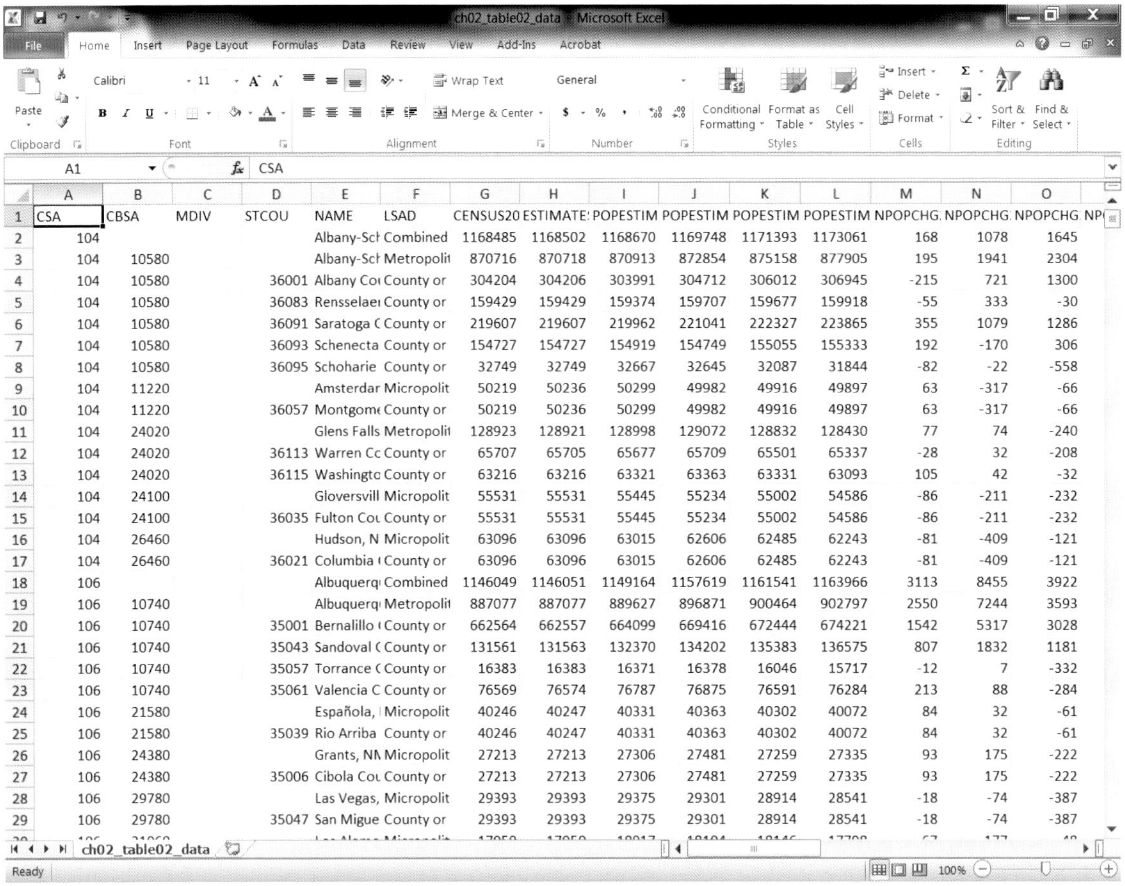

	A	B	C	D	E	F	G	H	I	J	K	L	M	N	O
1	CSA	CBSA	MDIV	STCOU	NAME	LSAD	CENSUS20	ESTIMATE:	POPESTIM	POPESTIM	POPESTIM	POPESTIM	NPOPCHG.	NPOPCHG.	NPOPCHG.
2	104				Albany-Sch	Combined	1168485	1168502	1168670	1169748	1171393	1173061	168	1078	1645
3	104	10580			Albany-Sch	Metropolit	870716	870718	870913	872854	875158	877905	195	1941	2304
4	104	10580		36001	Albany Co	County or	304204	304206	303991	304712	306012	306945	-215	721	1300
5	104	10580		36083	Rensselae	County or	159429	159429	159374	159707	159677	159918	-55	333	-30
6	104	10580		36091	Saratoga C	County or	219607	219607	219962	221041	222327	223865	355	1079	1286
7	104	10580		36093	Schenecta	County or	154727	154727	154919	154749	155055	155333	192	-170	306
8	104	10580		36095	Schoharie	County or	32749	32749	32667	32645	32087	31844	-82	-22	-558
9	104	11220			Amsterdar	Micropolit	50219	50236	50299	49982	49916	49897	63	-317	-66
10	104	11220		36057	Montgome	County or	50219	50236	50299	49982	49916	49897	63	-317	-66
11	104	24020			Glens Falls	Metropolit	128923	128921	128998	129072	128832	128430	77	74	-240
12	104	24020		36113	Warren Cc	County or	65707	65705	65677	65709	65501	65337	-28	32	-208
13	104	24020		36115	Washingtc	County or	63216	63216	63321	63363	63331	63093	105	42	-32
14	104	24100			Gloversvill	Micropolit	55531	55531	55445	55234	55002	54586	-86	-211	-232
15	104	24100		36035	Fulton Cou	County or	55531	55531	55445	55234	55002	54586	-86	-211	-232
16	104	26460			Hudson, N	Micropolit	63096	63096	63015	62606	62485	62243	-81	-409	-121
17	104	26460		36021	Columbia (	County or	63096	63096	63015	62606	62485	62243	-81	-409	-121
18	106				Albuquerq	Combined	1146049	1146051	1149164	1157619	1161541	1163966	3113	8455	3922
19	106	10740			Albuquerq	Metropolit	887077	887077	889627	896871	900464	902797	2550	7244	3593
20	106	10740		35001	Bernalillo (	County or	662564	662557	664099	669416	672444	674221	1542	5317	3028
21	106	10740		35043	Sandoval (	County or	131561	131563	132370	134202	135383	136575	807	1832	1181
22	106	10740		35057	Torrance (	County or	16383	16383	16371	16378	16046	15717	-12	7	-332
23	106	10740		35061	Valencia C	County or	76569	76574	76787	76875	76591	76284	213	88	-284
24	106	21580			Española,	Micropolit	40246	40247	40331	40363	40302	40072	84	32	-61
25	106	21580		35039	Rio Arriba	County or	40246	40247	40331	40363	40302	40072	84	32	-61
26	106	24380			Grants, NM	Micropolit	27213	27213	27306	27481	27259	27335	93	175	-222
27	106	24380		35006	Cibola Cou	County or	27213	27213	27306	27481	27259	27335	93	175	-222
28	106	29780			Las Vegas,	Micropolit	29393	29393	29375	29301	28914	28541	-18	-74	-387
29	106	29780		35047	San Migue	County or	29393	29393	29375	29301	28914	28541	-18	-74	-387

Figure 2.11 **An example of a dataset obtained in table format.** US Census.

Assuming that you can find and get access to data that will be appropriate to your study, locating useful data, especially if they are already in a GIS format, will be well worth the effort. Even if you find that none of the required data already exist and are available, that is still important information in planning your study. Knowing what you will need to collect and input yourself is important in adequately budgeting both time and money for your study.

Questions about location

The following are questions you should ask about your geographic location. These questions may appear basic, but they will help you gather information that is useful for integrating a GIS into your project. It is important to answer these questions before data collection so that you can arrange the categories and measurement devices for your project. Furthermore, thinking about the answers to these questions will help you identify important characteristics of your study location and will further assist you in the data collection process because you will have already established those aspects of your location that are important and determined which types of data variables you require. These issues are discussed in detail in later chapters. When evaluating the geographic aspects of your study, valuable considerations might include the following:

- Where does your research occur geographically?
- How would you describe the physical aspects of your study site or research area (i.e., what are the geographic boundaries of the study)?
- Which urban areas or cities are close to your study site? Are these relevant variables?
- Are there emotionally, spiritually, environmentally, physically, or culturally important units of analysis (neighborhoods, counties, valleys, mountains)?
- What are the different geographic locations associated with your independent variables?
- Do any unique, natural geographic features exist in your study site?
- Do any unique, manufactured geographic features or facilities exist within your study site?
- Do any geographic features of the area hold special meaning for the people who live there?
- Are there boundaries associated with your data? How are these defined?
- Are the data stationary through the timeframe of your analysis, or do they move? How will you account for these changes?

Questions for analysis

The following questions will aid in conducting your analysis. They focus on issues of boundaries, time, and place. All of these issues are discussed in greater detail in later chapters, but they are important to have in mind early on. A GIS is useful for conducting a comparative study of groups of people, for example, comparing two neighborhoods of a city. It's up to the researcher to determine the boundaries of a study. Boundary choice is essential, because changing the boundaries of an analysis can drastically alter the results obtained. Therefore the boundaries of your study are best dictated by the research question you are investigating and should be set in advance, as opposed to during the analysis phase.

Trying to determine whether geographic variables will be useful to your study can be tricky, but keep the following in mind: almost all units of analysis have an associated geographic location. Social scientists are often interested in measuring basic sociodemographic variables, such as age, income, gender, and ethnicity. These core variables can then be associated with geographic locations, such as neighborhoods, cities, and states. The possibilities are limitless. Anything that involves geographic boundaries can be studied using GIS. Sometimes boundaries are artificially created by the researcher, depending on the researcher's goals. At other times, the boundaries already exist and are based on manufactured geographic features (such as streets) or natural geographic features (such as rivers, lakes, and mountains).

The issue of boundaries is a very important one for GIS. One thing to keep in mind is that you are the boss when it comes to your data collection and the type of information that you want to include in your study. Therefore, the notion of what are appropriate boundaries in your study may require you to look back at your research questions or hypotheses. It is also important to consider whether you are dealing with socially constructed boundaries, such as political boundaries (e.g., city limits or state lines) or physical boundaries (e.g., mountain ranges, oceans, or rivers). Sometimes the two coincide, but often they do not. Sometimes we use boundaries that are defined more fluidly and conceptually (e.g., a traditional hunting ground, the heart of the community, or a gang's territory).

Questions for analysis include the following:

- What is the primary unit of analysis in your study?
- What are the physical boundaries for your unit of analysis?
- What are the social (conceptual) boundaries for your unit of analysis?
- Are your conceptual boundaries social, philosophical, or economic? Conceptual boundaries can be any parameters that reasonably organize people into groups relevant to the issues you are examining. Although conceptual boundaries are often physically defined, geographic boundaries (e.g., cities or neighborhoods), such boundaries may be self-defined, such as people who identify themselves as part of a particular group because of some perceived shared characteristic (e.g., a particular

ethnic or religious group or a group of people with a similar hobby or interest, such as gardening or bird watching). You can identify the boundaries of a group you are studying based on these conceptually defined parameters.
- What is the hypothesis or driving research question? What is the problem or issue your research addresses?
- What is the main geographic feature or variable you will examine in your study?
- How is this feature or issue related to your independent or dependent variable? (Also see the next question to help you answer this question.)
- Can you identify any geographic pattern by unit of analysis (e.g., group, neighborhood, country) relative to the topic or issue under study? For instance, if you are studying poverty, are certain neighborhoods in the study more poverty stricken than others?
- If your study involves a comparison of groups, are there differences or stratifications based on geographic location?
- If your study does not indicate any geographically based patterns, are there geographic variables worthy of exploration that you failed to consider?
- What themes emerge from your data? Do they emerge by unit of analysis?
- Are there themes that emerge specific to certain geographic locations?
- Do specific social, economic, and political data appear to cluster geographically?
- If this information is not easily available in a spatial database format, could such a database be easily created?

Steps in the research process

If you decide that you do want to employ a GIS in your project, you need to accomplish two nuts-and-bolts tasks right away: (1) identify the geographic region and (2) develop a data dictionary (figure 2.12).

First, you should identify the geographic region (or spatial extent) and important features of your study, including the categories, geographic features, and physical environmental features that you want to consider as part of the study. Essentially, this means drawing a study site boundary, along with having a good understanding of your topic and determining which geographic features you want to measure.

Period of military service
These periods represent officially recognized time divisions relating to wars or to legally-relevant peacetime eras. The data pertain to active-duty military service. In most tabulations of these data, people serving in combinations of wartime and peacetime periods are classified in their most recent wartime period.

Related term: Veteran status

PHC-1 tables
These are Census 2000 Summary Population and Housing Characteristics tables, a publication series which includes information on the 100-percent population and housing subjects. The data are available for the United States, regions, divisions, states, counties, county subdivisions, places, metropolitan areas, urbanized areas, American Indian and Alaska Native areas, and Hawaiian homelands. The series is comparable to the 1990 CPH-1 report series, Summary Population and Housing Characteristics. The series is available in printed form and on the Internet in PDF format.

Place
A concentration of population either legally bounded as an incorporated place, or identified as a Census Designated Place (CDP) including comunidades and zonas urbanas in Puerto Rico. Incorporated places have legal descriptions of borough (except in Alaska and New York), city, town (except in New England, New York, and Wisconsin), or village.

Related terms: Census designated place (CDP), City, Comunidad, Incorporated place, Town, Zona urbana

Place of birth
The U. S. state or foreign country where a person was born. Used in determining citizenship.

Related terms: Citizenship status, Foreign born, Native population

Plumbing facilities
The data on plumbing facilities were obtained from both occupied and vacant housing units. Complete plumbing facilities include: (1) hot and cold piped water; (2) a flush toilet; and (3) a bathtub or shower. All three facilities must be located in the housing unit.

Population
All people, male and female, child and adult, living in a given geographic area.

Related terms: Apportionment population, Resident population

Figure 2.12 An example of a section of a data dictionary for the US Census (the Census Bureau website uses the more public-friendly term *glossary*). Concepts included in the dataset, along with the associated terminology, definitions, and units of measurement, are described in detail. This information is critical to researchers when assessing a dataset to determine its relevance to their research needs and to interpret results obtained from its analysis. US Census.

Developing a data dictionary means developing a set of definitions and criteria for what constitutes your particular categories. In other words, you need to develop descriptions that clearly define each of the attributes that you would be recording in association with your variables. (We explore research design and considerations for incorporating GIS further in chapter 3.)

For example, you might be doing a study of urban communities, but how exactly will you define these? You might use a definition based on population density, the percentage of the landscape covered by concrete, or the number of roads per square mile. Developing and documenting clear definitions for each variable and attribute to be included in your study is an essential step before going forward with data collection and analysis. Doing this in advance will help to ensure that everyone involved in the study understands the data to be collected and that each item will be mutually exclusive, thus avoiding confusion later in the process.

Public health example

The best way to determine the purpose of your research is to think about the central question that you want to have answered. For instance, maybe you are interested in examining whether people who live in a certain section of town (e.g., the poorer section) suffer from respiratory problems. The different parts of the town may be described as being "rich," "middle class," or "poor" based on the income levels of residents who live there. The question of health connected to income could be addressed using a GIS. Following the earlier mentioned guidelines for integrating GIS into the study, you would want to determine the study boundaries and relevant geographic features that are going to be part of the study. In this case, you are interested in drawing your study boundaries based on income. In other words, you want to draw a boundary around the low-income section of town, the mid-economic area, and the upscale part of town. Where would you begin? You could begin by looking at US Census data for that town to determine the clusters of different parts of town based on household income. You could then draw your boundaries based on the clustering observed in the US Census data. Step 1 in the process also calls for identifying other geographic features that might be important to your study. In this example, that would include determining the locations of different factories, incinerators, or other production facilities (figures 2.13 and 2.14).

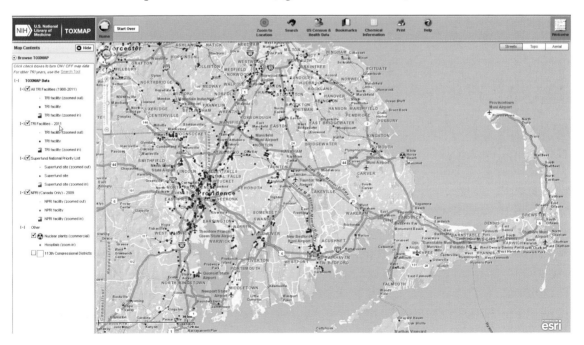

Figure 2.13 Mapped locations of sites listed in the US Environmental Protection (EPA) Agency Toxic Release Inventory (TRI). Sites are displayed here on an interactive map of eastern Massachusetts. These maps help users visually explore data from the EPA's TRI and Superfund Program. The underlying data are also available for **download from the EPA website.** Courtesy of the Division of Specialized Information Services of the US National Library of Medicine. Basemap data from Esri, DeLorme, USGS, NOAA, NGA, IFL.

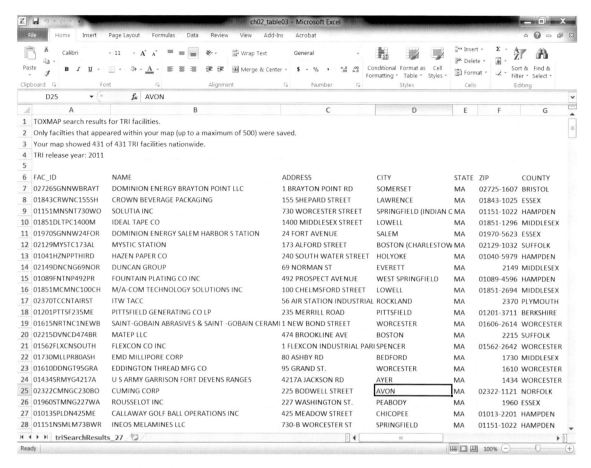

Figure 2.14   Data underlying the map shown in figure 2.13. Data from EPA.

Step 2 in the integration of GIS into your study calls for developing a data dictionary. You as the researcher would determine what level of household income would fit your categories, for example, rich (yearly income over $45,000), middle class (yearly income between $20,000 and $44,999), and poor (yearly income under $19,999). In your GIS, you could also gather information on the level of emissions from these different facilities by looking at the Environmental Protection Agency's website, where the EPA indicates industrial sites that emit beyond a certain specified level. This would be valuable information regarding air quality that could become a part of your GIS database.

You could interview people who live throughout the town and conduct a survey that inquires about their general health, income, and whether they suffer from any respiratory problems. As long as you know the geographic locations of respondents, you could enter this information, along with the survey answers, into a database. That way, when you conduct your analysis, you will be able to have geographic information for each unit of analysis (e.g., household). If you want to aggregate the data slightly to protect the privacy of the individuals, you could categorize respondents as living in a neighborhood versus at an actual street address. On further investigation, you may want to geographically locate various factories, incinerators, or other production-oriented facilities that could be emitting substances that affect people's respiratory health.

Another important thing to consider in determining your research purpose is the general theme that is a part of your research question. In the foregoing example, some of the potential themes might be environmental health, social inequality, and/or environmental justice.

After determining the general purpose of your research, you can then ask the question, How would GIS be helpful to the project? In other words, how would using a GIS enhance the study? A GIS is useful because it facilitates a more holistic and contextual view of a research problem or issue. It accomplishes this through bringing together a variety of different data types. Any study that you choose to develop will most likely include a variety of important variables. The trick in using the GIS is to identify which variables will best be studied using a GIS. This topic is discussed in greater detail in the following chapter.

Moving forward

In this chapter, you learned about strengths and challenges to each letter in the GIS. Additionally, you learned about conceptualization and the right questions to consider as you begin to frame a spatially based research project. Furthermore, you learned what questions to ask yourself about data and analysis to help move you forward with your project. You also learned about the conceptual framework, a useful tool for guiding conceptualization. Chapter 3 discusses research design, which is fundamental to the research process.

Review questions

1. What are some difficulties with the *G* in GIS, as discussed in the chapter?
2. How do you determine project goals?
3. What is the difference between a variable and an attribute?
4. How do you know as a researcher when it is appropriate to employ data aggregation techniques?
5. What are three questions you can ask about data?
6. What are some questions you can ask about location as you design your research project?
7. What are four useful questions to ask about analysis?
8. What is the difference between a logical data model and a conceptual model?

Additional readings and references

Bernhardsen, T. 2002. *Geographic Information Systems: An Introduction*. 3rd ed. New York, NY: Wiley.

Bolsted, P. 2008. *GIS Fundamentals: A First Text on Geographic Information Systems*. 3rd ed. White Bear Lake, MS: Eider Press.

Ormsby, T., E. J. Napolean, R. Burke, and C. Groessl. 2010. *Getting to Know ArcGIS Desktop*. Redlands, CA: Esri Press.

O'Sullivan, D., and D. Unwin. 2010. *Geographic Information Analysis*. 2nd ed. New York, NY: Wiley.

Steinberg, S. L., and S. J. Steinberg. 2008. "People, Place, and Health: A Sociospatial Perspective of Agricultural Workers and Their Environment." Humboldt State University. http://humboldt-dspace.calstate.edu/xmlui/handle/2148/428.

———. 2011. "Geospatial Analysis Technology and Social Science Research." In *Handbook of Emergent Technologies*, ed. S. Hesse-Biber, 563–91. Oxford: Oxford University Press.

Relevant websites

- **US Census Bureau, Geography (http://www.census.gov/geo/maps-data/data/tiger.html):** This website presents the TIGER (Topologically Integrated Geographic Encoding and Referencing) files. The US Census Bureau offers various file types on its website to map its geographic data.
- **Esri, ArcGIS Content (http://www.esri.com/data/find-data):** This website provides information on different types of maps, such as basemaps, reference maps, and specialty maps. It also discusses how demographic and lifestyle data can be used in Esri products.
- **US Environmental Protection Agency, Browse Data Dictionaries by Table (http://iaspub.epa.gov/sor_internet/registry/datareg/searchandretrieve/datadictionaries/browse.do):** This website presents the various elements that exist in data dictionaries and includes a complete list of all of the data dictionaries that the EPA possesses.
- **National Institutes of Health TOXMAP (http://toxmap.nlm.nih.gov/toxmap/flex/):** A GIS from the Division of Specialized Information Services of the US National Library of Medicine that uses maps of the United States to help users explore data from the EPA's TRI and Superfund Program.

Chapter 3

Research design

In this chapter, you will explore the different purposes of research. You will learn two different approaches to social research design: the deductive approach, which is underlain by the traditional scientific method, and the inductive approach, which is driven by the notion of grounded theory. You will learn a step-by-step process for designing spatial research using either a deductive or inductive approach. You will also learn key research concepts such as baseline data.

Learning objectives

- Understand the difference between inductive and deductive research
- Comprehend the important role that baseline data play in developing a GIS
- Differentiate between the purposes of exploratory and explanatory research
- Understand and learn how to determine your own unit of analysis for a research project
- Comprehend the value of triangulation to the research process
- Learn how ground truthing is central to any type of geospatial research
- Create and use a conceptual model in the research process

Key concepts

baseline data
conceptual model
deductive research
explanatory studies
exploratory research

ground truthing
grounded theory
hypothesis
inductive research
literature review

operationalization
theory
triangulation
units of analysis

What is the purpose of your research?

Research can have a variety of purposes. When doing applied research, you may often conduct **descriptive research** that shows the current state of things. Two additional purposes of research are to explore data and to attempt to explain why things are the way they are. Of course, the practical goal of many research projects is to solve a specific problem or suggest alternatives. A GIS is an excellent tool for all of these research goals, especially in situations where data from a variety of sources must be brought together. With the speed and analytical power of a GIS, it becomes feasible to explore many more analysis options and characteristics of data than might be initially expected.

Descriptive research

In a descriptive study, the main goal of research is to catalog and observe data. A descriptive study is useful for increasing understanding about something. An example of a descriptive study using GIS might be to look at the number and geographic distribution of ethnic populations in an urban area such as New York City. The goal of such a study might be to determine spatially where these different ethnic groups are located, and why (e.g., people of Italian descent are located in Little Italy, and a majority of those of Haitian descent reside in Brooklyn). Why would a researcher want to take this approach? The value of a descriptive study is that it provides detailed information that could be used as **baseline data**. Baseline data include any information used to establish a primary picture or understanding of the situation under study. Figure 3.1 presents an example of baseline data related to regional wildfires.

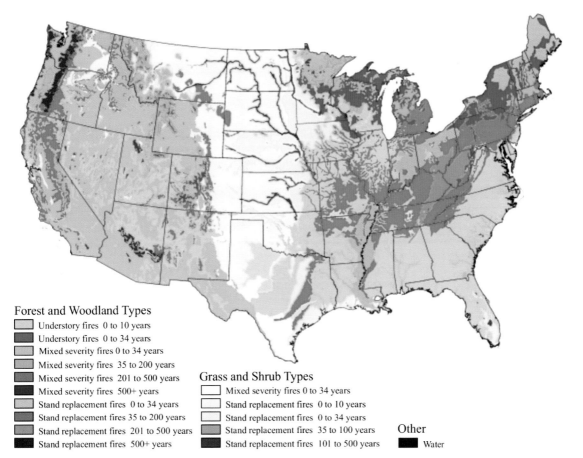

Figure 3.1 This map of North American fire regimes shows the typical period of return for wildfire. Such a dataset could be used as a baseline dataset for studies assessing changes to the frequency of fire in different regions owing either to natural or manufactured impacts to the landscape. US Department of Agriculture, Forest Service, Rocky Mountain Research Station.

Descriptive research, by nature, is informational and is often used to make policy decisions, provide services, analyze crime, track diseases, or answer a wide variety of other research-related questions. For example, by tracking the occurrences of cases of West Nile virus in each county in the country using GIS, you could determine where the virus is becoming a problem, how it is moving, and when it is likely to hit next. Using this information, you could alert local public health agencies, take action to reduce mosquitoes in at-risk locations, and develop appropriate policies for treating and preventing further spread of the disease. Figure 3.2 presents a spatial perspective of West Nile virus locations based on mosquito carriers that have been tracked and recorded. Such analysis leads to a better understanding of potential virus threats to the public.

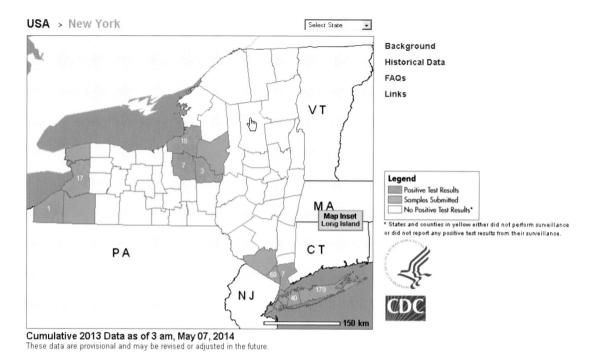

Figure 3.2 This map shows counties in New York State with mosquitoes confirmed to be carrying the West Nile virus. These data are tracked and updated on a regular basis throughout the year for all US states and territories. Data from Centers for Disease Control and Prevention.

Exploration

When a study is exploratory, there may be little known about the topic. The goal of exploratory research is to begin to develop an understanding of the topic so that you can begin to develop additional questions or hypotheses.

Exploratory research is research at the tip of the iceberg. In other words, the researcher is trying out something for the first time. An example of this would be if you lived in a community where there is a shortage of affordable housing. The word on the street might be that only "old people" can afford to live in the nice section of the city anymore. As a researcher who is conducting exploratory research, you could create a GIS map that portrays the age distribution of people in particular areas of the city.

Figure 3.3 presents a map showing the distribution of people aged older than sixty-four. To further investigate the question of whether only the older population is able to afford housing in a particular region, the next step would be to overlay these data of the elderly population with housing prices for the same region. It could be that the elderly population can afford their homes because they bought them a long time ago, when the prices were lower. You might accomplish this by using a GIS and adding US Census population data to the mix.

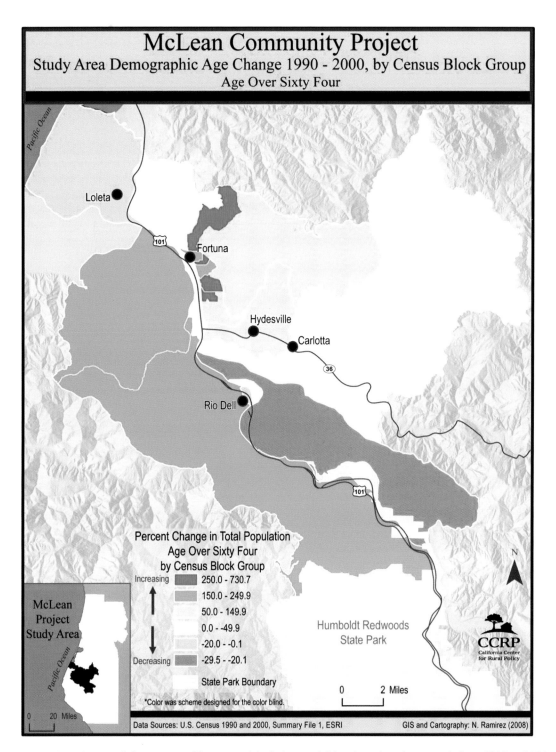

Figure 3.3 **A map of data grouped by census block, by age (older than sixty-four years), from 1990 to 2000, in the Eel River Valley near Fortuna, California.** Map courtesy of Sheila Steinberg and Nicholas Ramirez, Humboldt State University/California Center for Rural Policy, with funding by the McLean Foundation, Fortuna, CA. Data from US Census, Esri.

This type of exploratory research would be the starting point for a more in-depth study of housing issues that could be conducted in the future. Exploratory research is an appropriate place to start for the researcher who is becoming familiar with a particular topic or situation. It is also a good place to start for someone who is interested in making an empirical investigation of general knowledge held by members of the community. This provides a way for a researcher to further examine if the word on the street has any validity and, if so, to what degree the information is true.

The exploratory research process can be somewhat challenging, but also exciting, because it occurs early on in the creative investigational process. Figure 3.4 presents an exploratory process that a researcher can follow to take a research question from its conceptual form to a more operational form, where a researcher can specify the particular data he or she is going to examine.

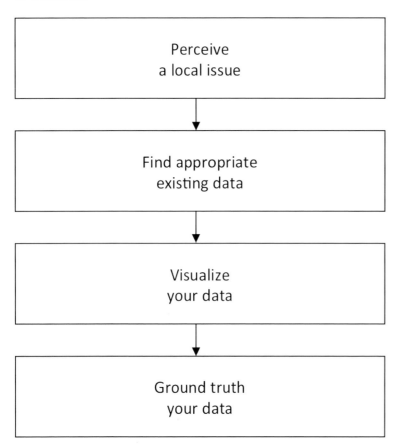

Figure 3.4 Exploratory process for taking a research question from concept to data in preparation for analysis.

Explanation

When explanation is the goal of your research, you are seeking to examine relationships between variables. A researcher conducting an explanatory study is interested in developing a scientifically based understanding of why things happen. **Explanatory studies** still involve description, but with a different goal in mind. One goal might be to develop an understanding of connections that explain why something occurs in some locations but not in others. How does GIS fit into this? It enables the researcher to consider spatial relationships as one potential component that helps to explain why an observed relationship occurs.

For example, let's say that you are interested in understanding why crime rates are higher in some neighborhoods than others. Is it due to socioeconomic issues, such as income and education? Are there spatial relationships, such as proximity to ATMs or convenience stores? Or are there negative correlations to locations of police stations, well-lit streets, or other physical factors? As part of your study, you could create a geographic picture of these variables within the areas that experience varied crime rates. The GIS will show you where the different variables intersect with higher or lower crime rates and quite possibly show that certain variable combinations have a stronger correlation with crime than others. To avoid committing the ecological fallacy (see discussion later in this chapter), you might wish to follow up the analysis by conducting interviews or surveys of residents and viewing trends in crime in the context of the national economy and crime in other areas to develop a complete understanding of why these higher crime rates occur in specific types of neighborhoods.

Of course, one must be cautious when using a GIS to avoid the temptation to draw conclusions based solely on appearance. Although visualization of the data is a powerful method for showing relationships, one must be careful to follow this up with sound analysis—be that spatial analysis in the GIS, spatial and nonspatial statistical analyses, or other means appropriate to the data. To help in this process, we provide step-by-step guidance for two different approaches to research: deductive and inductive.

Deductive versus inductive approach to research

A GIS can be useful for both inductive and deductive approaches to research. In a deductive approach, the researcher begins the research process by following a series of traditional steps. In this kind of research, the researcher begins with a review of literature, generates a conceptual framework, develops a **hypothesis**, and then tests this hypothesis by gathering data. Figure 3.5 visualizes the process involved in the deductive research approach, in which the researcher begins with an examination of the data to generate a **theory**.

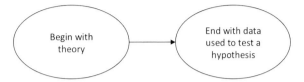

Figure 3.5 The deductive research approach.

By contrast, in an inductive approach, the researcher's understanding of the research topic and potential hypotheses emerge from the data. In other words, the researcher does not go into the study with any preconceived notions or hypotheses. Instead, the researcher begins the research process by collecting data and then seeks to develop an understanding of patterns observed. This ultimately leads the researcher to develop a theory to help explain the observed patterns. Figure 3.6 portrays the **inductive research** approach, in which the researcher begins with a theory and then collects and analyzes data to test a hypothesis related to that theory.

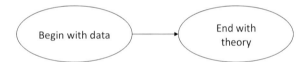

Figure 3.6 The inductive research approach.

Stages of sociospatial research for deductive research

In using a deductive research model, researchers have a clear idea or hypothesis that they want to investigate through the research process. We have developed ten steps in the research process, which is a variation of the eight steps associated with the scientific method as mentioned by Henslin (2003):

1. Choose a topic.
2. Define the problem.
3. Conduct a literature review.
4. Develop a hypothesis.
5. Develop a conceptual framework.

6. Choose research methods.
7. Collect and prepare the data.
8. Ground truth (verify) the data.
9. Analyze the data.
10. Share the results.

Choose a topic

The first step in conducting your research is to choose a topic. As a researcher, you may already have your topic chosen for you by an employer, an organization, a funding agency, or a foundation that hires you to conduct research. If you are independently choosing your topic, it is a good idea to select a subject or issue that you find personally interesting.

It is a good idea to keep a journal of potential research ideas. In this journal, you could write down research ideas or thoughts as they come to you. Keep a journal like this for a month, and then go back to it and look for themes that may have emerged. By reviewing your thoughts and ideas over an extended time period, you might find that one or two ideas surface along with developed options for further study. Of course, you may also need to consider time limitations. If you are a student, can you complete your study in the course of a semester, or will it be more like a graduate thesis, extending over a longer period of time?

If your project is related to your professional role or position, what is your time frame? Are you working within the confines of a budget or policy cycle? Is there a pressing need that you are trying to address, or is this research going to be used for long-term planning?

It is also good to pick a topic that is feasible to study. For example, you may find a comparison of nontraditional sexual practices around the world to be interesting but may encounter some difficulty collecting primary data on this topic. When evaluating the feasibility of a particular topic, consider what, if any, prior work has been done on the subject. Additionally, you should think about how to locate and collect the data necessary to complete the analysis. You can usually find some existing, relevant data related to your topic. This step is important for assessing feasibility, because you want to avoid doing research in a vacuum, without consulting past research practice, geographic locations where your topic has been studied, and prior research results related to your topic. Once you have determined a topic, you can think about choosing a study site, perhaps something close to home you can visit. To determine how a GIS would fit into your topic, think about which relevant variables can be collected in space and what boundaries of analysis are appropriate to exploring the data.

Define the problem

In defining the problem, you want to think more specifically about how you can narrow your topic. Having a good, solid definition of the problem will focus your research as you proceed with the project. We advise writing out a problem statement that includes the following four components:

1. An introduction to the topic
2. Established relevance or importance by citing the established literature
3. Background on your topic
4. A specific problem related to a particular issue or research hypothesis

To include a GIS in your definition of the problem, you need to think about what **units of analysis** you are interested in comparing. Are you interested in investigating an issue by looking at different groups who live in different places? Would it be more fitting to examine the same group over time? How you want to investigate your research question—meaning what groups, geographic locations, and issues your study involves—is important to know up front. Once you have determined these issues, you can begin collecting already existing GIS data that relate to your topic or creating data that pertain to your topic, or both.

Conduct a literature review

Conducting a literature review is an essential next step in the academic research process. Even if you are not conducting an academic research project, a review of some of the pertinent literature will help to inform and support the approaches to the topic that other researchers have used, which is important when you need to convince colleagues. The **literature review** should present relevant information from a variety of literature. A broad review is especially important if you are dealing with an interdisciplinary topic where one source does not adequately address all of the issues you are interested in studying in your analysis. Because GIS-based projects by nature draw on information from a variety of disciplines, a review of several areas of literature may be necessary.

For example, if you are using a GIS to analyze gaps in the public transportation system in a particular community, you will want to examine literature in fields as diverse as sociology, psychology, transportation, and GIS. Such sources might help in your understanding of which people use public transportation and in what ways, how public transportation systems are designed and routed, how GIS is used in modeling systems and behaviors, and perhaps other areas. Figure 3.7 illustrates how GIS can be used to portray public transportation in New York City.

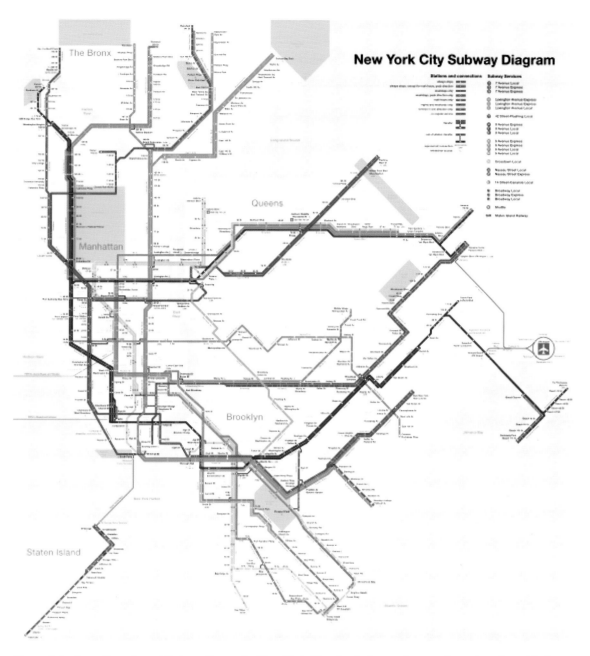

Figure 3.7 A portion of a public transit map for New York City. Analyzing such maps, in conjunction with data on demographics, ridership, crimes, or access to shopping, day care, or other services, could serve as an essential component of many studies. Questions that might be explored would be opportunities to add stops or adjust routes and schedules, increase the visibility of police on the beat, or site new social service agency offices to better serve the target population. Map courtesy of Jake Berman, http://maps.complutense.org. Subway service information courtesy of the MTA. CC-BY-SA-3.0.

Ultimately, by reviewing how others have thought about modeling or analyzing data in a spatial context—even if they never used a GIS—you will be better able to develop an appropriate model to implement in a GIS. (Note that researchers have, for many years, thought about questions spatially. The presentation of Dr. John Snow's research in London in the mid-1800s is one of the earliest published examples. Although the advent of GIS was more than a century off, the design of Snow's analysis would be easily carried out in a GIS today. For additional, classic examples of spatial concepts in social sciences, visit the Center for Spatial Integrated Social Science website's section on classic research at http://www.csiss.org/classics/.)

In writing a literature review, be sure to include a thorough review of the relevant information that pertains to your topic. This involves searching for a variety of sources of relevant information. It is always good to include many types of information in your literature review, ranging from journals to books, news stories, and websites. Check sources and credentials when relying on information from websites because it may not have been vetted for quality and credibility. It is up to you, the researcher, to critically assess the sources of information you find on the web, or anyplace else, before deciding which sources you will use.

When reviewing the literature, try to search for studies conducted on your topic using a GIS. This information may give you ideas about how to incorporate a GIS into your own study. Use the keywords "GIS" or "geographic information systems" when searching for these studies in the literature related to your study. This approach is generally more successful than going to journals focused on GIS and looking for information particular to the discipline you are analyzing; although, sometimes this approach can also yield good resources.

Your literature review should contain common themes and specific, relevant ideas and concepts and should also present relevant debates that pertain to your problem statement. By conducting a thorough literature review, you may gain insight into research methodologies, study designs, and variables that are directly relevant to your own study. You will also find that the literature can help to justify or legitimize the methods you will use in your own study.

Develop a hypothesis

Research is guided by a question or issue that a researcher or group of researchers thinks is worth investigating or needs to be addressed in some way. Regardless of your motivation, your research question is best guided by a hypothesis. A hypothesis is simply an idea about the research that seems reasonable and that you think explains the situation at hand, or it might be the underlying basis of debate about the topic. In other words, the hypothesis of your research is an educated guess about what you might find, but the validity of your hypothesis has yet to be proven.

When developing your hypothesis, think about the most important aspects of your proposed topic. A hypothesis is an educated guess about relationships you might find between variables. Once you can state your ideas about the research or analysis in one sentence, you will know that you have narrowed your topic enough to create a hypothesis. Your hypothesis should clearly state the main idea or the crux of your research project. In other words, a hypothesis is simply a statement that conveys what you might find or the state of some set of events. A hypothesis presents something that can be tested, explored, and further investigated.

The following hypotheses incorporate a spatial question and can be effectively analyzed in a GIS:

- A greater percentage of poor people than of wealthier people live in polluted environments.
- Newcomers who move to a rural area tend to cluster in areas with people of similar socioeconomic background.
- Rates of AIDS are lower among individuals with a college education.
- Individuals with strong social ties in their local communities are more likely to participate in local governance.

A final comment on the hypothesis is that it establishes what you believe to be important about your research project. As a researcher, you can think about how GIS might add to the understanding of your hypothesis by identifying key variables (figure 3.8).

Figure 3.8 **A chemical factory releases smoke into the atmosphere.** Nickolay Khoroshkov/Shutterstock.com.

Every hypothesis should have a single dependent and one or more independent variables. In the first hypothesis listed previously, a GIS could be used to take census data overlaid with pollution data from the Environmental Protection Agency's Toxic Release Inventory database. Socioeconomic information would come from the census data, and you could define (according to whatever criteria you deem appropriate) what areas are polluted (figure 3.9).

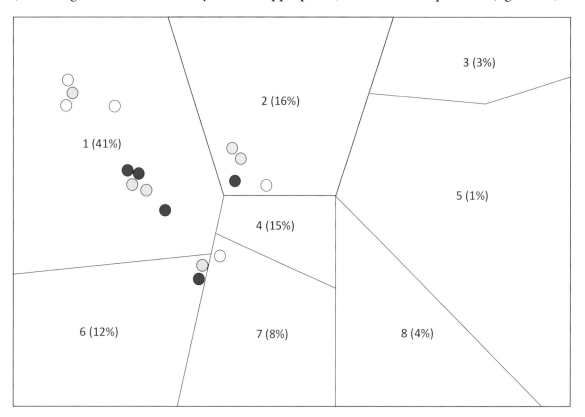

Figure 3.9 Facilities that are registered on the Environmental Protection Agency's Toxic Release Inventory are recorded as point locations in the database. Information regarding the specific pollutants released is also recorded. Here these are coded on an ordinal scale from low toxicity (white circles) to high toxicity (black circles). Assume that the polygons on the map represent the census tracts with a specified percentage of the population classified as poor (as defined from census data). We could use the GIS to determine if there is a statistical difference based solely on the point locations of facilities releasing toxic waste (do they fall in the census tract or not?) or based on other spatial concepts such as nearness or adjacency.

If you consider the spatial distribution of the pollution sources in figure 3.9, several things might be apparent. First, the majority of the mid- and high-pollution facilities are somewhat clustered to the left of center on the map in polygons 1, 2, and 7. It would be logical to assume that if the center of the map (polygon 4) were the historic city center, industrial facilities, particularly the older, higher-polluting facilities, would have grown up around the city center.

Over time, newer, cleaner facilities might be expected to grow on the edges of the city. Second, you will notice that facilities are generally located on the left side of the map, with no facilities sited on the right side of the map (polygons 3, 5, and 8). Last, we might note that although the central polygon (polygon 4) has no points inside it, and polygon 7 has only one light polluter, several other sources of pollution are just across the line in the adjacent polygons (polygons 1, 2, and 6). We will further discuss this example as we go through the remaining stages of research. As an exercise, you might also consider the other three hypotheses presented earlier or others you are interested in exploring. What considerations might be important as you operationalize this for your spatial analysis of the data in each of these situations?

Develop a conceptual model

What is a conceptual model? As discussed earlier, a conceptual model or framework is a working theoretical model that explains your view of the world. In essence, it is your chance to identify key variables in your study, to explain the links between these variables, and to explain the sequence and flow of relationships by using arrows. (We introduced the concept of creating a flowchart for GIS analysis in detail in chapter 2.)

Once you complete your literature review, you should have an easy time developing this framework. A conceptual framework is different from a literature review because it guides the research process of your specific project. The conceptual model establishes factors important to your study and indicates to all who read your study the predicted relationships between key variables.

Brainstorming is a good approach to developing a conceptual model or framework. You might start by writing down the important aspects of your study on a sheet of paper. Next, draw circles around these objects and use arrows to indicate relationships between these variables. As you identify your key variables, you can decide what type of contextual, geographic, or other information can facilitate a study of the relationships among these variables. When you identify relationships between the variables in your study, consider how geographic information might enhance your understanding of these relationships.

Consider our earlier example relating poverty and pollution. Given the spatial distribution of polluting facilities on the map, we would want to conceptualize what the actual locations really mean to our hypothesis: is presence of a facility inside a polygon all that matters, or do sites affect people for some distance (perhaps in a neighboring polygon)? If there is an effect over a distance, how does the mode of transport (through air, water, or solid waste) influence the distance and direction of the effect? What about the poverty data? Are all of the people in the census block distributed equally, or does the housing cluster in certain portions of the polygon? (This relates to issues of boundaries and the modifiable area unit problem discussed more fully in chapter 5.) The list could go on, but the important point is to think beyond what is directly visible, such as points falling inside particular polygons, to consider other factors or mechanisms that might be important in assessing the validity of your hypothesis.

A GIS is very useful for establishing how different variables relate to one another conceptually. Some GIS software programs provide a flowchart tool as a means to develop your analysis approach and, once populated with data and analysis functions, to run your model. Even when the flowchart is simply drawn on paper, it can guide you in developing a systematic or holistic approach to a particular issue under study.

For instance, suppose that you are conducting a study about an individual's attitudes about environmental issues. You may hypothesize that the nature of the community and its geographic location (e.g., proximity to unspoiled natural features, such as state or national parks or wilderness areas) influence community concern about environmental issues. Your conceptual framework would then incorporate the geographic variable of park and wilderness proximity into your model (figure 3.10). Using a GIS, you could map the locations of these natural features and overlay sociodemographic characteristics and residents' levels of environmental concern to see if there are differences.

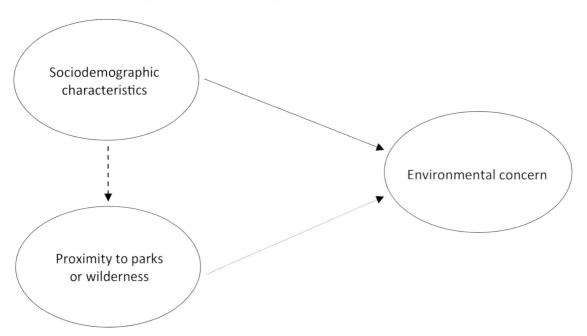

Figure 3.10   An example of a conceptual model or framework for environmental concern. Sociodemographic factors, such as age, race, gender, and income, might have some relationship to where individuals choose to live (dashed arrow) in addition to their environmental concern (solid arrow). Proximity to parks or wilderness is also expected to have an influence on environmental concern (dotted arrow).

Choose research methods

What are the factors to consider in choosing a research method? One should first consider the project goal and the sorts of data most appropriate in meeting it. Can you use existing data, or will you require new data (or both)? Should the data you use in your study be quantitative or qualitative (or both)? What will the boundaries of your study be? Research can be conducted at the local, regional, national, or global level or at any combination of the four levels. Will you be doing a descriptive study of one area or comparing multiple research locations? Are there one or several methods you might use in the collection and analysis of the data?

If time and money allow, it is sometimes beneficial to incorporate multiple research methods in studying your topic because a variety of methods adds a greater empirical angle to your study. One popular approach is **triangulation**, or cross-examination, which is simply studying the same phenomena using three different research methods. Triangulation gives the researcher greater choice in gathering information on the topic. When data collected using multiple methods all point to a similar result, confidence in your results is strengthened.

So you could approach your problem or issue using a variety of methods, such as surveys, key-informant interviews, external or participant observation, and a review of historic or archival data. You could incorporate a geographic component into some or all of these different methods.

AN ASIDE ON TRIANGULATION

Interestingly, the term *triangulation* is historically associated with geometry, not the social sciences. If you work primarily in the social sciences, you may be familiar with the term as it applies to using multiple sources or methods to arrive at a result or conclusion. However, if you do an Internet search on the term, you will find that the first several pages of resulting hits relate to mapping and geometry. The properties of triangles were well understood by the ancient Greeks and are the basis for many modern mapping and land surveying techniques. As we were writing this chapter, we realized that an unintentional relationship between the mapping and social sciences may have originated in the translation of this geometric mapping term into the social science lexicon. In mapping, triangulation refers to the process of calculating a distance to a location by knowing the length of one side of a triangle and the related angles. This is accomplished with an instrument known as a theodolite, or its modern equivalents, including the Global Positioning System or surveyor's total station (figure 3.11).

Figure 3.11 **A surveyor making assessments in the field.** CandyBox Images/Shutterstock.com.

Operationalization: Measurement

As you consider the selection of your research methods, it is important to refer back to your conceptual framework. You need to decide exactly what variables you will measure, and how. This requires that you define each of the concepts that you are using in your study and its associated level of measurement. Operationalizing a variable requires you to explain how you are going to measure the concepts mentioned in your hypothesis. Typically, the variety of data collection approaches that you employ is documented in the methods section of your study. When your methods relate directly to the creation of a GIS dataset, you need to be sure to record the data creation procedures in a metadata file, which should be included with each individual data layer.

In the example of the relationship between pollution and poverty, you might choose to operationalize pollution as shown in figure 3.9 into three simple categories: low, medium, and high. In doing so, you would need to define exactly what is meant by these terms. Doing this would be an example of translating data from a simple numerical form into a perhaps more palatable and easily understood frame of low, medium, and high. Of course, you would have to explain the data that are used as part of your categorization process for low, medium, and high. For instance, is "low pollution" something that is measured as a number of pounds or gallons released, or by a medical definition such as the odds of getting sick from exposure (e.g., one in 1 million as low; one in one hundred thousand as medium; and one in ten thousand as high)? Of course, you could come up with a variety of definitions, each viewed as appropriate or inappropriate, depending, for example, on legal definitions, regulatory recommendations, expert opinions, or local experience.

Coming up with a true definition is not necessarily as important (or possible) as explaining your definition so others can understand and interpret your results. One advantage of the GIS is that if you choose to change your definition or examine different scenarios, it is relatively easy to rerun your analysis with the adjusted definitions to determine how the results change.

Collect and prepare the data

When using a GIS as a tool for your research, collecting and preparing the data for analysis can be significant to the research process. For any kind of research, data analysis and organization are time consuming; for GIS-based projects, you can expect that 75 to 80 percent of the time and effort you expend will be used in collecting, creating, or converting data to ensure that everything is ready for the analysis. In later chapters, we extend our discussion of additional types of data and their collection and creation in GIS formats. When we say "creating data," we don't just mean making it up. Because GIS technology is fairly new to some disciplines, data may exist in hard-copy formats (paper maps, field notes, etc.) but not necessarily in a computerized, GIS-compatible format. For these reasons, the researcher may have to be persistent and a bit creative when seeking out potential sources of data. It is not uncommon to find multiple sources of what appear to be the same data, so it is essential to review and evaluate each prospective dataset before settling on the perfect source of information for your particular study.

Of course, not all data you may need will exist, so it is likely you could be collecting your own data for use in a GIS analysis. For many social scientists, data collection is one of the most enjoyable components of the research process. Why? This is the part of the research process in which researchers actively implement ideas that, until then, have only existed in their minds (and, ideally, during the two previous stages of the research process, have been committed to paper).

Data collection can come in a variety of forms. It may involve going into the field for face-to-face contact with the people in your study or examining in detail the secondary data.

This can involve distributing and collecting surveys or traveling to a faraway location to conduct a case study. Data collection is part of the research process that motivated many of us to go into natural or social sciences in the first place, and the part we have always found to be fun.

One should begin the data collection process with a good understanding of (1) what type of data are necessary for testing the hypothesis, (2) an idea of where to find this information (agencies, corporations, universities, etc.), and (3) clearly identified geographic components that best fit within the study. In this technological age, and given security concerns, some organizations and agencies are somewhat guarded about sharing detailed geographic information. However, a wealth of free, preexisting geographic information is available from a variety of data providers, libraries, government agencies, Internet resources, and others. Where you go looking for information depends on which geographic features you are looking for. (See chapter 5 for a more detailed discussion of measurement and GIS.)

Ground truth (verify) the data

Because a GIS involves technology that can use data acquired without ever leaving your office, it is theoretically possible to complete an entire analysis without ever leaving your desk. However, it would be foolish to believe such an analysis would be without flaws (especially considering the variable quality, scale, projections, data formats, etc., we discuss throughout this book). Even if all of your data can be acquired without fieldwork, it is a good idea to ground truth at key points in the process. Whenever you use GIS, you want to make sure that you ground truth, not only to ensure that the data you are using are appropriate to your question but also to validate that the results obtained in the GIS match what you find in the field. (Note that ground truthing of results, by necessity, comes after the analysis described in step 9, "analyze the data," but we discuss it here because the issues are similar for both input data and results.)

In most cases, **ground truthing** means actually traveling to the place where the study is located to get a visual on the data or results. In situations where physically visiting the site may be difficult, alternative sources might be used to cross-validate your data against a second, reliable source. For example, if you are studying urban development patterns, you might be able to use current aerial photography or satellite images to confirm that a particular location has or has not been developed. If you can go to the field, you would want to take along a map or maps that represent data you are going to use in your GIS analysis and make sure your maps are both valid and accurate. Of course, you cannot check everything; but simply spot-checking or sampling from the map can go a long way toward ensuring that you start with good data.

In our pollution example discussed earlier in the chapter, ground truthing could help to answer questions about the form of the pollution and where it travels. For example, observing a smoke plume from a factory may show that a prevailing wind carries the smoke in a particular direction most of the time. Or a drive through each census block could tell you if the residents are evenly distributed within the polygon or if they are clustered in specific portions.

Finally, ground truthing is essential at the conclusion of an analysis when you are evaluating the results. Most GIS-based studies will pinpoint certain locations as having met a set of criteria. Again, spot-checking in the field can go a long way toward telling you if the results look correct and if they make sense. If your study is large, you may want to spot-check a number of sites to assess the overall accuracy of the analysis (e.g., if nine of ten results are correct when checked on the ground, it would be reasonable to estimate accuracy at 90 percent). The topic of accuracy assessment in spatial data analysis could fill an entire book, so we only mention it here and encourage those who are interested to pursue this topic independently.

Analyze the data

In this step in the research process, you must again refer back to your primary research question and conceptual framework. The form of the analysis you select will in part be determined by the type of data you have collected and prepared in the GIS. Some analysis tools work exclusively on vector data and others exclusively on raster data. Of course, you can accomplish many operations with either. Furthermore, if you collected quantitative data, chances are you will use some form of statistical analysis. Many of the common, descriptive statistics are available within the GIS software. However, it is also common to enlist other statistical programs for some portions of the analysis. This is accomplished by extracting key information from the GIS using the spatial tools and exporting the raw numbers into a program such as SPSS, S-PLUS, R, or Excel to further analyze the information before returning it to the GIS to make maps of the results.

If you are working with qualitative data, you may require use of a data program designed specifically for qualitative analysis, such as HyperRESEARCH by Researchware, NVivo, or ATLAS.ti. In identifying your key variables, you can decide which types of contextual or geographic information might aid in studying the relationships between these variables. You can then analyze the geographic information you have collected as part of your study using a variety of methods.

Regardless of the type of data, the most important thing to keep in mind at this stage is to let your project dictate the analysis rather than let the software drive the analysis simply because the software contains a menu option or button. Retuning to your conceptual framework will be essential at this stage because the framework is your guide in the analysis. Chapter 11 provides more detailed discussion of completing the data analysis using GIS.

Share the results

For applied research, this is perhaps the most important step in the entire research process. You can share results through two basic avenues and for two different audiences: laypeople and members of the scientific community. If you are interested in sharing your findings with laypeople, you should package the results in a manner that is easily understood without a lot of scientific jargon.

We have found that producing a report that discusses the methods and highlights the main findings of a study to be most effective. You might also consider developing a visual presentation, which could be given in a variety of settings that highlight both your research methods and your main findings. If you plan to share your results with members of the scientific community, you will most likely want to write a paper for submission to a scientific journal and possibly also present your results at a professional conference.

GIS can be very useful in sharing your results because it allows for presenting data in a visual format. Most commonly, the visual output is a map, but GIS can also provide output as charts or graphs or, if done as an interactive presentation, as animated or interactive maps. Not surprisingly, any of these visualizations can have wide-ranging applicability in a variety of contexts—"a picture is worth a thousand words."

Grounded theory: GIS using an inductive approach

As a researcher, you also have the option of employing an inductive model in your research design. This type of approach begins with a different series of steps than those traditionally used for a deductive approach. An inductive approach begins with the data and proceeds to gleaning an understanding of themes and patterns. From this information, theory is then generated.

Grounded theory is an inductive research approach characterized by its sequencing: data collection followed by theory generation. It is called "grounded" because of its strong connection to the reality that the data represent. This inductive research approach is qualitative in nature. Grounded theory is an appropriate research method for assessing case studies, transcripts, oral histories, and archival data.

Glaser and Strauss (1967) first coined the term *grounded theory* in the late 1960s, in their seminal book, *The Discovery of Grounded Theory: Strategies for Qualitative Research*. Since that time, many other qualitative researchers have adopted and written about grounded theory. Grounded theory has become a popular approach embraced by a variety of disciplines, including public health, business, and criminology, just to name a few.

The key to determining if you will use grounded theory is to consider the purpose of your research. One of the primary attractions of grounded theory is that it provides the opportunity to "generate theory that will be relevant to [scientists'] research" (Glaser and Strauss 1967, vii), unlike verifying theory, which is used when following the traditional scientific method, which is deductive. Grounded theory is a good approach to employ when you are interested in the discovery phase of gathering information, because it is more appropriate for researchers whose goal is to generate information, themes, and patterns, not to prove theory.

The main premise of grounded theory is that theory emerges from an examination of the data. Rather than the researcher dictating themes and ideas that will be investigated, the data dictate what is relevant and important to study further. "Grounded Theory is based on the systematic generating of theory from data that, itself, is systematically obtained from social research" (Glaser 1978, 2). Thus, the grounded theory approach views research methods as part of the theory-generating process. The process is iterative; the researcher is constantly conducting analysis, looking for themes, and then conducting more analysis. It is a very hands-on approach to sorting through data.

The core of grounded theory is in analyzing and looking for patterns in the data. In the analysis, the researcher attempts to achieve *theoretical saturation* (Dey 1999). Theoretical saturation means no additional themes or concepts, categories, or relationships emerge from the data, which can only be achieved after the researcher has made a series of run-throughs with the data, identifying themes and looking for data that support the themes. Bernard (2000, 443) summarizes how grounded theory can be accomplished using the following series of steps:

1. Begin with a set of information (e.g., interviews, transcripts, or newspaper articles).
2. Identify potential themes in the data.
3. Pull data together as categories emerge.
4. Think about links between categories.
5. Construct theoretical models based on the links.
6. Present the results using exemplars.

Following these steps, you begin with whatever set of data or information you want to analyze. This information will most likely be of a qualitative nature. Identifying potential themes in your data can be done by hand or with the help of a qualitative data analysis program. As you sift through the data, certain words or phrases will begin to emerge consistently. You can then use the themes that you identify to develop a coding scheme (see Strauss and Corbin 1997) to complete the analysis of the themes. Step 3 calls for grouping, or categorizing, your information. In essence, you look for similarities, differences, and repetitions that occur in what has been stated. This is an iterative process that evolves as you analyze the data using your own specific coding process. (For more specifics on coding your data, see Dey 1999.)

Step 4 calls for thinking about the links between the grouped categories that you have seen emerge. This is akin to developing a conceptual framework or model. (See the section "Stages of sociospatial research for deductive research," step 5.) This leads to the next natural step, which is to construct a theoretical model based on the links that you observed. This is your best model of the relationships that you saw emerge between themes that you identified in the data. Finally, in step 6, you present the data using exemplars. These are nothing more than quotations or snippets from the data that illustrate the themes, concepts, or relationships that you are discussing. You can think of exemplars as examples (shared words, quotes, etc.) of concepts or themes that emerge from the data analysis process.

Sociospatial grounded theory using GIS

To date, GIS has rarely been incorporated into analyses that explicitly use grounded theory. We believe that the spatial information GIS provides can be an important additional component to research that adopts an inductive approach. The visual patterns in spatial data can prove a powerful indicator when you are exploring emergent themes drawn from existing data to develop theory. We have devised a series of steps that you can follow when using GIS as part of this approach to social research:

1. Determine a topic of interest.
2. Determine a geographic location of interest.
3. Collect the data (qualitative, spatially linked social data).
4. Geocode the data.
5. Ground truth the data.
6. Analyze the data and look for spatial and social patterns.
7. Generate theory (spatial and social).

These steps are similar to the ten steps of the deductive research process, except these are taken by incorporating a different approach. However, some of the initial steps are the same, and the same basic advice applies here.

Determine a topic of interest

In choosing your topic, you want to pick one that you find interesting. This will ensure you enjoy the research process and approach it with energy and enthusiasm. In considering your topic, you should also consider what might be feasible area of study for you considering time, money, and interest. Having a lot of time, money, or resources substantially influences your research method selection; for example, you might select detailed interviews. If you have less time, money, or resources, you may need to rely more heavily on available data.

Determine a geographic location of interest

Determining the geographic location of interest means that you identify a study location that is associated with your topic. This could be a neighborhood or county or something less defined, such as former residents of a community that no longer exists. In New Mexico, there was one such community, called Santa Rita. It was located next to an open pit mine, and when the mine expanded, the town site became part of a giant hole that was the mining pit. Many of the town's homes were moved to other surrounding towns, and ex–Santa Rita residents moved into them. A study today of these residents' perceptions of the town would be conducted about a place, the town of Santa Rita, that no longer physically exists (figure 3.12).

Figure 3.12   **Photograph of Santa Rita, New Mexico, circa 1945, prior to its relocation for expansion of the open pit mine.** Photographer unknown.

Assuming that you can track down the former residents of this town, you could conduct your interviews with these individuals regarding their perceptions of the town. You would also want to note where these individuals now live. Why? Because the geographic locations where former Santa Rita residents currently reside may be a factor that corresponds to their individual perceptions about the former town. For instance, one question you might ask would be, Do residents who live within five to ten miles of the old mining pit (Santa Rita town site) share different perceptions of the community than individuals who moved farther away from the town? Or do all residents, regardless of current geographic location, share the same perceptions of the town? The point we want to make is that the physical locations of a place can play an important role in the analytical process when using grounded theory.

Collect the data

When you collect your data, even when using a grounded theory approach, you can simultaneously collect information about the spatial surroundings. Why do we advocate this? Our reason for collecting both types of data comes from a philosophical belief that the social and physical environments interact with and affect one another. The degree to which a researcher employs a dual data collection process will be only determined by the researcher and her or his preferences.

It makes sense when employing a grounded theory approach to collect information on the geographic location and the natural environment related to your data. Why? Including this type of information could greatly enhance the emergence of themes, ideas, and relationships that exist in your model. In fact, it may lead to the inclusion of physical, social, and environmental features in your theoretical model, features you had not previously considered. That is where grounded theory and GIS are quite compatible. Grounded theory is flexible enough to allow for including and identifying a variety of different data types, including geographic information.

Geocode the data

When you spatially code the data, you are assigning a code that reflects the geographic location of your data. For example, say you are interested in analyzing different newspapers' coverage of the issue of immigration. In your content analysis, you choose to analyze newspaper articles from various locations around the country. As a part of your grounded theory analysis, you could code the location of each newspaper and note other attributes about the community in which it is published to see what kinds of patterns emerge in your data analysis. This would enable you to determine if the physical environment and population are related to perceptions of immigrants. If you fail to collect this information and treat all of the newspapers as the same, you may be missing a key explanatory element for your theoretical model. In conducting your analysis, it would be interesting to observe if differences in attitude toward immigration emerged between various newspapers' coverage of these issues. A content analysis of the data mentioned would reveal some themes and patterns that could then be crafted into a theoretical model. Figure 3.13 presents an example of content analysis done by categorizing relevant newspaper articles into themes, which were quantified and mapped to present a spatial analysis of where, geographically, certain topics are discussed.

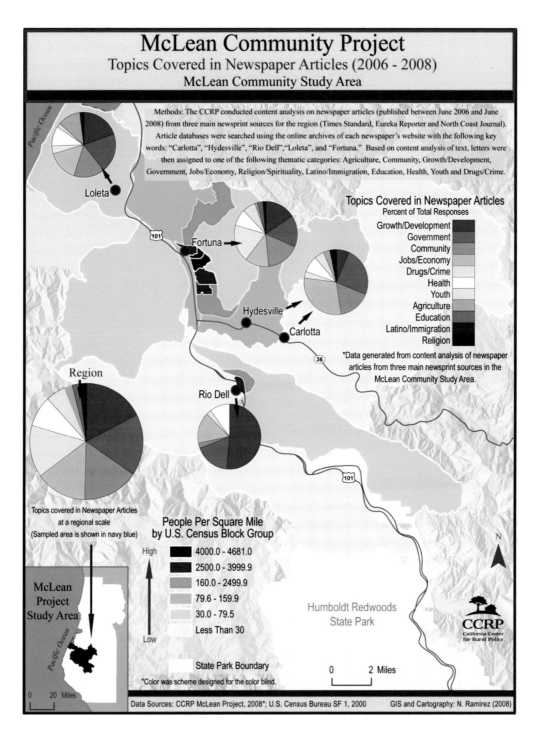

Figure 3.13 This map shows the results of a content analysis of newspaper articles for the Eel River Valley near Fortuna, California. Themes identified in each article are represented by pie charts associated with each town in the valley, overlaid with population density. Map courtesy of Sheila Steinberg and Nicholas Ramirez, Humboldt State University/California Center for Rural Policy, with funding by the McLean Foundation, Fortuna, CA. Data from US Census, Esri, Sheila Steinberg/McLean Community Project.

Ground truthing the data

What do we mean by ground truthing? Ground truthing involves checking to ensure that the computerized data you have are representative of what exists on the ground. How do you ground truth data? Most often, ground truthing is accomplished by physically visiting the location under study and field checking a subset of the data. In cases where this is not possible (inaccessible or unsafe location, historic data), alternative sources may be used as surrogates for ground truthing (e.g., phone books, property tax listings, historic records, and aerial photography). Ground truthing data is perhaps one of the most important steps of integrating GIS into social science research. Why? Anytime you are dealing with GIS data, although it may be tempting, you may not accept data at face value. You should check (at least a sample, if not all of) the data to ensure that they are without major errors and that they represent the information required at the level of detail necessary for your analysis. Most problematic are datasets that are not current or that were originally collected for a different purpose. For example, you might have a dataset from several years ago detailing the location of soup kitchens in a city. It would be wise to visit these locations to ensure they are still active or, if this isn't possible, to check the locations in a current phone book. Errors may also arise when data from multiple sources, scales, or projections are combined in your data analysis. Successfully employing grounded theory requires that the data informing the theory be free of major errors, because reliance on flawed observations or data can produce a theory that does not actually fit the situation under study. If your observations do not reflect reality as defined for your specific analysis, you risk a serious problem, especially when it comes to geographical data (e.g., the location of something).

Analyze the data and look for spatial and social patterns

As mentioned earlier, a major part of the grounded theory approach is the search for patterns in your data. Ideally, if you want to integrate a GIS into your information-gathering process, you connect each piece of data or information with its geographic location. For example, say you have a collection of historic diaries from the 1940s. As a researcher, you might be interested in understanding how World War II influenced people from that time period. You could employ grounded theory in your analysis of the diaries and simultaneously develop a coding system that notes the geographic locations where the diaries were recorded.

For instance, individuals living in different parts of the United States may have had very different experiences during the war, depending on myriad factors. For example, were military bases located nearby? Did the diaries' authors live in regions populated by particular ethnic groups from parts of the world viewed either positively or negatively because of the war?

Keeping track of the geographic information as you conduct your qualitative data analysis may reveal geographic patterns in the data indicating that location plays a role in the attitudes people express in their diaries. Certain geographic areas may have been harder hit by rationing, may have had a greater number of local men and women who went off to war, and so on. None of this would be obvious at the outset of a grounded theory analysis. However, if you keep track geographically of where the diaries were recorded, your analysis could produce some interesting results.

One might ask the logical question, How do you know when you have sufficiently analyzed the data? Dey (1999) provided a clear summary of the analysis process using grounded theory. He noted that researchers should conclude their research when they reach theoretical saturation, identify a core category or main story line, integrate the analysis around the main story line, and then use the coded information to modify the results, stopping the process with the emergence of a useful theoretical model (Dey 1999).

Generate theory (spatial and social)

Generating theories is the most creative part of the grounded theory approach. At this stage, you get to generate a theoretical model that reflects the patterns that you observed in your data. As mentioned in the previous step, the geography—or, rather, spatial location—associated with data may factor significantly into the theoretical model that you generate. In any of the examples provided in this section, the variable geography could potentially play an important role in the analysis. When you construct your model, you should indicate whether the physical, social, or environmental context, or all three, factored into grounded theory that was generated through the research process. Figure 3.14 illustrates how geography can play a role in people's perceptions of World War II, as evidenced in their diaries.

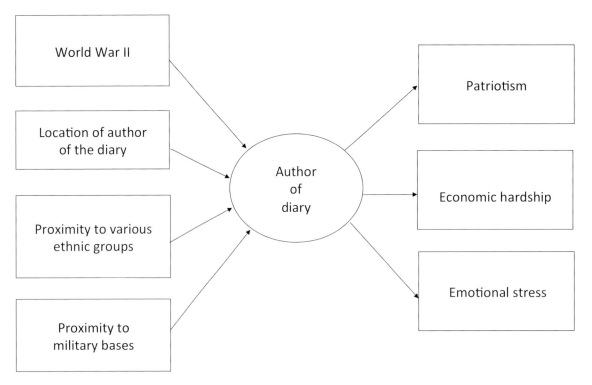

Figure 3.14 A variety of variables, including environmental, geographic, and social, are believed to affect the authors of the diaries assembled for an analysis using a grounded theory approach. In some combination, these might influence the themes that emerge from the diaries and help to inform the eventual theory suggested by this research.

Conducting a grounded theory analysis generates the following emergent themes: economic hardship, emotional stress, and patriotism. Perhaps those living in the heartland felt less stressed than those on the coasts, who feared attack by German submarines or Japanese kamikaze planes. Those near military bases might have felt a greater sense of stress because many of their friends and loved ones were directly involved in the war effort. Areas of the country with concentrations of ethnic groups tied to the Allies may have felt a greater sense of hardship or stress. This figure illustrates grounded theory in action from a historic perspective.

Moving forward

In this chapter, you learned the various purposes of research. You also learned two approaches to conducting research: deductive and inductive. Understanding various approaches to research further enhances your ability to tackle issues, problems, or situations effectively and with the most appropriate research methods. As part of your examination of inductive research, you learned about the notion of grounded theory. For the discussion of deductive research, you investigated the standard scientific method, which is commonly used across many disciplines. The information presented here has equipped you with a solid grounding in how to establish the framework for a research process. In the next chapter, you will learn about sampling and boundaries for research projects.

Review questions

1. If you are going to design a research project, how do you decide if you should use the inductive or deductive approach? What are the factors that will affect your decision?
2. What role do baseline data play in the research process? Do you have to have baseline data? Why or why not?
3. List the steps in sociospatial research for deductive research.
4. What is sociospatial grounded theory? Describe the steps in this process.
5. Why would you consider triangulation as part of your research process?
6. What role does a conceptual model play in the research process?

Additional readings and references

Babbie, E. 2013. *The Practice of Social Research*. 13th ed. Belmont, CA: Wadsworth Cengage Learning.

Bernard, H. R. 2000. *Social Research Methods: Qualitative and Quantitative Approaches*. Thousand Oaks, CA: Sage.

Brown, J. K., and J. K. Smith. 2000. *Wildland Fire in Ecosystems: Effects of Fire on Flora*. General Technical Report RMRS-GTR-42, vol. 2, 40, 56–68. Fort Collins, CO: Rocky Mountain Research Station, Forest Service, US Department of Agriculture.

Dey, S. 1999. *Grounding Grounded Theory: Guidelines for Qualitative Inquiry*. Bingley, UK: Emerald.

Glaser, B., and A. Strauss. 1967. *The Discovery of Grounded Theory: Strategies for Qualitative Research*. New York: Aldine Transaction Press.

Henslin, J. M. 2003. *Down to Earth Sociology: Introductory Readings*. New York: Free Press.

Steinberg, S. J., and S. L. Steinberg. 2011. "Geospatial Analysis Technology and Social Science Research." In *Handbook of Emergent Technologies*, ed. S. Hesse-Biber, 563–91. Oxford: Oxford University Press.

Steinberg, S. L., D. Zalarvis-Chase, M. Strong, and N. Yandell. 2008. *California Center for Rural Policy McLean Community Study Final Report*. Arcata, CA: Humboldt State University.

Strauss, A., and J. Corbin. 1997. *Grounded Theory in Practice*. Thousand Oaks, CA: Sage.

Relevant websites

- **Temple University Libraries: Qualitative Research: Grounded Theory** (http://guides.temple.edu/groundedtheory): This website explains grounded theory and discusses the advantages and disadvantages to using it.
- **Esri Video: Developing a Conceptual Framework for Geodesign** (http://video.esri.com/watch/193/developing-a-conceptual-framework-for-geodesign): This website links to a video that explores how to describe the concepts and relationships between the concepts that are part of your research project.
- **"Data Collection–Primary versus Secondary"** (http://dspace.library.uu.nl/handle/1874/23634): This article, by Joop J. Hox and Hennie R. Boije, University of the Netherlands, discusses the differences between primary and secondary data types and their collection.

Chapter 4

Research ethics and spatial inquiry

In this chapter, you will learn about research ethics and the unique opportunities and challenges that incorporating spatial research methods bring to a project. First, you will learn about the ethics involved in doing research in a spatial context. A number of social, cultural, and political considerations are important to keep in mind, particularly when data, which may be sensitive, are linked to a map. You will then learn about existing or secondary data sources and potential errors that can arise when using these owing to the data, processing operations within the software environment, or other human errors that may arise in the research process. Finally, you will learn approaches for maintaining confidentiality and anonymity of data and for masking data, as well as research data management considerations.

Learning objectives

- Understand research ethics and the central role it plays in spatial research
- Learn about the social, political, and cultural aspects of research ethics and spatial research
- Learn what the ecological fallacy is and how to avoid committing it in the research process
- Examine how ethics impacts data collection, sharing, and storage

Key concepts

accuracy issues
anonymity
confidentiality
cultural implications of research
data aggregation
data sharing

data storage
ecological fallacy
errors caused by analysis
errors in human inquiry
existing data source
inaccurate observations
overgeneralization

political implications of research
research ethics
selective observations
social implications of research

Research ethics and GIS

Research ethics relates to principles and practices that guide the way people carry out their research. Topics range from how a researcher recruits participants for a project to how data are gathered, stored, analyzed, and ultimately communicated and shared. For instance, with whom will you share your research data, and at what level of detail? Will your results be reported in your research findings? A variety of issues fall under the umbrella of research ethics, and in the context of spatially based research, several additional factors should be considered regarding the scale of analysis and detail of location-based data collected and reported.

You should consider research ethics anytime you perform research, especially when it involves potentially sensitive or personal information. A failure to consider research ethics may put people at risk. In some research contexts, documented procedures or professional practices may be available to help guide your work. Institutional review boards, ethics panels, or requirements of the research sponsor will often provide detailed guidance and review of a research plan. However, in some contexts, formal ethics protocols may not be in place, and even when they are, they may not address your specific research plan or its potential pitfalls. Although anyone can do research, not everyone does research in an ethically correct manner, and in a spatial context, powerful software and data collection tools may necessitate additional ethical considerations. Failure to incorporate research ethics may call into question the motives of your research project and the findings resulting from your work.

Many studies in social science include information that is personal and thus must be treated with care and confidentiality. When you store data in a GIS, this can be especially important because the information is linked to a specific location. You should consider how the data might impact particular groups or individuals relating to the issue or topic under study, or how details regarding locations of sensitive information on environmental or built infrastructure might negatively impact those resources. For example, specifically reporting the locations of individuals of a particular minority population could provide information to those who may seek to single out or oppress that population. Similarly, reporting locations of an endangered species of plant or animal, while potentially helping to protect the resource, could also provide poachers the information required to collect them for their own personal gain.

In traditional data collection, it is relatively easy to decouple personal data from an individual respondent by assigning a random ID code to the responses. Simply removing identifying information such as names and addresses can preserve privacy. However, when these same data are linked to a map, a respondent's address or other locational information can be obtained by looking at the map. Therefore, data scale or generalization of locations is often used as a means to obscure such details. For example, results could be mapped by city or postal code rather than at the scale of a specific street address or neighborhood.

As we discussed in chapter 3, research is built on collection and analysis of data to identify patterns. As part of this process, it is important to be thoughtful about how you go about collecting and analyzing data. Many research studies have the potential to result in positive outcomes, but an ethical researcher will also consider potentially unintended, negative outcomes and strive to ensure that work is conducted with an eye to ethics and professionalism.

Social implications

Essentially all research has potential for direct or indirect social implications stemming from the work itself or from the findings resulting from the study. The term **social implications of research** refers to the impact that research can have on various aspects of society. In other words, what does the research mean, or could it mean, for society? Whether it is a health study or research on the use of particular natural resources, research typically ties back to some societal question or interest, giving the work social relevance. Spatially based research brings an added potential for ethical questions because it may be used to document and highlight where social inequalities occur. Additionally, mapping where specific resources (human, capital, or natural) are located and who owns these resources is naturally important to people, and this information has the potential to be exploited in ways that may not be intended by the researcher. For example, identification of a valuable natural resource on land occupied by a group lacking political influence might lead to their displacement by individuals with financial or political power. Geographically based spatial research opens up new areas for discussion about how to meet the needs of society by bringing people and resources together. This theme runs throughout many different kinds of research, ranging across such varied disciplines as business, marketing, health, natural resources, and social services. We revisit these topics in more detail in chapters 8 and 9.

Cultural implications

The term **cultural implications of research** examines how the research can impact a particular culture. An example of a cultural study might involve trying to understand the boundaries and locations of sites that are important to a particular group of people, such as a tribe in the Amazon forest. This tribe claims that they have been using certain areas of the forest for many years to conduct their sacred rituals and ceremonies. Through asking the right questions using a survey or public participation GIS (PPGIS), you are able to determine where the tribal boundaries exist. In chapter 7, we present approaches for conducting spatial surveys, and in chapter 8, we explain how to engage in PPGIS. Through these processes, the researchers can interview members of the tribe. Ethical considerations factor in when you decide which details you will include and which you will leave out or obscure in the final report. For example, in a final report that might be given to other researchers or the public, you would not want to include a map showing the exact locations of tribal sacred sites, burial grounds, and special hunting grounds. If such a report were to go public, the result could be that poachers, grave robbers, and thieves interested in stealing tribal artifacts would gain knowledge of these locations through your map and, as a result, pillage the resources. This is an example of failing to consider sound spatial research ethics—because you failed to account for and protect the privacy of information about sites sacred and important to these forest dwellers, information that should not be shared publicly. Sharing information publicly which should have been kept private may not involve breaking any law but can certainly involve breaching the ethical boundaries of what is considered appropriate. Through making public something that culturally was supposed to be kept private—only known to the members of the tribe—you are putting an important aspect of the culture at risk. This is something you never want to do, because it can cause harm to the people who participate in and shared information as a part of your study.

Political implications

Data alone are not political. However, the ways that individuals or groups use data can be political. The term **political implications of research** refers to the possible political effects or impacts of the research. Sometimes people or groups will refute data that present facts simply because that information does not align with their point of view. A common example is when communities are assumed to possess specific political orientations. Various regions of the country have been long described as red states (Republican) or blue states (Democratic). This could lead someone to believe that everyone in California is a liberal and everyone in Texas a conservative. While these generalizations do relate to which party these states voted for in the past several elections, it is a drastic oversimplification to assume only Democrats live in California and only Republicans live in Texas. The reality is you can find people with a wide range of political views in every state, and digging into the details of the data at a local scale would help to clarify this fact. Nonetheless, this is an interesting example of the political aspect of data, because what we find is that even when data speak the truth about a situation, that truth may be challenged or even acted on based on what people believe rather than what the data actually show. This certainly plays out in national politics as presidential candidates choose to concede some states while focusing their campaigns on areas where they perceive demographics are more politically diverse.

Another issue to consider in terms of the political implications of data is the level at which politically charged data are collected. Incorporating a geographic variable relating individuals' responses to a specified geographic area can present certain challenges. Individuals may be less inclined to provide responses to a survey or other data collection attempt if they believe their privacy may be compromised. In other words, if a person is sharing detailed, personal data with you, such as level of income, the person might be afraid that if these data were to be shared publicly, the person could be negatively impacted.

There may be potential for marginalizing certain people in the town if community members learn about a person's opinions. For example, if you were studying individuals' opinions on gun control, you could use GIS to link responses to specific houses or neighborhoods, thus identifying which neighbor supports gun control and which does not. Because of the emotional nature of this topic, doing so could lead to unanticipated outcomes or conflict between neighbors. It would be preferable to mask responses at the household level to protect the privacy of individual respondents. You can accomplish this by visualizing the data at a sufficient level of generalization to mask individuals, for example, at the level of the ZIP Code, county, or state.

A consideration for political issues is that presenting this information publicly could have implications for the people who are tied to those data. In some cases, this is appropriate or required by law. For instance, the Federal Election Commission reports campaign contributions by name, occupation, and ZIP Code, making it simple to identify how much an individual has donated to a particular candidate's campaign (figure 4.1).

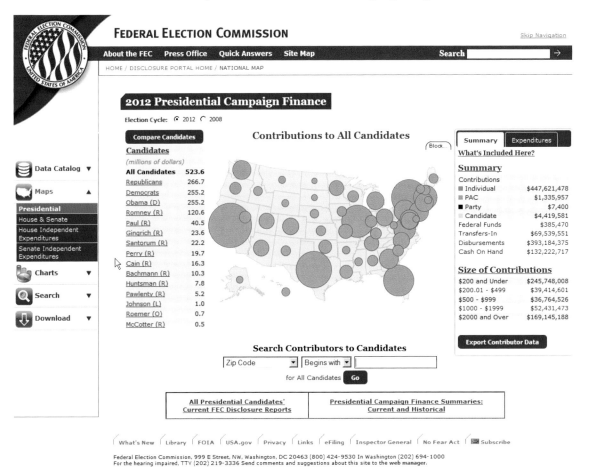

Figure 4.1 A web-based map of campaign contributions searchable by city, ZIP Code, or name. Data are also available for download. http://www.fec.gov/disclosurep/pnational.do. Federal Election Commission.

Similarly, specific geographic data tied to criminal behavior may be available via public website, for example, data regarding sex offenders are provided on Megan's Law websites in many states (figure 4.2).

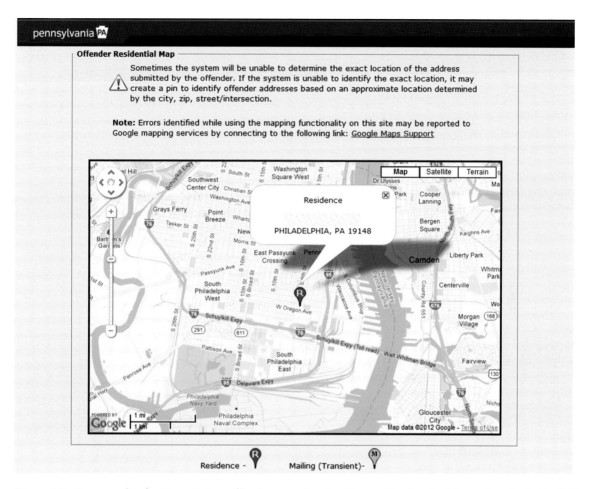

Figure 4.2   An example of a Megan's Law offender map showing the specific street address of a selected individual (hidden here). Such systems allow searches for offenders within a distance of a particular address or other features (e.g., a school, a park, or at a certain distance from a given street address). Courtesy of the Commonwealth of Pennsylvania, Pennsylvania State Police. Map data © 2012, Google.

The US Census collects geographic information with all of a person's personal information, but it does not publicly present this information; rather, it aggregates the data. However, we should note that after seventy-two years, these data connecting people and geography are made publicly available. In the case of the US Census, the underlying assumption is that after seventy-two years, enough time has passed that respondents are no longer living.

Errors caused by analysis

In the course of your data analysis, some errors may creep into your results. We use the term **errors caused by analysis** to refer to errors that emerge through the analysis process. Errors may be caused by a variety of issues ranging from major blunders in data entry and processing to subtle errors that go unnoticed. Typically, the major errors will be readily apparent and, in most cases, easier to identify and correct. These might be the result of an incorrect processing step or inadvertent use of the wrong data file. Such errors may result in empty output files or, if output is generated, results that don't make sense. For example, an intersection operation using data from two different counties would result in a blank or erroneous output because, by definition, counties would not be expected to share the same geographic space. In the following sections, we focus on errors that are more subtle in nature, errors resulting from the data sources.

Existing data sources

When working with existing, secondary data sources, you should pay attention to a number of factors relating to data format, quality, and utility. The term **existing data source** refers to a source of data that have already been collected and that already exist prior to you beginning the research process. Regardless of whether the data are spatial or tabular, it is important to review the metadata or other related data documentation for existing sources. Are the data you are using from the original, raw dataset, or have the data been summarized and aggregated in some manner? A good example of aggregated data is data obtained from government sources such as the decennial census. Data in the census are aggregated to the enumeration areas that represent groupings of individuals. Other data sources may provide data for similar geographies but develop data attributes from sampled data. Sampled data can be developed using a variety of approaches based on different sampling frames, sample draws, number of samples, and so on. Both complete census and sampled data types are valid.

Where you as a researcher must be careful is in selecting and understanding the data sources you incorporate into your research. Data that include complete, detailed metadata will allow you an opportunity to understand the nature and intended purpose of the dataset. Then, on the basis of your own research objectives, you can determine if and how combining these data into your work is appropriate, and equally important, how doing so will affect the interpretation of your results. Of course, some data you locate may not be well documented. Metadata may be missing, incomplete, or presented in a nonstandard format. Although metadata standards have improved and formalized over time, if you are working with historic data or data generated without these standards in mind, you may need to delve deeper to locate this information. For example, reviewing the methods section of a final report associated with such data may illuminate the sample design and data processing that led to the resulting dataset you are considering. In other cases, you may need to review agency documentation to locate the protocols and standards used for mapping or data collection. It is even possible that you will need to track down the individual responsible for the data in question and ask him or her directly for information about the study.

If efforts to obtain documentation of a secondary dataset fail, of course, this doesn't preclude you from using these data in your analysis; however, failure to secure the necessary information to validate the dataset should raise a red flag, leading you to stop and consider if using the data anyway is your best option. Could you find an alternative data source? Is it possible to collect or develop your own, new data? Although we cannot go back in time to recollect historic data, there are some possibilities for working around this problem. Incomplete maps of historic land cover and land use, for example, could be supplemented with new data, interpreted from historical aerial photography of the area. Other forms of data, such as surveyor's notebooks (which often predate aerial imagery) or oral histories passed down through generations of the local people, may also provide alternatives. Of course, these approaches are not a perfect fix and may require a significant investment of time and resources, so using an alternative to an imperfect or poorly documented dataset does come with its own issues.

Even when data are well documented and appear to perfectly meet your research objectives, they will, in fact, not be perfect. All datasets, whether the ones we develop ourselves or those we acquired from existing sources, have some error associated with them. Although careful researchers will do their best to eliminate or reduce errors, attributes may occasionally be miscoded or incorrectly calculated. To this end, it is reasonable to consider the source of the data. Do they come from a trusted and reputable source? Have the data been vetted by others in the field? In many fields of research, certain data sources (government agencies, universities, or other research organizations) are viewed as accepted and authoritative. Originating from such sources is a fairly good indication that the data will meet the needs of your study. In contrast, data from interest groups or individuals not affiliated with a known, reputable organization may require additional scrutiny to ensure they are not biased to support a particular perspective or agenda. Of course, there is no guarantee data from a specific organization are automatically more or less reputable than data from other sources. Ultimately it is up to you, the researcher, to carry out due diligence in selecting data and assessing their suitability for a given purpose.

Variations in data accuracy, scale, and temporal change

In chapter 2, we discussed a number of accuracy issues related to the spatial component of GIS data. The term **accuracy issues** refers to the accuracy of results in the project. Choices of mapping scale, projection, coordinates, and datums can all contribute errors in the mapping process. These issues can occur at several points in the data development work flow, starting from the original fieldwork or image interpretation from which the maps are made, through to various stages of data entry and processing in creating a GIS data file. As described in the previous section, reviewing the metadata of a given dataset to get an understanding of how the spatial data were acquired, prepared, and processed can provide important information that will help you determine the appropriateness of the dataset. Although the integrity of each individual data layer is an essential consideration, it is not the only important one. In a typical GIS analysis, we may use several data layers, which we compare, combine, and query in combination.

When data are combined in GIS, it is easy to assume that differences observed are real (figure 4.3). Although ArcGIS software will help us by aligning datasets to the same coordinates, projection, datum, and scale, this does not necessarily mean the underlying data were originally mapped that way. Overlaying a US Geological Survey quadrangle developed at 1:24,000 scale on a local city map at 1:1,000 may not be a good idea if your objective is to measure differences of a few meters. The quadrangle is not of a sufficient accuracy to accomplish an analysis at that scale, but nonetheless, we have observed many a novice GIS user unknowingly attempting to go down that path. Differences in map layers, which are artifacts of their development, may also come from differences in or changes to map projections, coordinate systems, and datums. Map data developed from digitized paper maps may also reflect cartographic alterations such as a displacement, or offsetting, of some features to facilitate map reading on the paper medium. While such cartographic conventions are appropriate to printed maps, their translation into an interactive and dynamic GIS realm means some features may not be located where they actually belong.

Another category of errors can come into play through the data processing operations carried out in your GIS software. For example, there may be computational differences in how an actual algorithm is implemented in the software you are using, or in the software the originator of a dataset used, that affect the accuracy of your results. Many of the computational differences come from rounding or other tolerance settings in the GIS software. Having an understanding of what software and what tolerance settings or algorithms were used can help you better assess the accuracy of data you are using in your analysis and the interpretation of results you obtain.

It is worth remembering that even though GIS is based on computer analysis, a variety of assumptions lie behind the processing methods and algorithms in these programs. Human errors, discussed earlier, that may affect data collection and quality remain when that data are entered into the computer. Be careful not to trust the computer too much. Be logical in the conclusions you draw, and be clear about your assumptions and potential error when you report your results (figure 4.3).

Figure 4.3   Consider what appears as a visually small error in the alignment of the two circles in the figure. If these circles represent the overlap between two variables in your study, it might suggest that they are highly correlated. However, is this correlation statistically significant?

If you were conducting a study that required 95 percent correlation and the computer reported a value of 94.3 percent, what would you do? What if there was 95.2 percent correlation? These differences, especially when small, may just as likely come from errors in mapping or other stages of the data collection process. It might be unethical to cancel a new project or program to assist a community if it falls just short of a predetermined (and perhaps arbitrary) cutoff value, especially if the shortfall is just as likely an artifact of the analysis process.

For the most part, social science research does not require detail on the order of centimeters, or even a few meters, so many of these more subtle errors may be irrelevant. The key point is that although ArcGIS may report results of measurements to several decimal places, most likely you should not.

Errors in human inquiry

Engaging in social research is not a trivial matter. There are several things to be aware of in the early design stages of any social science research project. The term **errors in human inquiry** refers to the various human-induced mistakes that can occur through the research process. In his book *The Practice of Social Research*, Earl Babbie (2013) points out some common errors made in social research that should be addressed: (1) inaccurate observation, (2) overgeneralization, (3) selective observation, and (4) illogical reasoning. With a little forethought, these pitfalls can be avoided.

Inaccurate observations

To ensure that you avoid **inaccurate observations**, it is important to have a well-defined plan for data collection. Most important, whenever possible, you should record data as they are observed rather than rely on your memory to properly and precisely recollect what was observed. Data collection forms, tape recording devices, and clear definitions of criteria and terminology relevant to your observations are essential.

If you will have more than one person involved in data collection, it is essential that everyone understands and uses the same definitions when making observations. If your study allows for it, it helps to get your data collection staff recording the same information through practice. Have them observe together in the same place for several data collection runs to ensure that all observers get acceptably similar results. Once you have everyone synchronized to seeing the same things, it is more likely the individual results from multiple observers will be consistent across your study. Approaching a study with a defined plan for collecting information will result in a higher likelihood of engaging in accurate observation.

Overgeneralization

Overgeneralization is another common error in research that means you've made conclusions as a researcher that go beyond your results in the research project. This error is easy to commit because it results from the limitations (time, money, or other resources) common to most studies. Overgeneralization of study data can occur in two forms. First, it can occur when your sample size is too small or is collected with an invalid sampling method and thus does not accurately represent the intended study population. Second, overgeneralization can result when the results of one study are inaccurately extended to other situations where they may not apply.

If you noticed that several houses in your neighborhood go up for sale in the same month, you might make an overgeneralization that the neighborhood is going downhill and all the neighbors are fleeing. An even greater overgeneralization would be to assume that the entire city is going downhill without checking the status of other neighborhoods. It is entirely possible that this hypothesis is correct, but this is only one possible explanation. You would need to collect more information to determine the real reasons behind people moving out of your neighborhood.

Selective observation

Following on the concept of overgeneralization is a natural inclination to see what we believe. **Selective observations** might be described as extending a pattern of observation that is likely to support previous findings, in effect limiting your ability to objectively observe. In other words, we see what we expect to see and miss things that do not fit our perspectives. Allowing preconceived notions to influence data collection can be a self-fulfilling prophecy and is an easy trap for any researcher to fall into when he or she wants to see a hypothesis or ideas supported.

To avoid this pitfall, it is important to develop your research design in advance, specifying the number and type of observations to be collected. For example, to understand why people decide to remain in or leave various neighborhoods, you might survey a sample of houses from each neighborhood in the city rather than only those whose houses are for sale in your own neighborhood. Stratifying your sample design by one or more sociodemographic or physiographic characteristics is one approach. Alternatively, you may prefer to randomly or systematically sample the population. In any case, having an advance plan is the best way to avoid the problem of selective observation.

Illogical reasoning

Finally, the problem of **illogical reasoning** is another potential pitfall of which to be aware when engaging in good scientific research. Whenever you conduct research, especially research that is aided by computer analysis, it is essential to consider the logic behind the data and results of the analysis. If the results obtained in a study seem strange to you, there is a good chance they are. In the next section, we discuss the concept of **ecological fallacy** as allows people to identify "whose" data they might be? These two issues can best be addressed by focusing on two key concepts: anonymity and confidentiality.

Anonymity

Anonymity means that when you conduct the research, you do not ask a person's name or for any other identifiable traits that could be used to tell who that person is. You can ensure anonymity by choosing not to collect such identifiable information. Many companies and universities that conduct research prefer that research be conducted anonymously so as not to risk sharing information about people. When you conduct anonymous research, you are protecting the research subjects in your study. For example, a survey is described as having been done anonymously if no individually identifiable information is collected as part of the survey.

Confidentiality

Confidentiality has to do with protecting information about people. When collecting data where individuals could potentially be identified via specific details contained in the data, it is important to keep that information protected. Confidentiality does not disallow you from collecting and analyzing detailed data from your study subjects, it simply means you will take appropriate steps to keep the raw data secret and not share them with the larger public. In most research situations, individual responses would be maintained in a secure setting, for example, in a secure research lab, locked filing cabinet, or password-protected computer. When results are reported, individual responses should be masked, or aggregated in some manner, to protect the identities of the individuals who shared their information with you. At the conclusion of a study in which confidentiality is important, it may be appropriate to destroy the underlying data, maintaining only the final, aggregated results. As a researcher, you should be aware of the research protocols of your particular organization and develop a plan for how you will keep information.

Data aggregation

Data aggregation is one potential solution to dealing with some of the privacy issues that inevitably arise as a result of mapping social variables. In **data aggregation**, the researcher simply chooses another, more general spatial level at which to view the data. Although the researcher may have the capability to connect individual responses with specific households, by aggregating data, the researcher may elect to analyze the data at the neighborhood or community level (figure 4.4). This would ensure the ethical privacy of responses related to individuals.

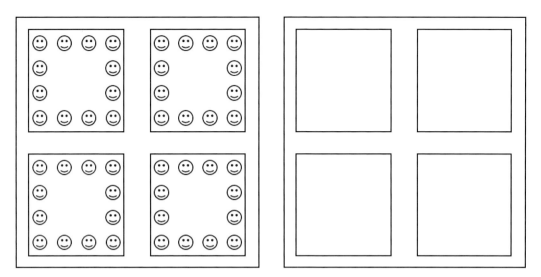

Figure 4.4 Data may be collected by individual households, as shown on the left. Although it is possible, and perhaps important, to retain the household-level information, many studies do not require this detail. If not, it might be preferable to aggregate data into larger units of analysis. The map on the right would be an example of aggregating data to each block or neighborhood. Other units of aggregation might also be used, for example, street, census block, ZIP Code, telephone calling prefix, or city or county.

Other, creative solutions to the privacy problem can decouple individuals from the data, depending on the study's specific requirements. For example, if you need to maintain individual responses, you can map the data to a false coordinate system so that the true geographic location on the ground is not stored in the GIS. This would be accomplished by mapping locations relative to a false origin located someplace in or near the study area. By using a false origin, it is possible to maintain spatial information such as distances and directions without producing a map that could be used to navigate back to a specific respondent's home.

For example, a true geographic location might be described in UTM coordinates as (479688, 4600488). UTM coordinates for the northern hemisphere are based on an origin located to the west of the UTM zone (for the *x* values) and at the equator (for the *y* values) measured in meters. You can simply alter the coordinate values by some set amount. In other words, simply by adding or subtracting a fixed value to or from all of the x,y geographic coordinates in the study, the true locations are no longer known, but the spatial positions would be maintained. In your GIS software, this sort of adjustment can be accomplished en masse for an entire dataset by setting the values to be applied with the tools designed for the reprojection of data.

Figure 4.5 shows an example of the calculations necessary to adjust the UTM coordinates given in the preceding paragraph as well as a simplified example of adjusted numbers that shows that the spatial relationships of distance and direction are maintained.

Original coordinate values (UTM coordinates)	Arbitrarily selected adjustment value (-284359)	Adjusted coordinate values (UTM)
X = 479688	X = 479688 - 284359	X = 195329
Y = 4600488	Y = 4600488 - 284359	Y = 4316129

Original coordinate value pairs (UTM coordinates)	Arbitrarily selected adjustment value (+3)	Adjusted coordinate values
(4, 6)	(4+3, 6+3)	(7, 9)
(5, 2)	(5+3, 2+3)	(8, 5)

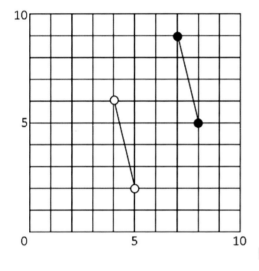

Pythagorean Theorem:

$$\sqrt{a^2 + b^2} = c$$

For purposes of mapping, we assign 'a' and 'b' to the 'x' and 'y' coordinates and the hypotenuse of the triangle 'c' (shown as a diagonal line on the plot)

Relative distances between the original points and the adjusted points do not change so long an identical adjustment is applied to both the X and Y values. In both cases, the difference between the X values equals one and the distance between Y values equals four. Therefore, the resulting distance between the original points and the adjusted points stays the same (4.12) when solving the Pythagorean calculation.

Figure 4.5 The first calculation shows the adjustment of true UTM coordinates using an arbitrary adjustment value to be added to or subtracted from the true coordinates. The second set of calculations uses smaller values to make the example easier to follow. This calculation and the associated plot show that the mapped (Euclidean) distance between points, as calculated by the Pythagorean theorem, does not change when these adjustments are made. The original points are shown in white and the altered points in black. It is important to note that when mapping geographic locations with latitude and longitude (which are spherical coordinates), this process is not valid. You must first project your map data into a planar coordinate system such as UTM.

Masking

One other consideration when masking spatial data is to remember that even if you do adjust values so that mapped coordinates are no longer apparent, you must be careful about the information included on printed maps related to your study. For example, including common cultural features such as roads and political boundaries, or natural features such as rivers or lakes, may provide an informed viewer enough information to recognize the location of the study (figure 4.6). For studies that require confidentiality, you will want to use map figures that provide minimal information.

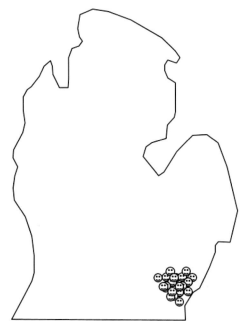

Figure 4.6 In this figure, without telling the reader anything about the location of the study, most people with a general knowledge of geography would find it simple to determine several things about the location of this study. Ask yourself the following: Can you determine what country this study was completed in? What about the part of the country? Can you make an educated guess about the city that was the focus? Odds are you could answer most or all of these questions correctly given a simplistic map that only gives an outline of the major region where the study occurred.

In the example in figure 4.6, these data might be easily masked by leaving off the outline of Michigan's Lower Peninsula. In that case, you would be hard-pressed to determine where the study was done. Leaving out even basic geographic features and details can go a long way in providing confidentiality to your study.

Primary and secondary data

Data aggregation is also a consideration when you are deciding whether to use existing data (secondary data) or to collect your own, new data (primary data). Sometimes the detail provided in data available to you is insufficient to serve the needs of your particular research study. For example, if you are really interested in conducting a study at the small-town level and the data for your study are only available at the county level, it would be difficult to draw any conclusions about the small towns, or even to separate the small-town data from those of the large towns that exist within the county. If the information you require is not available, your only option is to create your own data at the appropriate level of detail to conduct your research. (For more detailed discussion of primary and secondary data, refer to chapter 5.)

Keep in mind that just as we have discussed data aggregation, others will also aggregate data. Before going to the trouble of collecting all new, more detailed data for a study, it is worthwhile to contact the agency or institution that produced the data you are considering to determine if more detailed information was retained and could be obtained for the purpose of research. For example, the US Census publishes only aggregated data for official release. However, the individual records are maintained and do become available after seventy-two years. Thus, at the present time, Census data from individual respondents collected in 1940 or before can be accessed, providing an incredibly rich dataset compared to the aggregated Census data.

Of course, a variety of factors may influence who can be provided access to any particular dataset. If your research requires more detail than is readily available, it may be worthwhile to look into the possibility of obtaining the extra detail. This is also worth keeping in mind when you do collect and create your own datasets. If the policies and procedures of your organization allow for the unaltered, individual data to be retained for future researchers to use, that original data may be worth maintaining even if the data will not be distributed to the general public as part of the final report. All issues related to procedures used for data collection, input, aggregation, or other processes should be documented in a metadata file (data about the data) so that it is clear what the data are to others who may be considering consulting them in the future.

Although complete metadata will contain substantial information about the topic, sources, and methods associated with the collection and preparation of your data, key components of metadata should always address the following seven topics (US Geological Survey 2012):

1. What does the dataset describe?
2. Who produced the dataset?
3. Why was the dataset created?
4. How was the dataset created?

5. How reliable are the data; what problems remain in the dataset?
6. How can someone get a copy of the dataset?
7. Who wrote the metadata?

A complete metadata document following national or international metadata standards can run multiple pages. ArcGIS provides tools to assist you in developing compliant metadata files in the ArcCatalog application (figure 4.7). It is worthwhile to note that ArcCatalog does not use the term "metadata" on the interface but instead labels the tab Description. When viewing this tab, metadata may be created, edited, or imported via the buttons along the upper edge of the Description window.

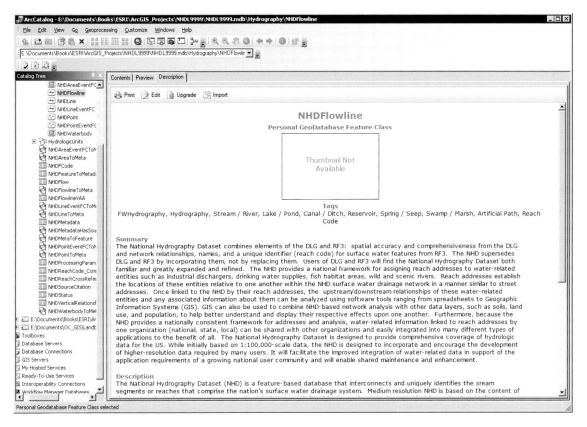

Figure 4.7   A view of ArcCatalog showing metadata for a data layer selected on the table of contents at the left side of the window. Clicking the Description tab shows metadata already associated with the dataset. If metadata are missing, they may be imported or created using the Edit, Upgrade, and Import buttons along the top of the metadata window. Metadata for NHDFlowline data from USGS.

Unfortunately, complete and compliant metadata are not always included with datasets you may obtain, or such information may be available in formats not directly associated with the GIS data structures. In such instances, you may need to find supporting information in separate documents, on websites, or even by talking with individuals who are familiar with the data in question. Of course, as a researcher, it is important that you understand the data sources you use in your work to ensure you are using the most appropriate and best information available.

Ethics and data sharing

The term **data sharing** refers to the sharing of data from a research project. The issue of sharing data is a key consideration that needs to be addressed at the beginning of a research project. Why? Because as you collect data, you want to be able to explain to people participating in your study how the data will be used, with whom the data are going to be shared, and how participants' confidentiality will be protected. Your responses to these issues may affect whether people will want to participate in your study.

Spatial data projects often involve the use of existing or what we call secondary data. But organizations and agencies are not always so forthcoming and willing to share these data. Therefore, it becomes important to be able to negotiate with the holders of such information to be able to use it. This may involve promising to share your data or results with the secondary data holders at the conclusion of your research. Another option aside from sharing the entire dataset is to share the output of the data, such as a PDF of a map.

Data sharing can often be a part of a research project. For example, during your research, someone may share information with you that you analyze and use to produce a report that you will provide to other researchers, your funder, or even the general public or media. With the Internet, many research projects develop websites for distributing data files, reports, and other project information, making your handling of data and research results even more important to consider.

Ethics and data storage

If you have data that are in a digital format, the data must be stored on a computer that has certain safeguards in place to keep those data safe. The term **data storage** refers to where and how you keep the data. As we have indicated in this chapter, any research project could have data or information that could impact people if it were to become public. Considerations for data storage may include the location where digital files are kept and who has access to them. In networked computer environments, there may be opportunities for confidential data to be accessed or left vulnerable to the outside. Care should be taken to ensure computers with sensitive data are secured with passwords and access controls adequate to maintain confidentiality of data. For highly sensitive data, it may be preferable to use computers that are isolated from public computer networks or in secured facilities. Other data security issues may arise owing to a lack of procedures for properly controlling data access and copying, for example, data copied from the secure computers to a laptop or external storage device by a well-meaning individual wanting to get extra work done at home. Even the best laid data security plans can be scuttled by a lost laptop or USB drive.

At the conclusion of a research project, you should consider appropriate methods to delete your digital data. Depending on your research protocols, it may be appropriate to retain a backup of your raw data for a period of time. If this is the case, data should be copied to removable media and stored in a secure location. It is important to recognize that deleting data from your hard drive does not erase the data in a secure manner. Even casual computer users realize that deleted data often are retrievable from the computer's trash can. Although emptying the trash on your computer provides a slightly better level of security, an experienced individual can still retrieve deleted data from the system. For truly secure data deletion, you should use software specifically designed to securely delete data files, especially if your data are particularly sensitive.

In the Microsoft Windows environment, where most ArcGIS users work, there are a number of options for securely deleting data, including SDelete, available from the Microsoft Technet website (http://technet.microsoft.com). If you work on another computing platform, you should seek out a similar tool. Keep in mind that data may be stored in more than one location. For example, if field computers were used for data collection, or if project staff copied data to removable media, those should be properly tracked and erased as well. As a conscientious researcher, you should identify any working files or data backups (including offsite backups) that may need to be deleted. It is also important to check temporary file locations that may be set in the configuration for ArcGIS or other analytical software you used in entering and processing the data. It is a good idea to retain copies of log files generated by SDelete or whichever tool you use as proof that all data were deleted according to your research protocols.

Review questions

1. What are some of the ethical issues you should consider when collecting primary spatial research?
2. As a researcher, what is your recourse if someone challenges your data on political grounds?
3. What are some tactics you could employ to minimize errors in observation?
4. What are some cultural implications of spatial research?
5. What are some errors that can occur in analysis?

6. What are some ways to avoid errors in accuracy?
7. What is data masking, and how might it be useful in spatial inquiry?
8. What is the difference between primary and secondary data?
9. What are the ethical considerations of using primary data?
10. How do you ensure the protection of groups when you are mapping information that is important to their survival and could be dangerous if shared with a larger audience?
11. What are the ethical considerations of using secondary or existing data? How do you navigate the challenges?

Additional reading and references

Babbie, E. 2013. *The Practice of Social Research*. 13th ed. Belmont, CA: Wadsworth Cengage Learning.

Creswell, J. W. 1998. *Qualitative Inquiry and Research Design: Choosing among Five Traditions*. Thousand Oaks, CA: Sage.

———. 2003. *Research Design: Qualitative, Quantitative, and Mixed Methods Approaches*. 2nd ed. Thousand Oaks, CA: Sage.

Panel on Confidentiality Issues Arising from the Integration of Remotely Sensed and Self-Identifying Data, Board on Environmental Change and Society, Division of Behavioral and Social Sciences and Education, and National Research Council. 2007. *Putting People on the Map: Protecting Confidentiality with Linked Social-Spatial Data*. Washington, DC: National Academies Press.

Stewart, C. N., Jr. 2011. *Research Ethics for Scientists: A Companion for Students*. New York: Wiley.

Taylor, D. F., and T. Lauriault, eds. 2013. *Developments in the Theory and Practice of Cybercartography: Applications and Indigenous Mapping*. 2nd ed. Cambridge, MA: Elsevier Science.

US Geological Survey. 2012. "Metadata in Plain Language." http://geology.usgs.gov/tools/metadata/tools/doc/ctc/.

Relevant websites

- **Primary Data Collection Methods (http://www.preservearticles.com/201104125345/methods-of-collecting-primary-data-in-statistics.html)**: A brief summary article discussing several methods of data collection.
- **The Influence of Data Aggregation on the Stability of Location Model Solutions (http://onlinelibrary.wiley.com/doi/10.1111/j.1538-4632.1997.tb00957.x/pdf)**: A paper discussing the influence of data aggregation on census block group data.

- **Metadata Examples (http://www.state.nj.us/dep/gis/metaexamples.htm)**: The New Jersey Department of Environmental Protection provides several examples of Federal Geographic Data Committee (FGDC) compliant metadata.
- **Federal Geographic Data Committee (FGDC) (http://www.fgdc.gov/):** An interagency committee that promotes the coordinated development, use, sharing, and dissemination of geospatial data on a national basis.
- **International Standards Organization (http://www.iso.org/iso/catalogue_detail.htm?csnumber=26020)**: Metadata schema (ISO 19115:2003) for describing geographic information and services.

Chapter 5

Measurement, sampling, and boundaries

In this chapter, you will learn about measurement, sampling, and boundaries. You will learn how to select a sampling method for spatial analysis; the differences between primary and secondary data; and about concepts, variables, and attributes. From a sampling standpoint, this chapter focuses on the use of probability and nonprobability sampling and spatial sampling considerations for stratification, data interpolation, and modeling. The crux of this chapter is how to incorporate spatial elements into sampling design.

Learning objectives

- Understand the difference between primary and secondary data sources
- Learn how to operationalize concepts in GIS
- Understand different types of sampling
- Learn which type of sampling is most appropriate for your particular research study
- Understand the spatial aspects of the study area and how to define that boundary

Key concepts

attributes
concept
conceptualization
discrete data
edge effects
interval variable
modifiable area unit problem (MAUP)
nominal variable
nondiscrete data
nonprobability sampling
ordinal variable
physical entities
primary data
probability sampling
purposive sampling
quota sampling
ratio value
secondary data
snowball sampling
variable

Moving beyond your personal experience

Although GIS analysis is generally an office-based activity, the questions we use it to explore are not. To be an effective analyst, it is always helpful to experience the environment you are studying. Along with subject area expertise relevant to the topic of your study, your own personal experience can go a long way toward enhancing your ability to collect, analyze, and ultimately understand your data and the interactions within them that are important. Attempting to analyze data for locations and topics outside your personal experience, though possible, may be more difficult. The classic parable of the blind men and an elephant provides an excellent analogy (figure 5.1). Lacking knowledge of the elephant, the blind men each draw different conclusions about the elephant based on a small sample of data. In research, you are often limited in both the amount and detail of available data.

Figure 5.1 *Blind Monks Examining an Elephant*, Hanabusa, Itchö, 1652–1724. Ukiyo-e print illustration from Buddhist parable showing blind monks examining an elephant. Each man reaches a different conclusion based on which part of the elephant he has examined. US Library of Congress, reproduction LC-USZC4-8725.

When you get out into the field, you have an opportunity to gather data from a number of perspectives. Perhaps most important, the in-person experience can help you to observe the environment you are interested in and help you determine what data may be relevant to answering your questions. Consider, for example, attempting to develop a community analysis for a particular neighborhood or city. While you can certainly obtain GIS data from existing sources, including parcel and zoning data from a county planning office or demographic information from the US Census Bureau, you may miss important details easily observed on the ground. Are the homes and businesses well kept, or is the neighborhood run down? Are the sidewalks cracked and parks overgrown and strewn with trash, or is the infrastructure clean and well maintained? Annotating existing sources with additional details observed in the field or developing new thematic layers can help you understand the environment you are investigating.

In natural environments, the same may be true; for example, although GIS layers showing mapped locations of water, vegetation, or geologic features are often available, the scale and detail included in such data may not be sufficient to tell the complete story. A visit to the location may quickly inform you of the quality of the environment, for example, is the vegetation in the area diverse with both overstory canopy species and understory layers, or is it a manicured timber plantation with uniformly spaced trees and no understory vegetation? One may be an excellent habitat for the bird you are studying and the other quite poor, yet this may not be apparent on a map simply labeling the polygon as forest.

In chapter 9, we explore a qualitative ethnographic type of field research. By contrast, in this chapter, we focus on the type of field engagement that you would use to collect quantitative or categorical data for built and/or natural environments. The built environment can be described as any sort of human-made structure or artifact. Examples of this would be buildings and roads. The natural environment comprises those features that are not constructed by humans. Of course, there are many examples where the line between the built and natural environments is somewhat vague: a river that has been channelized and constrained by levies, a park where hills may have been placed by a bulldozer years earlier, or a lake formed by damming. Although for some analyses, these distinctions will be important, distinguishing these cases is not always necessary. The important message is that with a bit of knowledge and on-the-ground experience, you will be better prepared to observe and document those characteristics of the data that are important to your objectives.

In field research, the physical environment becomes a lab of sorts. To conduct effective field research, it's important to be aware of what you are looking for out in the field. It is also important to consider how you plan to gather this information. In other words, as a scientist or researcher, you want to have done your homework in advance and researched the area before you go into the field to collect your data so that you can be efficient and effective when you get there. For example, consider how you might plan for a fishing trip. You would conduct background research to determine what sort of gear, bait, and other equipment will be most useful. You might determine some of this by reading fishing guidebooks or conducting research on the Internet. You might also talk to others who have been there or contract with a local outfitter who can guide you to the best fishing spots. Additionally, you would want to gather some temporal data—information such as when the best time to catch fish at a particular location is and what types of fish can be caught in different places.

Before you go out in the field, you can gather a lot of background information on your topic (see chapter 6) through using existing or secondary data sources. As a starting point, you probably want to begin by gathering any preliminary data or background information about your topic and the physical environment where you plan to conduct your research. As a researcher, you could begin by thinking about your goals and the topic that you hope to better understand once you get into the field, and the physical environment or place where your team is going to gather their data.

Because field data collection often involves some type of quantitative measurement, you will need to consider several aspects of data collection. What do you wish to document or measure, and what tools or technologies will you need to accomplish this? Quantitative survey measurements will likely require you to take equipment for measuring distances, directions, heights, and so forth. Depending on the desired accuracy, you may require a measuring tape, a Global Positioning System device, or a highly accurate laser range finder. For a less detailed analysis, perhaps a hand compass and pacing will provide sufficient accuracy. A study of community safety may be well served by a binary classification of streetlights that work versus those that do not, whereas another researcher may desire the added quantification provided by taking a light meter into the field to measure illumination at intervals along the sidewalk.

When one thinks about conducting field research for the natural or built environment, one of the first questions that emerges is how you will physically define your study area. Sometimes the definition will be straightforward, such as boundaries defined by a legal parcel or jurisdictional unit. In the simplest of cases, those boundaries may even be clearly defined on the ground as streets bounding the community or fences surrounding a ranch. In others, the boundaries may be more abstract, for example, an ecological boundary defined by a combination of factors, including climate, vegetation, and geology. In traditional mapping, we are comfortable with the concept of polygons defining clear boundaries between units of analysis, but in many natural and social systems, these boundaries are less obvious.

When the study area is large or access is difficult, it may be essential to consider sampling strategies that provide you with appropriate coverage. Although from a statistical perspective, it might seem best to randomly sample, in practice, this is not always possible. In a natural setting, randomly selected points may lie kilometers from the nearest road, perhaps requiring you to traverse dangerous terrain to get to them. Keeping safety in mind, it is often best to limit your sample selection to locations that are both representative and accessible. Using preliminary data and leveraging the power of GIS, you can assess the environment and develop a sample selection that is more manageable, limiting the time and effort required to collect field data, while ensuring a representative sample.

It is not our intent to discuss all possible sampling schemes in this chapter but rather to provide a few examples as a means to stimulate your thinking. For example, one method might be to select random points along a road or trail network and then, at each point, run a hundred-meter transect at a specific (or perhaps a random) bearing off the road into the environment under study. A variation on this approach would be to travel a specified distance off the road to establish a circular plot of a specified radius, which then will become the unit of analysis. Constraining either of these to a reasonable distance from the road or trail can help to optimize the travel time to each sample location. If there are known categories that your study is considering, such an approach can be used in combination with the various strata to ensure that a sufficient number of samples are taken in each.

This sort of approach is particularly useful when your research occurs in remote locations where you may have limited time and resources available to conduct your fieldwork. While the random point twenty-five kilometers off the trail may provide for a truly random sample selection, if it takes you three days to get there to collect the data, your study will either consume an exceptional amount of time and resources or, more likely, will run out of the resources needed to collect an adequate number of samples.

In this chapter, you will learn about the various factors that influence where, when, and how you draw those field research boundaries. What are different types of study boundaries you should consider? How do you determine what such boundaries are, and where do you indicate them physically on a map? Determining the answers to these questions is very important before heading into the field.

Choosing a sampling method for your spatial analysis

You need to consider a number of factors before choosing the type of data you will collect and deciding how to sample for those data. The most important question to address is whether you plan to use primary or secondary data.

Type of data source: Primary or secondary

When beginning a study in GIS, it is important to select and understand the data to be used in the analysis. Generally speaking, we refer to data collected directly by the research staff for the specific project as **primary data**. By contrast, data collected by someone else, for a different purpose, are referred to as **secondary data**. Although both data types have their place in any study, there are pros and cons to using each type.

The advantage of primary data is that you know exactly what you are working with, and you have complete control over the data creation process. This can be extremely beneficial in obtaining an end result that meets the specific goals of the study. However, collecting all of your data from scratch for every analysis you do can be quite time consuming and costly. Herein lies the value of secondary data sources.

Secondary data, although not collected with your specific research question in mind, are often appropriate, and their use can certainly save a lot of time, money, and effort when you need to get your analysis done quickly and efficiently. As discussed in chapter 6, a wide variety of GIS data sources are available from various government agencies, private firms, and other sources. Many of these secondary data sources are well documented, with metadata meeting the Federal Geographic Data Committee (FGDC) standards and with specific criteria for accuracy and completeness. When data are well documented with FGDC metadata, you can more easily and accurately determine the dataset's appropriateness to your study. In short, consider the source of and purpose for the data.

Generally, we assume that when data are collected for research purposes, care will be taken to ensure that the data are unbiased and complete, but this is not necessarily so. Every dataset is collected with some purpose in mind and, potentially, some bias based on this intended use. This is not to imply that data are intentionally biased; rather, data collected for a purpose other than the one you may be using them for may have built-in bias. For example, you may obtain a dataset that was collected by surveying shoppers at a local mall. Although those data may have been perfectly appropriate for a study of shopping behavior among people in that specific mall, the data may not be appropriate for a study of the purchasing habits of the entire community. Therefore, it is important to be sure that when you obtain secondary data, the dataset is appropriate for your purposes and meets your specific research needs.

Secondary data can raise concerns in two general situations: first, when it is similar to, but not exactly, what you need; and second, when it is unclear how and why the data were originally collected, raising questions about how and why particular variables were selected and measured. In many cases, using secondary data will require some effort on your part to prepare the data for use in your analysis. In the case of existing GIS data, this could range from a simple map reprojection, taking a few moments, to a more extensive reorganization of the database attributes to facilitate linkage to the basemap or other spatial information. (Please refer to chapter 7 for a discussion of database structure and compatibility of attributes.)

Some considerations that will influence your decision to create your own data or rely on secondary data include the following:

1. *Your study question*. What is your study question? Your data selection should serve the needs of your analysis. If existing, secondary data can do this, you may be able to save the time and effort required for collecting new data.
2. *Data requirements*. What type of data are you looking at? Are the available data up to date? If you are looking for base data (infrastructure, built environments, etc.), the data may not have changed significantly since available secondary sources were collected. However, if the data involve people or natural environments, things can change rapidly, and available secondary sources may not be appropriate.
3. *Combined approaches*. Sometimes you can use a hybrid approach. Perhaps you can work with an existing source and make updates to bring it up to date or add the missing attributes that you require.

In the end, realities of time and money will typically limit how much new data you collect and how much existing data you include or edit to use. In the remaining sections, we focus largely on how data are measured in preparing your GIS database. Even though you may use secondary sources exclusively, it is important to understand how and why those datasets were collected and organized in the way they were. Without that knowledge, you run the risk of incorporating inappropriate data into your analysis, which can lead to incorrect or misleading results.

As you evaluate the fitness of any data for use in a particular analysis, it is essential that you have a firm grip on the goals and design of your study. The following sections detail the process of conceptualizing your project, measuring variables, and designing the analysis framework. With these things in mind, you will be better prepared to evaluate when secondary data fit the needs of your analysis and, if they do not, how best to go about collecting your own data.

Concepts, variables, and attributes

A **concept** refers to a body of ideas taken together as a unit. As part of the research process, you will go through a process called **conceptualization**, in which you clearly define and develop a label to name the main ideas that are part of your study's theoretical model. Examples of concepts include poverty, community strength, and environmentalism. Each of these concepts reflects an area that could be studied in great detail. It is important to recognize that one simple concept may comprise multiple attributes.

AN ASIDE ON DISCRETE AND NONDISCRETE DATA

One of the fundamental issues when mapping information, whether for use in GIS or elsewhere, is the form of the data relative to the tools used to map them. Traditional maps and their computerized counterparts typically use lines to define borders. Consider the sketch on the left in figure 5.2. If you are told that the areas defined on the map represent parcel boundaries, those are a **discrete data** type. Property boundaries can be precisely surveyed and mapped such that the map accurately represents them.

However, if you were told that this was a map of annual rainfall in millimeters, you would rightfully assume that the borders in the figure on the left must be generalized. That is, the amount of rain falling on the land does not change abruptly as you cross a line from one mapped polygon to another. More realistically, you would expect rain to vary gradually across the region of the map, with some areas getting more and others less.

The figure on the right is a more realistic rendition of a **nondiscrete variable** changing gradually across the map. Whereas raster mapping can reasonably simulate nondiscrete data, many existing data sources are structured in discrete, vector formats. The trick for the analyst is to accept this simplification of reality and treat these variables as though they are discrete or, alternatively, find ways to compensate for maps made in this way by converting the data to a raster format.

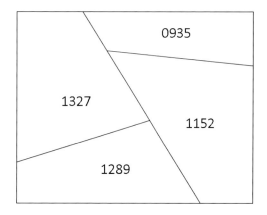 

Figure 5.2 Discrete data (left) and continuous, nondiscrete data (right) give very different representations. Choosing an appropriate format for data collection and visualization is important.

Operationalization of concepts in GIS

Once you define the concepts that will inform your analysis, you must determine how you will measure each one, a process termed *operationalization*. This is typically accomplished by identifying the specific variable or variables to measure. A **variable** can be described as an indicator for the measurement of concepts. Often in the GIS context, these variables are **physical entities** such as a person, a facility, or another feature that can be located in space and about which data can be collected for use in the analysis.

For example, if you were going to identify variables related to the concept of poverty, one indicator you could measure would be income. Of course, there are other variables that could be measured as components of the concept of poverty, for example, nutrition, education, number of children, and employment. A variable can be anything that provides you an indication or measurement of your concept. Depending on the concept being measured, you may choose to use just one or perhaps several variables as indicators. Each of these variables may include one or more attributes. When considering the variables that you want to include in your study, you should choose ones that relate back to the main concepts you are studying. In other words, the concept behind the research question must always drive the selection of variables and attributes.

The database structure behind a GIS system dovetails well with this approach. As discussed in chapter 7, GIS organizes the data in database tables that are linked to geographic locations. With this structure, individual variables are often stored in the first or second column of the table using a user-defined naming system. (These may be real names or codes assigned to the individual items.) The **attributes**, or specific descriptive characteristics of the variables, exist in subsequent columns, as shown in table 5.1.

Table 5.1 A simple data table containing entity and attribute data collected about four individual study subjects

ID	Variable/entity	Attribute_1	Attribute_2	Attribute_3	Attribute_4	...	Attribute_n
1	Jane	53	Female	Hispanic	College	...	45,000
2	Chris	27	Male	Caucasian	HS	...	35,000
3	Bob	61	Male	Asian	College	...	55,000
4	Sue	36	Female	Caucasian	Graduate	...	65,000

Note. The ID may be assigned automatically by your GIS or database software.

An attribute is a descriptive characteristic recorded for each variable. Attributes of a variable can exist in numeric or nonnumeric form. The GIS can account for both numeric and nonnumeric data. (See chapter 7 for a further discussion of this concept.) For instance, if you are interested in the concept of rural community wellness, what are some variables that could be examined as indicators of this concept? You might start by examining the health facilities that are available in these rural areas. You would consider questions such as what type of facility is available—is it a clinic or hospital? During what hours is the facility available? Is it a twenty-four-hour facility or open only during business hours? Is it located in a rural or urban area? What is the bed capacity of the facility? Given the preceding questions, using a GIS, our primary variable would be health facility. Attributes might include names such as TYPE, HOURS, LOCATION, BEDS, and so on, which would be identified and used to develop the actual data table within the GIS software or another data entry package.

These are all descriptive characteristics of the variable facility, which could be integrated into a GIS to help investigate your hypothesis. As a result, one of the main variables or entity types in your study would be health facility. All of the descriptive attributes that accompany your variable could be classified as attributes. How these entities are ultimately represented in the GIS is another consideration. Will you represent them as point locations, or is it more useful to measure and map the actual shape of the building footprint? In this example, the former would likely be sufficient, whereas the latter would create a significant amount of additional work with very little analytical payoff, because the physical shape of the facility is unlikely to have much to do with the level of care provided.

Different data types: Matching geographic and social variables

Data come in a variety of formats, often including variables of different types. Variables are typically categorized as being nominal, ordinal, interval, or ratio.

A **nominal variable** consists of data representing the information about variables where the attribute of the variable does not have a quantitative basis. Nominal variables typically are used to name the data without giving them any numeric value. For example, if we reconsider table 5.2 (now with the columns labeled according to their data type), we can see that for the nominal attributes shown, there is nothing inherent to the data we can use to judge quantitative information based on their names.

Table 5.2 A simple data table containing entity and attribute data collected about four individual study subjects

ID	Variable/entity	Attribute_1	Attribute_2	Attribute_3	Attribute_4	...	Attribute_n
1	Jane	53	Female	Hispanic	College	...	45,000
2	Chris	27	Male	Caucasian	HS	...	35,000
3	Bob	61	Male	Asian	College	...	55,000
4	Sue	36	Female	Caucasian	Graduate	...	65,000
		Nominal	Ratio	Nominal	Nominal	Ordinal	Interval

Note. Columns are marked according to the level of data measurement: nominal, ordinal, interval, or ratio.

Because the nominal data have no quantity associated with them, we cannot make quantitative judgments about relationships between them. For example, it would make no sense to make statements such as the following:

Chris > Bob
Hispanic = Caucasian
2 × Male < Female ÷ 7

For simplicity in data entry, you might opt to assign a numeric code to nominal data. For example, in a study of communities, you might choose to describe a location as urban, suburban, or rural. These three categories, or attributes of the variable location, tell us something about the variable but do not have an inherent quantitative basis. Even if you choose to code them as city = 1, suburban = 2, and rural = 3, the numbers assigned are simply nominal and hold no meaning in a quantitative sense.

A dichotomous nominal variable (e.g., male–female) might also be coded using numbers to allow for quantitative analysis using a system called *dummy coding*. In this situation, the variables are assigned values of 0 and 1. You could use a system of dummy coding to highlight the differences in the variable gender, between male and female. The number 1 = female and 0 = male can then be employed in a quantitative analysis to differentiate between two categories for a particular variable. Using dummy codes can provide a means to use the variables in an analysis that requires numbers.

In a GIS context, the process of coding zeros and ones can be especially useful when conducting an analysis with yes–no or acceptable–not acceptable type data. By using dummy coding where 0 = not acceptable and 1 = acceptable, raster data layers can be multiplied to ensure that any location that is unacceptable for one of the criteria will result in a "0" in the output. Similarly, if all criteria are acceptable, the result of a multiplication of raster layers will be "1" (figure 5.3).

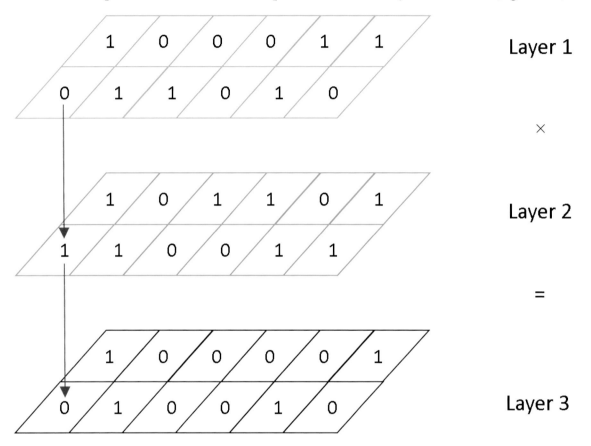

Figure 5.3 Coding raster layers with only zeros (unacceptable) and ones (acceptable) provides a means to mathematically identify pixel locations in the resulting data layer meeting all of your desired analysis criteria. By multiplying these two layers (layer 1 × layer 2) using map algebra (see chapter 11), we obtain a new dataset (layer 3) where zeros indicate locations where one or both criteria were unacceptable and ones indicate where both criteria were met. This same approach works equally well for studies with any number of layers coded in the preceding manner.

Ordinal variables are best described as variables or attributes that can be logically rank ordered but with quantities that are still not defined. These are the types of variables that could be measured along an ordered scale. The Likert scale, an example of which is shown in figure 5.4, is commonly used in survey research and presents an excellent example of an ordered scale in which the specific quantity between each level is undefined.

Strongly Disagree	Somewhat Disagree	Neutral	Somewhat Agree	Strongly Agree
1	2	3	4	5

Figure 5.4   An example of a 5-point Likert scale. The numbers indicate a respondent's level of agreement. Using an ordinal variable of this type leaves the relative distance between values undefined. It is up to each respondent to interpret their meaning.

Although data on such a scale are clearly ordered, there is no defined spacing between categories. In other words, we know a respondent who selects "2" has shown a lower level of agreement than a respondent who selects "4," but we cannot state that the difference of two has any real meaning in a quantitative sense, nor can we say that the gap between these individuals is identical to the gap between individuals selecting "1" and "3."

For the variable education (Attribute_4 in table 5.2), you could measure it by the following: (1) less than high school, (2) high school graduate, (3) some college, (4) college graduate, or (5) graduate or professional degree. Again, in this example, the numbers indicate something in terms of the rank order of these variables (a level of education, where a higher number indicates more years of formal education), but the distances between the measures do not have specific quantitative meaning. The goal with ordinal values is to obtain a rank order of responses.

In a GIS context, the process of coding ordinal data is again useful in some quantitative analysis applications. Because there is an order to data of this type, queries of the GIS data can appropriately incorporate mathematical statements that would not have been sensible for nominal data. For example, statements such as "college > high school" or "neutral < somewhat agree" make logical sense. When presented in word form, they may be more difficult to code in the GIS. Therefore, it is helpful to code ordinal data with numbers rather than character strings for the purpose of analysis in GIS.

An **interval variable** incorporates a true distance between the recorded measures, and therefore quantitative analysis becomes much more practical. In table 5.2, we see that Attribute_n represents annual income in dollars. We can examine these numbers and determine that Chris makes $10,000 less than Jane and that Bob makes $10,000 less than Sue. In both cases, the difference of $10,000 is the same, even though the locations along the income scale where each of these salaries falls are different. Interval data can be thought of in terms of a number line where the distances along the line are at a fixed scale, much as the scale on a map would be expected to remain consistent for the entire map.

In the GIS context, interval data offer an analyst the ability to conduct detailed quantitative analysis of the data because typical mathematical and statistical operations are valid and meaningful with these numbers. Distances between values are meaningful, including geographic intervals (e.g., distance between locations) or intervals of other attributes, as in the example of income.

The last of the four variable types used in data collection is the ratio measurement. The distinct feature of a **ratio value** is the presence of a true zero. This concept is sometimes confusing because some uses of zero are arbitrary. A good example of the use of a false zero is the description of time. The year zero on most calendars is not truly the beginning of time. It is widely accepted that regardless of the human calendar used, time existed before the point humans selected as the year zero. Many cultures through time have devised calendars to meet their own needs, each one having a different zero point (e.g., Roman, Mayan, Hebrew, Hindu).

In table 5.2, Attribute_1 refers to age. As far as any individual is concerned, before the individual is born, he or she has no age. Zero in this case is an absolute, or true, zero. One cannot be a negative number of years old. This might raise the question as to why we classify income as an interval variable rather than an ordinal variable. The simple answer is that it is possible to have negative income. Anyone who has a home loan or credit card realizes that dollars can be positive or negative. Having zero dollars at any particular point in time will not necessarily get you out of paying your bills.

The reason the distinction in the zero is so important is because in quantitative analyses where the zero is arbitrary, things can get confused, especially if the zeroes end up causing other data to incorrectly zero out in a multiplicative process, or perhaps worse, the zero causes your computer to crash because of a division-by-zero error.

Data sampling and GIS

Sample design is, of course, a substantial area of inquiry that we can only briefly cover in this text. We present an overview of commonly used sampling approaches and suggestions about their implementation in a geographic context. If you are unfamiliar with sample design, you might want to explore this topic in more detail by referring to sources specifically related to sample design referenced at the end of this chapter.

Probability and nonprobability sampling

Sampling is an important part of the research process. The type of sampling you engage in is dictated by your research question. Sampling can be divided into two basic categories: probability and nonprobability. The type of sampling that you choose will be based on the question that you are trying to answer as part of your research project as well as the resources you have available. Generally speaking, probability sampling approaches take more time and effort than nonprobability approaches, especially when working in GIS with a geographic component to the study.

When working with spatial data, your sampling frame is the entire area of the study, and you must be prepared to collect data at any randomly selected location on the landscape. For some studies, the ability to physically access a randomly selected location may raise concerns of cost, safety, or privacy that can make random sampling in a geographic context a bit more complex.

Probability sampling can best be described as sampling that incorporates some form of random sampling. In these sample designs, every element that is part of your study has the same chance of being included in the sample. A sample is merely a subset of the entire population used in an analysis when collecting data for the entire population (census) is not possible. The population is defined as the entire group of elements you are interested in studying. For instance, the opinion polls that rank public attitudes of the president cannot sample the entire population of the United States. Instead, they randomly select individuals throughout the country and question these individuals about their opinion of the president. This group of people is representative of the larger US population. The reason we can say that the opinions of this smaller group of people reflect the opinions of the entire US population is because they have been randomly selected.

The advantage of probability sampling is that random sampling allows for generalizing your findings to a larger population. It is a more representative sample type because it avoids biases inherent to other forms of sampling. The disadvantage of conducting a probability sample is that it sometimes is difficult to get a sampling frame that is complete enough to include all of the potential elements.

Nonprobability sampling

Nonprobability sampling can be explained as sampling that is not based on the random selection of elements to be a part of the study. In certain instances in social research, a nonprobability sample would be preferred over a random sample, for example, for topics for which the researcher has a targeted population that needs to be interviewed as part of the process. For example, say you want to study what it is like to be a member of the Hells Angels; you could not take a random sample of the general population but rather you must target individuals who are known members of that particular population. Studying such a group would involve taking a nonprobability sampling approach. There are four different types of nonprobability sampling: (1) purposive sampling, (2) reliance on available subjects, (3) snowball sampling, and (4) quota sampling (Babbie 2012).

Purposive sampling

Purposive sampling means that specific characteristics are integral to your study to justify sampling a population from a specific and perhaps somewhat limited group. Purposive sampling is a type of nonprobability sampling that occurs when the researcher deliberately chooses the population that he or she wants to include as part of a study. It is a type of sampling often employed by people who have a clear idea based on the research question about whom they want to interview. Earl Babbie (2012) notes one characteristic of purposive sampling is that it can be difficult to enumerate the entire population you wish to study. A researcher has a reason for choosing the type of people he or she interviews. The researcher may become aware of these people through contacts, their outward visibility, or their membership in a particular organization.

For instance, if you are interested in studying successful entrepreneurship in your hometown, you first need to identify the business entrepreneurs who have been successful, perhaps by reviewing the local business section of the newspaper. To carry out your study, you would select and interview a number of business leaders who fit your definition of success. Even if you don't identify every successful entrepreneur in the community, a purposive sample could provide an appropriate set of individuals to give you a sense of what it takes to be successful in the community.

Available subjects

This type of sampling can be problematic but is commonly employed in the social sciences. When collecting data from available subjects, individuals are interviewed at some particular location. This can be very useful if the locations are spatially relevant to the study question. For example, if available subjects are sampled to find out why they visit that particular location at that particular time, it could be a powerful dataset tied to a point in space. However, if sampling of available subjects is done for a study unrelated to location, for example, to determine presidential favorability ratings, the location may introduce an unanticipated bias into the sample. For example, if your sampling location is an upscale shopping mall, you may find very different responses than if you had sampled outside the local ballpark.

In other words, this approach severely limits the representativeness of the sample to the overall population. An advantage of available subjects is that it provides a means for rapid data collection without going through the difficulty of first defining the population and then developing a sampling frame from which to randomly select respondents. There also might be times when this method would be appropriate, for instance, if you were interested in studying people's perceptions of some local or one-time event.

Snowball sampling

In this type of sampling, you rely on already interviewed subjects to provide information on further potential research subjects. An example of this would be the use of key-informant interviews. Say you are interested in interviewing military veterans who are antiwar. It might be easy to obtain a list of veterans in your hometown, but you would not necessarily have knowledge about the political feelings of these individuals until you interviewed them. Therefore, you might engage in **snowball sampling**.

In snowball sampling, you begin by interviewing people in the community who are knowledgeable about a topic. These people can usually be identified by asking around to find out who might be good to talk to about the subject of interest. Additionally, you could begin your snowball sample by talking to individuals who have been featured in the local newspaper as being related to your topic. As the initial interviews are completed, the subjects are asked if they can recommend other individuals who could be interviewed about your topic. In this way, you benefit from the individual knowledge of the respondents, who are more likely to be aware of other individuals who fall into your target population.

An example of a question to ask to elicit others to interview would be, "Who else do you recommend that I talk to about this topic?" Over time, this technique builds a list of potential research subjects using information that early respondents provide along the way. The way to double-check that you are on the right track is to see which names appear repeatedly as numerous different interviews are conducted. Certain individuals being recommended repeatedly indicates that you are interviewing appropriate people, given that others in the community recognize and confirm their expertise in the particular area that you are researching.

Quota sampling

Quota sampling is a stratified sampling approach in which the researcher selects the sample based on characteristics that are defined in advance. These are typically characteristics that exist in the larger population that the researcher wishes to study. Quotas can be used to identify respondents based on information obtained as part of the sampling frame. In social science contexts, you may determine that you want to sample equal or proportionately relevant numbers of individuals from different community service organizations. If this were your objective, you might obtain membership lists from each club (e.g., Rotary, Lions, Elks, Kiwanis) and, from each of these, randomly select a certain number of individuals to survey. In a geographically based study, the criteria for stratification might be spatial, for example, you might have a certain quota of respondents from each neighborhood in the city.

Random sampling

In **random sampling**, all elements or individuals in the study population have an equal chance of being selected. If you choose to use random sampling, you can approach the task using a number of methods. For example, if you were developing a sampling frame for a community, the sample might be generated from a list of addresses, phone numbers, membership lists, and so forth. If sampling spatially, one could generate a random list of locations (x,y coordinates) either across the entire study area or based on subset (stratified) areas within the overall study region. Using coordinates with GIS is particularly convenient for spatial studies because you can easily generate the coordinates using random number tools in a spreadsheet, statistical package, or the GIS. Armed with x,y coordinates and a handheld Global Positioning System receiver, it is a relatively simple process to navigate to the selected locations for the purpose of collecting physical, environmental, or sociodemographic data. For instance, it could be useful as a means of selecting locations to visit for the purpose of survey data collection.

Study area and sample unit boundaries

We have alluded to the need to define study and sample unit areas geographically when conducting a study with GIS. On the surface, this might seem straightforward; however, you should review a number of considerations before assuming you know the proper boundaries to use for your analysis. Before discussing the methods to go about selecting your boundaries, we should point out two important considerations with regard to the eventual GIS analysis. These are the related issues of edge effects and the **modifiable area unit problem** (**MAUP**).

Edge effects

In any spatially defined study, a problem can arise in analyzing data near the edge of the study boundary, especially when the analysis involves quantitative procedures; this is known as **edge effects**. Consider the simple analysis question of population density in a city full of happy people illustrated in figure 5.5.

The upper and lower halves of the figure show the same population in space. The placement of the study boundary, represented by the box, varies. In the upper portion, we would obtain twice the population density (ten people per unit area) as we would obtain in the lower portion of the figure (five people per unit area). Of course, there are infinite possible placements of this study boundary, each one potentially giving us a different result.

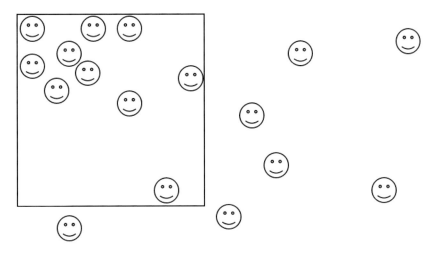

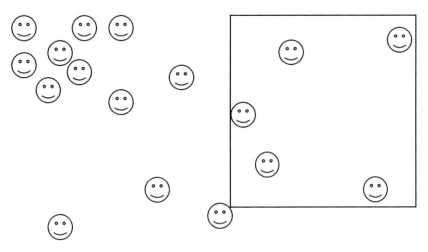

Figure 5.5 For a given population (represented by the happy faces in the upper portion of the figure), a study boundary can be placed in an infinite number of locations. The upper and lower portions of the figure represent just two of these possibilities. Each placement of the study boundary provides a potentially different result for whatever attribute or variable is to be measured based on the selected population (or sample from the population).

Without seeing the map or knowing something about the placement of the sample boundary, we have no way of knowing what is occurring just beyond the edge of the study area. For example, in the lower portion of figure 5.5, we may not capture information about the community that is common to the large number of individuals on the left side of the figure. Thoughtful selection of your study boundary can be an important step in any study design, whether or not it is done in a GIS.

Of course, one advantage of the GIS is that if we have data about all of the people in this town, we can run the analysis multiple times (multiple realizations) using a Monte Carlo simulation approach. For example, if we were to run the analysis one hundred times, with one hundred random placements of the boundary, we could determine what the most common (and thus most likely) result is, while also obtaining statistics regarding the distribution and standard deviation of those results.

Another example might be if we were looking at cancer rates in a particular population without knowing about an important source of pollution upriver or upwind of the study site. Knowing something about the relative geographic positions of important variables is essential to determining where and what to include within your study boundary. Conversely, having a sense of these issues can improve your ability to critically assess the meaning of the results of your own work and the work of others. You can think of a study boundary much like a cookie cutter; whatever falls outside the boundary gets thrown out and is no longer part of the data (or potential data, in the case of a sample).

The modifiable area unit problem (MAUP)

A related issue is the actual size of the sampling frame or study boundary used. As we alter the geographic area of the study boundary, the results obtained from the analysis can change. Again, consider the happy residents in figure 5.6. The area of the lower sample boundary is exactly twice the area of the original upper boundary. This results in a population density of ten people per unit area in the upper figure, whereas the area of the lower boundary contains fourteen people. Because the lower boundary is twice as large, the resulting density is actually 14 ÷ 2, or seven people per unit area.

Of course, the placement of the boundary will also influence the results. Thus, these two issues work in concert, either offsetting or potentially compounding the variability in the results. As mentioned previously, if you have the ability to run multiple realizations in a Monte Carlo simulation, you can determine the answer with the highest probability of resulting from the data.

Unfortunately, a large portion of the spatial data collected do not take these issues into account. Being careful to consider how and why your study boundaries are defined can be essential to getting results that will be accepted as valid. Interestingly, this concept is fairly new to the analysis of spatial data, having only been identified in the last twenty-five years. Most notably, Openshaw and Taylor (1979) originated much of the work on this issue, with its relevance to the social sciences being recognized in the last decade or so.

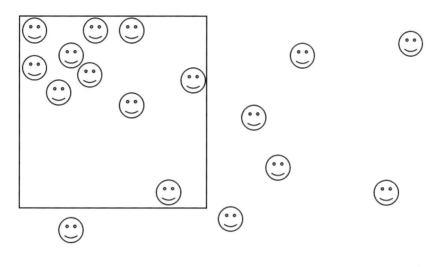

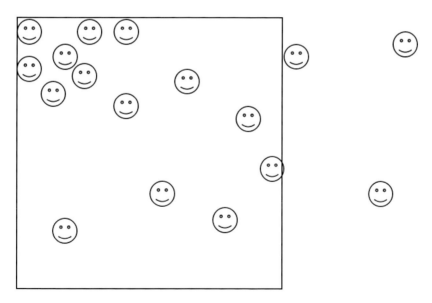

Figure 5.6   An example of the modifiable area unit problem. For a given population, represented by the happy faces, the size of the sample unit, represented by the box, can significantly alter the results obtained. In the lower portion of the figure, the area of the sample unit is exactly twice that of the sample unit in the upper portion. With identical populations, this change results in two distinct population density measures: ten people per unit area in the upper figure and seven people per unit area in the lower figure.

Selecting boundaries

In the preceding examples, the researcher artificially selects the boundaries simply by dropping a square frame onto the landscape. This is a valid approach and may be done randomly or systematically with the use of a grid of sample areas. However, in many studies, other types of boundaries may be associated with the data. For example, you might select from census tracts, ZIP Codes, county or state borders, or a wide variety of other options. As you consider one or more of these as potential study boundaries, give due consideration to the goals of your research and the influence your choice of boundary might have on the resulting sample design and analysis.

One approach is to consider your study from a geographic standpoint—that is, consider the geographic scale. At what level does your research question focus? Perhaps you are interested in local-level differences between blocks or neighborhoods. If so, the use of census blocks might be an excellent choice for stratification, sampling, and analysis. If you are doing a statewide or national study, county or state boundaries may be more appropriate. Are there important features that would influence the data just outside one of these boundaries (e.g., a large industrial facility or a military base)? How was boundary placement determined and for what purpose?

If your study is longitudinal and involves time series data about a particular place, you will need to determine how that place is defined. Where are you going to draw the boundaries? Presumably you would choose to draw the boundaries with consideration of the research question being investigated. However, this can be a bit complicated, because often the boundaries change over time—cities expand, political districts are redefined, and so on. If you are interested in looking at the same geographic area over time, you may need to establish a fixed study boundary, especially if the other boundaries are inconsistent over time.

Conversely, you may be interested in the change in the boundary. If that is the case, your study data will need to allow for the largest geographic area necessitated by the input data in your analysis. Possibly this will necessitate having access to or creating geographic data for the same location over different time periods so you can determine the outer limits, or geographic extent, of your overall analysis. Examples of boundaries that may change over time include almost any politically or jurisdictionally defined boundaries (e.g., school districts, voting precincts, or city limits). Boundaries based on sociodemographic factors can also change over time. For example, a lower-income neighborhood may be gentrified and become an upper-income neighborhood.

If comparing data at one time for different locations, you may need to address consistency between areas. Do geographical commonalities or differences exist between your research locations? If so, this may inform decisions about the stratification of study areas.

You may choose to weight or compute ratios of the data for each location to normalize for the differences. For example, a comparison of population between a large city and a rural town would be better achieved by taking a census of the entire population and determining a density for the entire area of each location, as opposed to using a fixed sampling frame as shown in figure 5.5. If you were to sample fixed areas of each community, it is possible that the placement of the samples could produce biased, misleading results.

Finally, there is the issue of how boundaries are conceptualized. Political or legal boundaries are typically well defined. However, if we consider a concept like "wilderness," different individuals will tend to have different conceptualizations of what the term means, even though a legal definition may exist. Similarly, even the definition of a city varies by individual. Consider how people refer to their hometown when visiting another place. Although you may be from a suburb of a major urban area, for example, Naperville, Illinois, you might say you are from Chicago when visiting Europe or even Los Angeles. Conversely, if you are in downtown Chicago, you would more likely specify your hometown as Naperville. How an individual defines space is rather personal and often contextual. In defining boundaries, you would want to do two things: (1) clearly define the concept being studied and (2) clearly define the boundaries of your study area based on the concepts you are examining. Thus, when analyzing data geographically, it is essential to clearly define your boundaries or to define them based on local knowledge or spatial conceptualization as appropriate to your study.

In some studies, you may be interested in examining a social or physical system. If that is the case, you might prefer an ecological or structural approach to boundary definition. Many systems have boundaries associated with their definitions, which would be appropriate in developing a geographic extent for your data in the GIS.

Review questions

1. Is it better to use primary or secondary data in a research project? Please explain your answer and define the difference between these two types of data.
2. What is the difference between probability sampling and nonprobability sampling?
3. Are there times when purposive sampling is preferable over random sampling?
4. What are some problems that could occur when using the available research subjects approach to sampling?
5. What is the modifiable area unit problem? Please provide an example.
6. What factors do you need to consider when you select your geographic boundary for a research project?

Additional readings and references

Babbie, E. 2013. *The Practice of Social Research*. 13th ed. Belmont, CA: Wadsworth Cengage Learning.

Bolsted, P. 2008. *GIS Fundamentals: A First Text on Geographic Information Systems*. 3rd ed. White Bear Lake, MS: Eider Press.

Openshaw, S., and P. Taylor. 1979. "A Million or So Correlation Coefficients: Three Experiments on the Modifiable Area Unit Problem." In *Statistical Applications in the Spatial Sciences*, ed. N. Wrigley, 127–44. London: Pion Press.

Ormsby, T., E. Napolean, R. Burke, C. Groessl, and L. Bowden. 2010. *Getting to Know ArcGIS Desktop*. Redlands, CA: Esri Press.

O'Sullivan, D., and D. Unwin. 2010. *Geographic Information Analysis*. 2nd ed. New York: Wiley.

Relevant websites

- **Electronic Statistics Textbook (http://www.statsoft.com/textbook/):** This site, from StatSoft, is a complete introductory statistics textbook online, including sections on measurement scales, sampling, and analysis. StatSoft, Inc. 2013. *Electronic Statistics Textbook* (electronic version). Tulsa, OK: StatSoft.
- **US Census—Sampling (http://www.census.gov/cps/methodology/sampling.html):** This website provides a description of the sampling method that the US Census Bureau used for the Current Population Survey.
- **Columbia CNMTL-QMSS e-Lessons: Quantitative Methods in Social Sciences (http://ccnmtl.columbia.edu/projects/qmss/samples_and_sampling/types_of_sampling.html):** This website presents and examines different types of sampling, such as probability sampling, simple random sampling, snowball sampling, and cluster sampling.

Chapter 6

Using secondary digital and nondigital data sources in research

In GIS-based research, a significant amount of time and effort goes into data acquisition and preparation. Data may come from a variety of existing sources or could originate from new collection. With the widespread use of GIS, a growing number of digital data sources are becoming available. Therefore, it is common for many research projects to incorporate a substantial amount of existing, secondary data. Examples may include GIS layers or tabular data representing demographic, health, economic, or environmental information. Data may be acquired from a variety of sources, including local, regional, statewide, national, and international governments; nonprofits; and private organizations. In this chapter, you will learn how to locate, assess, and gain access to existing or secondary data relevant to your project.

Learning objectives

- Understand how to find appropriate secondary data relevant to your project
- Learn how to evaluate data suitability
- Learn how to collect data from online and offline sources that can be easily incorporated into a GIS, including the use of Global Positioning System technology
- Learn the role that validity and reliability play in research
- Learn how to use news as a source of data

Key concepts

data acquisition
data from offline sources
data from the Internet

data suitability
news as a source of data

reliability
validity

Evaluating data sources

In chapter 5 we discussed the differences between primary and secondary data. Collecting new data can be time consuming and expensive, so it is valuable to consider if data are available before embarking on a new data collection effort. In this chapter, you will learn what factors to consider when evaluating and using secondary data. When existing data are not appropriate or available, it will be necessary to collect new, primary data, as discussed in detail in chapters 7 through 10. At the beginning of any research study, it is important to select and understand the data to be used in the analysis. When working with spatial data in GIS, there are additional considerations, such as the mapping scale and accuracy and cartographic representation of the variables of importance. Generally speaking, we refer to data that the researcher collects directly as **primary data**. By contrast, data that someone else has collected, for a different purpose, are referred to as **secondary data**. Although both data types have potential value in a given study, there are pros and cons to using each type.

Searching for secondary data

Countless sources of data exist, and we can in no way cover them all here. Many useful datasets are available via the Internet, especially those distributed by government agencies in the United States and around the world. Such data are often provided in formats compatible with popular GIS software packages. As one searches for data at the local level, the Internet tends to be less helpful. Although many local-level organizations and agencies use GIS and produce data, they may not have the need or the resources to provide online access to their data. Many smaller, local-level organizations, such as nonprofits, do not have the legal mandate to make their data available. It is just as likely that you will find yourself visiting local organizational or agency offices or making phone calls to find out who, if anyone, has the data you need.

In other words, searching for data to use in a GIS analysis is no different than searching for any type of data necessary in conducting a research project. It would be foolhardy to assume that all of the data you will require for a GIS-based study will be ready and waiting for you to download from the Internet. This may seem obvious, but surprisingly, the majority of people with whom we interact around the issue of data make this assumption. Getting out on the ground, talking to people who have the data, and collecting the data yourself can still be essential components of the process and should not be discounted. In fact, for a typical GIS project, as in any study, a substantial proportion of time and effort goes into data collection and entry prior to carrying out the actual data analysis.

We should mention one other item of note at this point. Many data are not readily available, even data collected by government agencies, which are considered public. There are a variety of reasons for this, often related to data incompatibilities, budgets, individual personalities, or agency rivalries. Regardless of the reasons, you may find that it takes some effort to acquire data from some organizations.

Evaluating data suitability

In any of the preceding cases, finding existing data online or through direct contact with an organization or individual can be time consuming, and there is no guarantee that when you do locate such data, they will precisely meet your needs. This process is called **evaluating data suitability**. Careful evaluation of available data is important to any research project. Spending time and effort to secure an existing dataset aggregated at a county level will not serve the requirements of a neighborhood-level study. Similarly, obtaining a dataset that is poorly documented (e.g., doesn't include information on how and when it was collected and entered into the computer) may not serve you well; time would be better spent locating or collecting data of appropriate quality. In short, this is where the computer acronym "garbage in, garbage out" (GIGO) comes into play.

In short, when considering data from any existing source, there are five key considerations:

1. Do the data contain the information you need for your study?
2. Are the data appropriately documented (metadata) such that you understand how and why they were collected and coded in the way provided?
3. Is the format of the data appropriate for your study, and if not, will you be able to convert it?
4. Do the data contain a geographic element to link them to the GIS, or if not, could one be added easily?
5. If there are multiple versions, is this the best one for your purpose?

Obtaining GIS data from the Internet

Vast amounts of GIS data are now available via the Internet. A search on the World Wide Web for "GIS data" returns millions of websites. The issue of locating data online is not so much one of finding data as it is one of finding appropriate data for your project. To this end, there are several important considerations when searching for data online.

Perhaps the most important consideration relates to the source or provider of data. With thousands of providers online, it is not unusual to locate multiple versions of a dataset that appears to meet your needs. In very broad terms, data will be available from four unique types of providers: (1) government agencies, (2) universities and research organizations, (3) nonprofit organizations, and (4) private firms. What may be less apparent is that in many instances, these varied data sources may all be based on the same, original government data.

Much of the existing, available GIS data originate in government agencies such as the US Census Bureau, the US Geological Survey (USGS), the Environmental Protection Agency, the National Aeronautics and Space Administration (NASA), and others. In many cases, data you obtain from private firms fall into the category of *value-added data*. In simple terms, this refers to government data that have been somehow enhanced by a private firm and packaged in a user-friendly manner. For example, the USGS quadrangle maps, a commonly used data source, are publicly available; however, a number of private companies have taken the time to collect and organize these maps on CD-ROM or website interfaces to facilitate their easy use. Some include additional information for specific purposes, such as planning a backpacking trip or navigating with a Global Positioning System unit in your car.

Of course, these companies charge a fee for these products. When you purchase such products, you are not paying for the data per se but rather for the convenient access and interface the companies provide for working with the data. The advantage of data that have been preprocessed in this fashion is that they tend to be easier for the novice user to access, may already be packaged for a popular GIS software package, or provide other benefits.

Similarly, nonprofits, especially those with limited budgets, may obtain freely available government data and add value to the information as well. Where we live, a number of nonprofit environmental organizations monitor natural resource issues related to logging and salmon. These organizations tend to be suspicious of the private industries that they are monitoring, so they will obtain public data and supplement them with their own field data. Because these nonprofits tend to run on very limited budgets, they often share data among themselves and rely on volunteer labor to compile the GIS data.

In situations where similar data are available from multiple sources, the value of one version over another may come through the added information, through the time and effort saved by using data someone else has packaged, or a variety of other perceived benefits. The question for you, the researcher, is which data meet your needs the best. If time is of the essence, the cost of commercial data may well be worthwhile compared to the time and cost of in-house data preparation and processing. If, however, time and processing are less significant concerns (perhaps you have student interns from the local university), obtaining original data directly from the agency that produced the data could be preferable.

This all sounds great, but what other issues should you consider when searching for secondary data? Everything we discussed in chapter 2 is a consideration. That is, once you think through your project, you should have a sense of the **scale** of the analysis, the **data model** that is most appropriate (vector or raster), and the map **coordinates** and **datums** with which you wish to work. All else being equal, avoiding conversions in any of these four areas will reduce the potential for error in the spatial component of the data. If you must make conversions, changes to the coordinate system or datums are generally reliable in the GIS software.

Scale conversion is a first consideration. A map collected at small scale (e.g., 1:100,000) would be a poor choice for use in a GIS analysis where you need to know on which side of a road something is located. Although you can zoom in on the screen and see this detail, the error associated with the lines on a map of 1:100,000 scale permits the line representing the road to be misplaced by well over a hundred feet, more than enough for a particular object to appear on the wrong side of the street. A good rule of thumb is to obtain maps at a larger scale than necessary; it is OK for you to zoom out. The opposite is not true—zooming in can lead to errors in your analysis.

When converting data models (raster to vector, or vice versa), a similar scale problem can occur. For example, when converting raster pixels that were originally derived at thirty by thirty meters (nine hundred meters squared) into a vector map, you should be careful not to analyze the resulting vector map at accuracies greater than the original accuracy. Figure 6.1 shows the origins of this conversion problem and how an unknowing end user of the data might not have the detail that was anticipated.

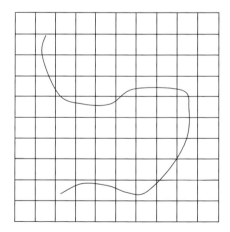 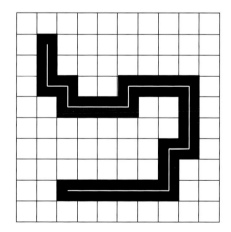

Figure 6.1 The graphic on the left represents the true shape and position of a linear feature in space. On the right we see the pixels (represented as black squares) that would be mapped in a raster format, for example, a land use, land cover map provided by the US Geological Survey. If you were to take the raster map and convert it into a vector dataset, the resulting line (in white) would be placed at the center of the raster pixels and no longer accurately represent the original feature.

An equally important issue relates to the data attributes. The attributes are the information about features represented on the map and contained in a database that accompanies the map information. If you were looking at the attributes for a dataset showing local businesses, there might be information about the number of employees, types of industry, gross sales, and so on. As in any data collection process, the data coding process will have a predetermined level of measurement (nominal, ordinal, interval, or ratio) incorporated. Regardless of the spatial accuracy of GIS data, if the information you require for your study is not present in the attribute database, your analysis will not proceed.

In some cases, the data attributes may be decoupled from the map information. That is, you may locate some of the datasets you require as plain database files, which then must be connected to the spatial component of the map. For example, you might have a US Postal Service map showing the locations of the ZIP Codes in a community. Separately, you could obtain membership lists (these being databases in some form) from a variety of organizations of interest in your study. Although these datasets may not have originally been associated, or even created by a single organization, they can be linked to create a spatial dataset, as discussed in detail later in the chapter.

Of course, it is possible that the data you need will not be found on the Internet or that the data you find online are insufficiently documented or from an unknown source so that you are uncomfortable including the data in your study. It is important to remember that just because data are available on the Internet does not necessarily mean they are of high quality. In many cases, you may find that searching for data elsewhere or creating your own, new data is preferable.

Choosing GIS variables

One of the most powerful things about using a GIS as part of the research process is that GIS allows for the holistic study of a question or issue. We can describe using GIS as providing a socioecological model of whatever issue or problem you are interested in studying. Why do we say this? A GIS allows for the simultaneous integration of many different variables and of information from a variety of sources and disciplines in examining and answering questions. Furthermore, the visual output of GIS provides a compelling means for communicating complex information using easy-to-understand outputs. Although similar research results could be summarized and communicated with statistical tables, regression graphics, or lengthy reports, all of which remain important tools for the researcher, map-based outputs are perhaps the most readily accessible for communicating research results quickly and efficiently.

As mentioned earlier, the GIS can be part of both the data collection process as well as the analysis. A GIS gives the researcher the option to communicate with the research subjects on another, visual level. This could be invaluable in a study or culture where perhaps people are illiterate or don't speak the same language and where a visual representation of data could make more sense to the subjects of the study. Beyond this, a GIS allows for the creation of models that mirror real life. Using the GIS allows the researcher to contextualize variables in space and time, thus incorporating a variety of social, environmental, physical, or even conceptual environments into a study. It provides what we define as a holistic systems approach, which is socioecological in nature. The researcher who studies almost any social problem or issue could find such a tool useful.

Questions to consider about data

As you consider which forms of data and variables are appropriate to your study, you should simultaneously think about (1) how GIS data will be integrated into your study and (2) how GIS (or, more generally, space) relates to your chosen variables. These are not always simple questions to answer, but their answers can be informed by considering a series of questions relating to the "GIS fit" for your project:

1. What is the main goal of your study (predictive, comparative, or descriptive)?
2. Does your study involve a variety of concepts and variables, and do their locations matter?
3. What data exist for your study location and/or variables?
4. Does your study involve a comparison of different locations?
5. Does distance play a role in your study?
6. What are your project resources?

As you answer these questions, it will help you to determine if available secondary data sources are appropriate for use in your research. If such data are not available, you may need to consider options for the collection of new data specifically designed to meet your research requirements. It is important to recognize that many researchers use both secondary and primary data in their research projects. In the following sections, we explore each of these questions in more detail.

1. What is the main goal of your study (predictive, comparative, or descriptive)?

Research projects can have a variety of goals. Although GIS has a potential role in many research studies and projects, its use will vary depending on the goal of a particular study and the types of data. A study that is descriptive focuses on documenting and portraying information conditions or other information about a particular study location, time, and topic or theme. A GIS would be useful here in recording the status of various features, facts, and information about a particular place in a spatial sense. For example, the decennial census in the United States provides a snapshot of the nation every ten years. Combined with data representing multiple themes or variables, a study based in part on these data can provide a picture of the population and its relationships with a variety of other information. Such studies provide a descriptive snapshot for a particular place and time (figure 6.2).

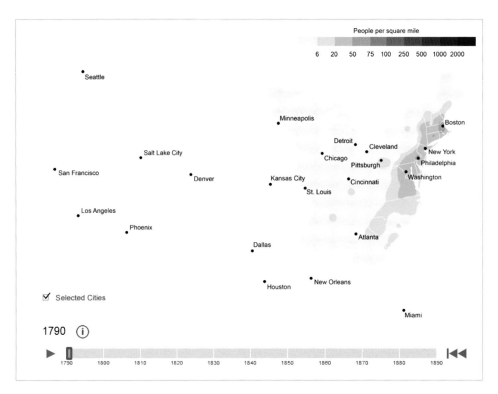

Figure 6.2 These three visualizations from the US Census Bureau show the westward population migration from 1790 to 1890. US Census Bureau, https://www.census.gov/dataviz/visualizations/001/.

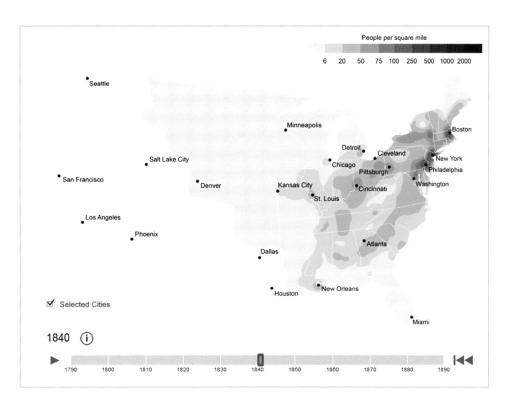

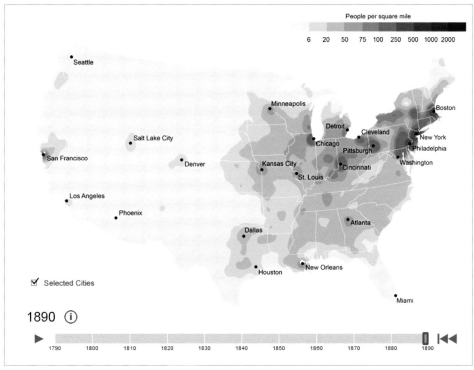

Figure 6.2 (cont'd)

In a predictive study, the goal is to develop an analysis of one or several possible outcomes based on known or existing information. Many scientific studies use sampled data representing a limited set of parameters to develop predictive models representing natural or behavioral scenarios. For instance, one might look at relationships of socioeconomic status relative to demand for particular products or services to optimize the selection of locations for a new or growing business franchise, or examine relationships between rainfall and infiltration rates to predict the likelihood of flooding in a particular region before approving land use or the development of new water control structures.

In a comparative study, one seeks to assess similarities and differences between two different sets of data. The comparison could be based on two different geographic settings or two different groups. For example, you might want to assess differences between members of a particular ethnic group in both rural and urban settings, or might be interested in the outcomes for a particular ethnic group in two similar urban areas in different parts of a state. A GIS could be used to track and document characteristics related to these groups in each setting, allowing for comparison of group-based information in different environments. Even though decennial census data are valuable in many studies, they may not be sufficiently detailed or current to answer your research questions.

2. Does your study involve a variety of concepts and variables, and do their locations matter?

To answer this question, you need to be able to think about the different components of the question and the role that spatial relationships might play in answering it. For example, does your research project focus on one area, or does it consist of multiple areas? If it is focused on one area, does location play a role in answering your research question? For example, you might be interested in the effects of particulate matter on community health and have a hypothesis that those living upwind and downwind of a suspected source would show statistically different levels of respiratory disease. In this situation, there are clearly spatial relationships that you might want to explore. Examples could include locations of populations with particular health issues, information on wind speed and direction, data on the content and chemical characteristics of relevant particulates, and so forth. Such factors would require knowledge of models on how these factors affect heath and additional supporting information. As is true with many research questions, some, and potentially many, data types will have a spatial component that would benefit from a GIS-based analysis. Although existing data may be available, if spatial relationships such as wind direction were not considerations at the time those data were collected, the necessary location information to distinguish which neighborhoods are represented in the data may be missing.

3. What data exist for your study location and/or variables?

No matter what topic you choose to study, there is a good chance some existing data will be available. Your first goal as a researcher is to locate available data and assess how well it matches up with the needs of your study. For example, does an existing dataset corresponding geographically to your area of study contain attribute information relevant to your topic? If so, do these data also correspond with the temporal needs of your study? An issue that surrounds data use is whether the data fit the problem. While it is common for data to exist in some form (e.g., digital, printed formats such as newspaper articles, data sheets, or written reports), these data may be outdated or collected using methods or measurement levels inappropriate for your research.

4. Does your study involve a comparison of different locations?

If your study focuses on answering a question by looking at data from two or more locations, there is a clear value in exploring the data spatially. A variety of factors may affect the comparison sites differentially, be they environmental, infrastructural, or otherwise. Using the GIS, you can extract spatial variables to compare and contrast information between localities, which may provide important insights that are not otherwise apparent. Using secondary data requires the information available for each location being studied and compared. If all of the data are from a single, consistent source, the secondary data could be very useful; however, if you are limited to data collected by multiple sources (perhaps locally for each area considered in your study), the manner and time of collection must be similar to make comparisons valid. Data collected by different organizations or using different methods may not provide the needed information to answer your questions.

5. Does distance play a role in your study?

You can determine if space or distance plays a role in your study by understanding if proximity and location are important to your research question. For example, where is the study location situated relative to natural resources, access to infrastructure, markets, or other important factors? If there may be a question relative to access or exposure to particular factors in space and time, using the GIS to quantify and assess these characteristics can reveal important information. For example, although two locations may, on the surface, appear similar based on their socioeconomic descriptions, they may demonstrate very different mechanisms that lead to this circumstance. For example, two seemingly identical upper-income neighborhoods might have developed for completely different reasons, which would be missed looking only at their socioeconomic data. Perhaps one capitalized on the local resources (logging, mining, fishing, etc.), whereas the other may be an upper-income suburb populated by highly educated professionals (doctors, lawyers, businesspeople, etc.). These differences in access to particular resources, markets, or other factors may not show up in socioeconomic descriptions limited to ethnicity and median household income.

6. What are your project resources?

Carrying out any project takes resources, including time and money. For some questions, available data may not meet your needs. When resources allow, it may be appropriate to obtain new, primary data specific to the needs of the question at hand. In other cases, you may opt to explore using the next best thing, perhaps a dataset that mostly meets your needs and that can be updated, or supplemented, to serve your objectives. In other cases, a surrogate dataset could perhaps be used as a stand-in for the data you require. For example, a study requiring traffic counts might substitute a GIS layer indicating road type (highway, arterial, collector, and local) to develop estimates of the number of vehicles rather than investing in a new count.

In recent years, access to individuals with GIS experience has grown tremendously. Similarly, access to powerful GIS software and analytical tools has greatly expanded as the computing hardware to support it becomes more affordable. Software technologies such as ArcGIS are now available through a wide variety of channels, including cloud-based options, making it possible to support sophisticated analysis with a minimum of infrastructure within an organization. Simultaneously, spatial data have proliferated both online and from a wide variety of organizations, making access to secondary data much easier than it was just a few years ago. In short, there's plenty of hardware, software, data, and people to support your spatial analysis needs. However, what may be missing is the specific knowledge and expertise to bring these together in service of answering your specific research question.

Validity and reliability

When preparing to collect data for a study, whether secondary or primary, two important considerations related to your measurement of the variables and their attributes are **validity** and **reliability**. A good empirical study cannot proceed without data that are both valid and reliable.

Validity

The term **validity** means making sure that there is congruency between the concepts that you want to measure and your measurement techniques. In other words, does your research project or study really measure what you are setting out to measure? This is relevant not only in assessing available secondary data, but also in collecting your own primary data. For instance, let's suppose that you are interested in studying the popularity of smoking cigarettes in a community. You could collect primary data; for example, you could employ unobtrusive measures, observing how many people purchase cigarettes at a local store at different times during the day. Another approach might be to go to different public places, for example, restaurants and bars in certain states, to observe how crowded the smoking sections are.

Consider, though, that some concepts are strongly correlated, in which case it may be easier to observe one over the other through using secondary data sources. For example, assume there is reliable information regarding the relationship between the annual number of medical visits per capita and cigarette use. If this were true, perhaps obtaining medical data from a secondary source would serve as a reasonable surrogate for cigarette use and could save time and effort in the field.

Validity of secondary data and GIS

In the GIS context, validity is important for accurate analysis. Earlier, we provided an example of surrogate data in GIS, substituting a road layer coded with road classes in lieu of actual traffic counts. For this substitution to retain validity in our study, we would need to feel confident that assumptions for traffic levels, the concept of interest, hold true (or reasonably so) for the area and road types available in the surrogate data. One means for doing this might be to apply a sample of recent traffic counts in the study region to all of the identical road classifications across the study area.

As another example, if you are interested in studying urbanization of a rural area, you would need to understand the amount of land that has been converted from a natural or agricultural designation into a developed designation over time. Developed land is a category that would include houses, roads, and other built features. These features indicate human development via the presence of manufactured structures.

One could use a number of data sources to get at this concept, for example, census data, building permits, or agricultural sales records for the specified time period. To conduct a valid study, you would want to choose variables and employ methods that reflect the actual concept that you are studying—urbanization. A likely surrogate that already incorporates the spatial components might be based on historic maps or aerial photos. You could interpret features visible on these sources and compare them to features on more recent maps or photos to compare acreage in particular land use categories through time. A review of the county assessor's records might show zoning designations and ownership changes that indicate when and how land was converted from one use to another.

A simple way to accomplish this would be to derive a change map by overlaying datasets in GIS to determine the percentage of land in the county that has been converted from natural categories to the category of developed land. It might be an invalid measure to make assumptions about the phenomenon of urbanization by looking at the amount of agricultural land in production. Some of these areas may simply be idle, serving as fallow or abandoned fields, conservation easements, parks, or serving other nondeveloped, nonurbanized uses.

Reliability

Reliability is the second major consideration in the process of measurement. Reliability can be defined as quality and consistency of measurement: Do the research methods that you employ in your study provide an accurate and consistent measure over time? As an example, imagine you were checking your weight on a scale at home. If you stepped on the scale in the morning and it gave your weight as 130 pounds, and then in the evening after work you came home and the same scale gave your weight as 160 pounds, chances are there is something wrong with the measurements reported by your scale. In other words, your scale is not reliable. It is up to the researcher to determine the reliability of the measurement devices being used in the field.

For field-based research, this could include everything from the instruments used to measure and record data to the consistency of the field staff in using tools or asking survey questions. In theory, if the same entity is measured several times (over a reasonably short period), those measurements should remain consistent. This should be true if the same item is measured with different instruments (e.g., all thermometers should give the same air temperature) or by different people (e.g., Bob and Judy should both obtain the same air temperature when each is using a given thermometer). Of course, we expect the values for some measurements to change over time, for example, the temperature may change several degrees between morning and afternoon. Others, such as the position of a fire hydrant, should remain stable over many years. Reliability of secondary data sources is also important. Descriptions of the instruments used, data collection methods, and quality control for data input or transcription should be present in the metadata. If this information is not available, you may need to make a determination about the reliability of the data based on other less rigorous factors (e.g., Did a reputable organization collect the data? Does the data come with other documented sources?).

If you are collecting primary data, be certain that all procedures related to data collection, measurement, and data entry are well defined and consistent. Staff should be trained in collection and documentation protocols, and instruments used to collect and manage the data should be calibrated to ensure accuracy and consistency. Simply put, it is a good idea to check your methods and equipment to anticipate and correct, in advance, any issues that could affect the reliability of the data.

Reliability and GIS

The California Redwoods are an environment that attracts recreational users and preservationists who enjoy interacting with the forest in different ways. If you were interested in measuring the impact that mountain bikers have had on erosion along trails in a park, how might you measure this in a GIS? One possibility might be to walk the trails, marking a map at locations where you observe erosion along the trail (perhaps how an erosion map from twenty years ago would have been created). These data could then be **digitized** into a GIS data layer. Another approach would be to use a **Global Positioning System (GPS)** receiver to go into the field to record locations along the trail where erosion is present and download these data directly into the GIS.

If, upon review of these two datasets, you notice significant differences in the amount and location of the erosion recorded along the trail, there are several plausible explanations. Differences could simply be due to the reliability of the technology used to map erosion. Your GPS unit could be misreading locations because of a dense canopy, or perhaps some locations were mismarked on the map while you were hand recording the data. If data included in the study are coming from historical, secondary sources, there might be reliability issues relating to how the data were originally collected. Sometimes it is best to use more than one measurement technique or compare multiple sources of data to ensure that data collection is reliable and, when discrepancies occur, that you can determine which technique is best in your particular application.

When data are entered into the GIS, the reliability of this process should also be assessed. Field data marked on a map or collected via a GPS receiver need to accurately line up to the proper locations on the basemap being used. Mismatches of datums, projections, coordinate systems, and so forth can lead to misalignments in the geographic position of the data. If these mismatches are severe, they could invalidate your study results. If secondary data derived from nondigital sources are part of your study, the methods for how these data were transcribed into the computer and converted to digital formats for use in GIS may affect the reliability of the resulting data. For example, data hand entered or scanned from paper files may be entered incorrectly or misinterpreted by the scanning software.

Furthermore, it is important to note that as spatial data are entered into any GIS software, in particular by hand, they may be entered using a different set of tolerances or settings related to positional accuracy. Regardless of the specific technique used (scanning, digitizing, etc.), any data entry processes for maps and images involve a rasterization process (even when creating vector data). Fortunately, modern GIS software makes it relatively easy to set and view tolerances associated with data entry, editing, and transformations, such as changes to projection, datums, and so on (figure 6.3). It is important to understand what accuracy was associated with the collection and entry of any secondary dataset under consideration for use in your analysis as well as any effects that may come with conversion of data from one format or platform to another.

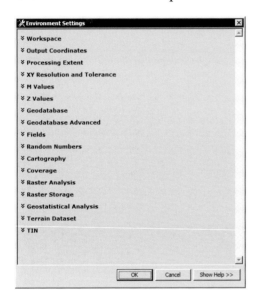

Figure 6.3 Environment settings, such as those defined in ArcGIS, provide control over how the software handles items like x,y resolution and tolerance globally. Similar settings are available during data editing processes. Esri.

Figure 6.4 illustrates why this is the case. All computer screens, scanners, and digitizing tablets are composed of a set of pixels, CCDs, or gridded wires. When a mapped location is entered into the computer, there is a good chance that the actual coordinate on the ground will not align perfectly to the grid system used internally by the computer for data storage and display. This results in the point being shifted to the nearest grid intersection when stored in the computer, and although these shifts are typically extremely small and do not significantly affect map accuracy, they do exist. In most cases, this issue can be ignored; we mention it here both for completeness and also so that if you see minor shifts in spatial data you have entered carefully, you will know where these originate and, if necessary, adjust the tolerance settings to minimize the issue.

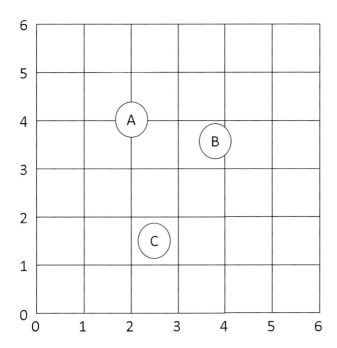

Figure 6.4 Points that should, in reality, fall in between grid intersections are recorded in the computer's memory in whichever grid intersection is closest. Point A falls at intersection (4, 2) and would be properly mapped at that location. Point B lies off of any intersection but clearly lies nearest to (4, 4) and would snap to that location. Point C presents the most complex case. Because it lies equidistant from four different intersections, the software may map it to any one of them, and as the end user, you probably do not have an easy means to control this decision.

In GIS and other computer graphics programs, a **snap distance** will be defined (figure 6.5). This describes how close two points entered into the computer must be to be considered unique. Conversely, if two points are within the defined snap distance, they will be automatically connected by the computer. Of course, if the spacing used for the grid is large, these errors will increase proportionately. Therefore, choices made for scanning resolution, the megapixel rating of a digital camera or sensor, snap distance, or other data tolerance settings can significantly affect data quality. Keep in mind that although you can adjust software settings to some degree, these errors cannot be completely eliminated and are a normal part of the data development process in GIS. It is important that you have a basic understanding of how the computer stores data on a grid so you can minimize errors to a point that they will not affect your study results.

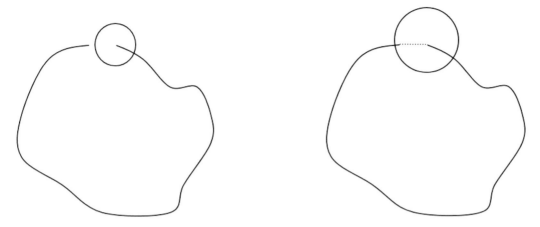

Figure 6.5 Snap distance is best viewed as a circle of a given radius. In the left-hand figure, with a smaller snap distance set, the gap in the polygon would remain open. However, the same figure prepared with a larger snap distance setting would result in the GIS automatically closing the gap. Depending on the type of data, and the accuracy of entry, errors can exist in two forms: (1) gaps can exist where they should not and (2) gaps can be closed inappropriately where they should remain open. Neither is desirable, so finding an appropriate balance for these tolerance settings is important.

Obtaining data from offline sources

Quite often the data you require will come from sources that do not have an online presence. Examples include local agencies and organizations, professionals, students, or archival sources. The same data quality considerations hold true for data obtained through offline channels; however, when you are obtaining data from offline sources, several additional issues are worth considering.

In the case of historic or archival data, it is highly likely that the data exist only in printed or other physical formats rather than as digital computer files. Examples of such sources include anything from a printed map sheet, a figure in a text or journal, or even the actual data sheets or notes residing in someone's file cabinet. Other sources of data might include photographs, audiotaped interviews, or video footage.

Any data that exist in one of these physical formats will require conversion to a digital format if they are to be integrated into a GIS. When we refer to a digital format, we simply mean that the data are in a format that a computer can directly manipulate. When you obtain data in physical formats, you will need to enter them into the computer as a first step in the data development process. Therefore, if there is an option to obtain similar data in digital form, you can avoid a substantial amount of data entry. The time and cost of entering nondigital data into the computer for integration into the GIS analysis should be weighed against the benefit of doing so.

For many questions, including data sources collected in the past is obvious. The primary consideration should lie in the reliability and consistency of the data, especially if they will be used as baseline information for determining trends or changes. Data collected in the same way (both spatial and nonspatial) by the same organization over time, for example, the US Census, can prove extremely valuable in an analysis. However, data collected using a variety of different methods, mapping scales, and technologies may simply lead to an erroneous analysis.

Let's illustrate this point with a simple example. Imagine using a map of North America drawn by explorers in the 1500s in conjunction with one collected using satellite imaging to determine changes in the location and extent of populated places in the eastern United States. Of course, we would expect that the accuracy of maps made in the 1500s would be substantially different than those made today (figure 6.6). Thus, differences observed between these two datasets are due to differences in the data quality more so than they are due to actual differences in the sizes and locations of the places under study.

Figure 6.6   An explorer's map from the 1500s depicting the Americas as mapped at that time. *Americae sive qvartae orbis partis nova et exactissima descriptio/avctore Diego Gvtiero Philippi Regis Hisp. etc. Cosmographo; Hiero. Cock excvde. 1562; Hieronymus Cock excude cum gratia et priuilegio 1562.* (Gutiérrez, Diego, fl. 1554–69.) US Library of Congress.

Although the preceding example presents an obvious difference in data types, there are often more subtle differences. For example, what differences in data quality, accuracy, or methodology would you expect to be present in data collected just a few decades apart? What about data collected a few years apart? What about data collected on a single topic but by different organizations? What about other information that may not be included in a particular dataset?

Consider data on occurrences of lung cancer. Would you feel confident that lung cancer is decreasing if you were to see a drop in the number of cases between 1970 and 1990? What if you knew that the 1970 data came from a government public health agency and the 1990 data came from an industry report? The more likely explanation would be that the differences are not due to a real decrease but rather to a difference in data sources and methods of analysis. What if all the data came from a single, reliable source, but then you discovered that a large, polluting industrial facility in the study area closed down in 1975?

One of the great benefits of a GIS is its ability to integrate data from a variety of sources. In the preceding example, even if you didn't know the relevance of pollution, socioeconomic factors, medical care, and so on, if data are available, the GIS is an effective tool for exploring potential relationships between and among various datasets, often from different disciplines and sources. So long as the data you incorporate are of similar mapping quality and scale, they can be integrated with other sources using location as a common link. However, when the detail of one or more datasets is of substantially lower resolution (i.e., one dataset is state level and another is county level), meaningful spatial analysis may be possible only at the level of the least detailed data.

Statistical interpolation methods (discussed further in chapter 11), used to estimate data values in unsampled locations, can sometimes provide a workaround but may not be optimal. Deciding when data are appropriate to include in an analysis and when new or different data may be required will depend on the kinds of questions and answers you require. If local detail is necessary, estimating from state or national data may be inappropriate; however, for broad-based studies, that same national dataset may be perfectly acceptable.

Using news as a source of data

News can provide an excellent source of data. In addition to being a good source of background data for your research, the news may actually be the data you are studying. Some researchers may choose to investigate a research question that involves the news as a data source. The news has a lot of information that could be useful for a researcher who is interested in conducting content analysis within a GIS. Almost all articles in a newspaper will list a location for the story in the opening line. This alone could be the basis for linking stories to geographic locations. It might also be important to know the location associated with the newspaper or the author of the article. Similar spatial information is included in some form in magazine, television, and online news formats.

News analysis might be done using a content analysis approach, among other methods. A researcher would search through the news stories in an attempt to find patterns related to a particular topic or subject. Let's say, for instance, that you are interested in conducting a content analysis of newspaper articles related to immigrants (see chapter 12 for details on how to conduct content analysis). To incorporate a GIS into your study, you could note the location where the newspaper is produced and see if that relates to the national origin of the immigrants being discussed. For instance, would the articles referring to immigrants from Canada be published from locations in the northern United States? Or would articles that focus on Latino immigrants appear more often in West Coast or East Coast newspapers? These are questions that could be answered by analyzing the content of articles in newspapers from different cities.

Another example using news as a source of data and integrating this into GIS might be if you worked for a transportation agency that wanted to identify patterns of car chases, which have become a regularly reported event in the media. You could integrate the GIS to help identify patterns in your analysis. For example, you might be interested in determining if there are differences based on the police jurisdictions in which the chases occur.

Let's say that, in your analysis, you were going to go back through a review of TV news segments over the last ten years for various parts of the country. You could get a measure of the dates on which car chases occurred, the times that the chases occurred, what types of road the chases occurred on (freeway or city street), where these streets were located geographically, and whether the drivers were charged with some other crime in addition to the car chases. The GIS would be an important part of the process of identifying patterns of car chases. Identifying these patterns could, it is hoped, provide information to the authorities about how to better intercept and prevent such chases from occurring in the future.

Review questions

1. How does one determine the GIS fit for a research project?
2. What do we mean by the terms *validity* and *reliability*?

3. How does ArcGIS tie in with validity and reliability?
4. What are secondary data?
5. What are the advantages and disadvantages to using secondary data?
6. How does a researcher decide which secondary data are the best to use for a particular project?

Additional readings and references

Elsley, M., and W. Cartwright. 2011. "Issues Related to the Use of Non-traditional, Digital Data Sources for Enhancing Park Management Data." Paper presented at the Surveying and Spatial Sciences Biennial Conference, Wellington, New Zealand.

Ghose, D. T., and D. K. Mohanta. 2013. "Reliability Analysis of a Geographic Information System–Aided Optimal Phasor Measurement Unit Location for Small Grid Operation." *Journal of Risk and Reliability* 27, no. 4: 450–58.

Guion, R. M. 2004. "Validity and Reliability." In *Handbook of Research Methods in Industrial and Organizational Psychology*, ed. S. G. Rogelberg, 57–76. Malden, MA: Blackwell.

Kirk, J., and M. Miller. 1985. *Reliability and Validity and Qualitative Research*. Thousand Oaks, CA: Sage.

Murray, A. T., and T. Grubesic, eds. 2013. *Critical Infrastructure: Reliability and Vulnerability—Advances in Spatial Science*. New York: Springer.

Steinberg, S. L., M. Strong, N. Yandell, and A. Guzman. 2008. *California Center for Rural Policy Rural Latino Project Final Report*. Report for the California Endowment. Humboldt State University, Arcata, CA.

Relevant websites

- **Atlas of Canada (http://atlas.gc.ca/):** This site is an example of an interactive GIS website.
- **Centers for Disease Control and Prevention (http://www.cdc.gov/):** This agency provides an array of national databases related to public health, disease, births, and deaths.
- **Esri (http://www.esri.com/data/find-data):** Esri provides ready-to-use, high-quality geospatial data for GIS visualization and analysis projects.

- **Harvard University Graduate School of Design (http://www.gsd.harvard.edu/gis/manual/data/):** By Paul Cote, Geographic Information Systems Specialist, cultivating spatial intelligence and sources of spatial data.
- **National Aeronautics and Space Administration (http://www.nasa.gov):** Provides a variety of imagery and monitoring data for the world or portions thereof.
- **Natural Resources Canada (http://www.nrcan.gc.ca/home):** Provides a variety of national datasets, including aerial images, geographic places, maps, and topographic information.
- **US Census Bureau (http://www.census.gov):** The official site for the US Census, demographic data, TIGER files, and other related data.
- **US Environmental Protection Agency (http://www.epa.gov):** Provides a variety of national databases related to environmental quality.
- **US Geological Survey (http://www.usgs.gov):** Provides a variety of national datasets including aerial images, geographic places, maps, and topographic information.

Chapter 7

Survey and interview spatial data collection and databases

In this chapter, you will learn about data collection via development of survey and interview instruments for use in a spatial analysis framework. You will run through a series of steps and questions in the interview creation process that will allow you, no matter what your background or discipline, to develop useful and spatially based interviews and surveys and the resulting databases. Additionally, you will learn how to approach data collection using GIS in the field with and without a computer. You will also learn about various data collection considerations, units of analysis, and factors to consider in creating a spatial database.

Learning objectives

- Learn how to incorporate the spatial component into surveys and interviews
- Learn how to use GIS both with and without a field computer
- Learn how to choose a basemap to which to tie your research data
- Learn how to collect data that can be easily incorporated into a GIS, including the use of Global Positioning System technology
- Learn options for importing and exporting data between a GIS database and other sources

Key concepts

database concepts
data collection considerations
developing your data
interview-based data collection and sampling
spatializing questions
survey-based data collection and sampling
unit of analysis
using GIS in the field

Developing your own data

At the beginning of any research project, you will face the following question: Where am I going to get the data necessary to answer my question? Depending on your topic, you may be able to draw on existing data, either within your own organization or from a number of public and private sources (secondary data). Sometimes the data you require are specific to the question and need to be collected from scratch (primary data). In many studies you will use a combination of secondary and primary data.

One key advantage of primary data is that you, as the researcher, have control over the method and form of data collection and storage. This helps to ensure you have the best available data for your analysis. You could collect such data in many different ways. In this chapter, we focus on two common approaches: the survey and the interview.

Survey-based data collection

A survey is a useful tool for collecting information from a variety of people. It can be a quick way to generate data for your project. Surveys are good for collecting data on people's attitudes, opinions, and practices. Of course, for use in a spatial analysis, responses must be linked to some location, for example, a neighborhood, city, or county, to allow for comparison across the geography of interest.

When is it appropriate to use a survey? The best time to use a survey is when you want to gather specific information quickly and efficiently. The survey is the perfect tool to use to obtain targeted information in a relatively short time frame. You have likely seen the results of surveys in your everyday experience. The media commonly use mapped results to express public opinions regarding current events, consumer confidence, or even which team the public thinks will win the Super Bowl. Surveys may be administered at one point in time, a process known as the cross-sectional study, or at multiple points in time, for a longitudinal (across-time) study.

Though most survey research is done using similar approaches, there are a variety of methods for collecting data. A survey can be done face-to-face or via phone, Internet, or postal mail. Surveys typically consist of a number of questions related to the topic under study and can involve both open-ended and close-ended questions. An **open-ended** question gives the interviewee or respondent the opportunity to craft his or her answer; there are no bounded answers. On the other hand, a **close-ended question** presents a query along with a series of answers from which the respondent (the person who completes the study) can select.

A survey involves collecting data in a written or oral format and may be self-administered or given in a manner where the researcher reads the survey questions to the **respondent**—the person completing the survey—and records the answers. One commonly used set of survey data for spatial analysis is the decennial census. Although these data are not specifically collected for your study question, they can provide a significant, spatially coded set of social, economic, and demographic information at a national level. The US Census uses a combination of self-reported answers and answers solicited by staff. Census takers are sent out into the community to help administer and collect the survey data. Sometimes people who are older or who have special needs or have difficulty reading or filling out the forms may require additional assistance completing the survey. Survey questions are always numbered.

The key to designing a good survey is to first have a topic in mind that you want to explore. Next, you want to be clear about the questions you need to answer relative to your topic. So, for instance, if the research topic is access to healthy food, the overarching questions that we might want to investigate as part of this process are as follows: (1) Do people have access to healthy food in their neighborhoods? (2) What sorts of healthy foods are available? and (3) Where are these healthy foods available?

A survey should consist only of questions for which you really need answers. Because it may be tempting to collect answers to lots of questions when conducting your own survey, it is a challenge to determine how many questions to include on a survey. Often researchers err on the side of asking too many questions, sometimes including information that is not central to the project. The danger of this approach is that too many questions may lead to the respondent opting not to complete the survey.

You can use the two types of survey questions, open-ended and closed-ended, when designing the survey. A close-ended question can be as simple as a yes or no question: Do you live in the Northwood neighborhood? Yes or no. Anytime you have questions that are categorical, you can assign numbers to the possible answers. This allows you to *precode* the answers within the survey. Then, when it comes time to enter these data into your database, you will have numbers to associate with the different possible answers.

Using the previous example of a study assessing people's access to healthy foods across a particular region, you decide to use a survey because it is a quick and easy way to gather this information. Is healthy food available in your neighborhood?

1. Always
2. Sometimes
3. Not usually
4. Never
5. Don't know

The person completing the survey would have the opportunity to choose from these five categories to assess whether healthy food is available in the community. By contrast, an open-ended question allows the respondent the option of providing his or her own answer without any bounds. An example of an open-ended question might be as follows: Where in your neighborhood do you shop for healthy food? This would allow the respondent to answer without being influenced by the presence of specific answers or categories to choose from.

Survey sampling

A key topic to think about before you actually administer the survey is how you are going to sample. Are you interested in people from a particular geographic region? How are you going to determine the boundaries for this region? A smart approach before conducting a survey is to develop a sampling frame—essentially, a list of all of the possible people from the population who could be included in your study. If you were to randomly sample from this list, you would avoid any bias that might come in choosing survey participants. By engaging in random sampling, you gain the advantage of generalizing your findings to the larger population. If, however, you engage in nonprobability sampling or nonrandom sampling, your results may not be reliably generalized to the larger population. There are times when you may be seeking out a particular group or population that warrants a purposive sampling approach, but this should only be adopted when it is appropriate to the purpose of the study. For example, you may be interested in studying individuals who suffer from a particular medical condition. Even in this situation, if you have the ability to generate a list or sampling frame of potential respondents, it is then preferable to randomly select from this list to produce a random sample for the survey.

In figure 7.1, we provide some example questions taken from a survey instrument for collecting data from businesses, including a geographic component. The survey was designed to examine the flow of goods and services on local and nonlocal bases. The full survey was more than forty questions; however, a number of specific questions were included to get both qualitative and specific (e.g., mappable) spatial information from respondents. If you are able to ask the specific addresses of survey respondents, that would be the most direct way to gather location information; however, because many surveys collect potentially sensitive, personal information or opinions, it is not always feasible to request a specific address or location. In such cases, you should consider a geographic identifier that provides an appropriate spatial scale for your analysis, while simultaneously maintaining a reasonable level of confidentiality for the respondent. For example, you might ask for the nearest major cross street, the ZIP Code, or the name of the neighborhood. For studies examining larger geographic regions, it may be sufficient simply to know the city, county, or even state from which the responses come.

What percentage of the products that you use in your business come from outside your community?
1. none – everything we use is from our community
2. 1 to 20%
3. 21% to 60%
4. More than 60%

What percentage of your business sales are to people from outside of your community?
1. none – everything we use is from our community
2. 1 to 20%
3. 21% to 60%
4. More than 60%

Why did you choose this geographic location for your business?

How many minutes does it take you to drive from home to work?
_____ minutes

Where were you born (city & state, or country)?
_____

What is the closest major intersection to your business? (within 1 mile) please list the cross streets

Cross streets _____ and _____

What is your ZIP code? _____

Figure 7.1 Example survey questions used to obtain both qualitative and mappable spatial information as part of a survey of businesses in an economic study.

Interview-based data collection

Interviews provide another method that can be used to collect data. The main difference between an interview and a survey is that an interview is normally conducted by one individual (the interviewer) asking questions of another person (the interviewee). Another distinction is that the interview tends to involve more open-ended questions than are typically found in a survey. Interviews can be conducted with anyone you choose to sample as part of your research project. Conducting key-informant interviews is an approach commonly used at the beginning of a research project. A key-informant interview is usually conducted with someone who is knowledgeable about the topic under study. This person does not need to be a technical expert but instead could be someone who is involved with and knows about the issue being studied.

When is it appropriate to use an interview?

Interviews are usually the instrument of choice when the researcher wants to go more in depth on a topic or issue. The interview is an excellent research method in part because of the interviewer–interviewee dynamic. This enables the researcher to approach the interview in a formalized way using a predetermined set of questions. You want to get the person being interviewed talking about the topic under study, without perhaps being too rigid in approach, as would occur with a survey.

Interviews may involve either a formal approach or be more loosely organized. The latter, "on-the-fly" approach allows the researcher the ability to insert probing questions throughout the interview as the flow of the interview warrants. The danger with this approach is that it could introduce a level of inconsistency to the data collection process, particularly if the interviewer is not consistent in how he or she approaches the interviews. To minimize this, it is advisable to follow a well-developed interview schedule. Adding in questions such as, "Is there anything else that you would like to share with me about this topic?" at strategic points can facilitate producing additional valuable material.

Sometimes open-ended questions allow the respondent to address topics that you failed to ask questions about but that may be salient to the person and the topic at hand. The sample interview schedule in figure 7.2 was designed to collect information about a Latino community in a small northern California town. You may note that this interview contains a significant number of open-ended questions. Although open-ended questions can be more difficult to analyze, their use was a conscious decision in the design of this interview schedule. We employed this approach because the Latino population in this region was somewhat insular and difficult to discover information about. The more conversational nature of the open-ended questions was deemed culturally appropriate and was expected to be more effective than a set of more narrowly defined interview questions. We wanted to employ a data-gathering approach that would facilitate engagement and interaction with the population. In the sample interview, you will note that many questions include a spatial component to facilitate coding into a GIS database for analysis.

> Organization/Agency: _____   Title/Position: _____
>
> Location of Interview: _____   Date: _____
>
> Where do *members of the study group* typically go to meet their health needs?
>
> Why do *members of the study group* go there versus somewhere else? (i.e., close location, accept Medical/insurance, word-of-mouth, provider-patient relationships, etc.)?
>
> How far do *members of the study group* in your community typically travel (in time or miles) to meet their health needs?
>
> Are there places that *members of the study group* gather in the community?
>
> What place do you consider to be the heart of the community? Where is that located?
>
> What are the names of the neighborhoods or communities where *members of the study group* live? How far are they from here/where you work?
>
> Who else would you recommend that I interview about health issues of the local *members of the study group*?

Figure 7.2   Sample key-informant interview questions that incorporate spatial information for a GIS analysis.

Interview sampling

As we have pointed out, an interview differs from a survey in that it is conducted by an interviewer, so it usually involves some form of interaction between the respondent and the researcher. The choice of whom to interview for a particular research project is really up to the researcher. This choice will depend on the goals of a project. For instance, if the goal of your project is to provide an oral history of significant issues that occurred where you live, you would want to conduct interviews most likely with people who have some knowledge and experience with the history of that community or place. The interview can be either randomly sampled or nonrandomly sampled, depending on the purpose of your project. If you randomly sample as a part of your study, then you will be able to generalize to the larger population. However, if you use a nonrandom sample, then you will not be able to generalize to the larger population. It should be stated here that with interviews, there can be very good justification for nonrandomly sampling your interviewees, especially if you employ a snowball sample approach. You can achieve a diversified sample if you look to interview people within your study area who are, for example, from different occupations. So if you want to diversify your sample to try to include a breadth of different types of people, you might want to try this approach. Projects sometimes seek a certain type of person who meets a certain set of criteria, and in those instances, you may not want to use a random sample method.

Spatializing your survey or interview questions

Spatializing your survey or interview to facilitate its analysis in a GIS environment is a relatively simple process. The key to spatializing your data collection instrument is to think about what role space plays in your research question. Do particular geographic boundaries delineate groups or participants in your study? For instance, are you interested in the opinions of people from different neighborhoods or communities within a particular town? Also consider what level of spatial detail you require for your analysis. For instance, is it enough to know the ZIP Codes where respondents live, or would it be better to gather data on the nearest major cross streets to the respondents' homes? Major intersections are a familiar landmark for most people and provide an approach that we have used often in our own socioenvironmental data collection when we have been interested in knowing where people are proximate to other natural or built environmental features, for instance, an environmental hazard, food pantry, hospital, or bus route. Being able to examine the role of space, place, and proximity is important to spatializing your survey or interview.

Using GIS in the field, with and without a computer

Without a field computer

One common question that may come up for the scientist who is interested in using GIS is as follows: Do I need a computer to gather data that could be analyzed using a GIS? The answer is no, not necessarily. Your data gathering in the field may involve having people fill out surveys, collecting oral histories, or conducting ethnographies (discussed further in chapter 9). Or you may be directly recording measurements or observations about the environment in your study area. The important thing to note is that if you do not have a computerized GIS database that you are using in the field, you need to be observant and take careful hard-copy notes about the geographic locations that pertain to your key variables. This information can be coded into the GIS upon your return to the office.

This might be as simple as marking data collection locations on a map or noting a geographically identifiable attribute such as the street address or the name of the town or watershed in which the researcher was located when the particular data were collected. Of course, if there are issues of confidentiality in your study, you would want to be careful only to record at a level of detail consistent with the needs of the study and according to whatever institutional human subjects research process is appropriate given your organization and funding source.

With a field computer

If your study allows for you to take a laptop or field computer with a GIS system on it, you may be able to code and enter your data directly into the GIS system as you go, minimizing the need for later data entry and/or transcription. Many low-cost devices are now available that can facilitate spatial data collection directly in the field. These may be as commonplace as a Global Positioning System (GPS)–enabled smartphone to more sophisticated, custom field computers with products such as Esri ArcPad software or other similar software on board. In addition to providing for direct data entry in the field, often including direct access to GPS coordinates, such systems can be preconfigured with the data collection forms and answer options to ensure accurate coding of data directly in the field. Of course, having GIS technology in the field can afford you, the researcher, an evolving look at the geographic relationships you are exploring in your study as you go. Of course, in some research settings, this type of data exploration in the midst of the study may be considered inappropriate; however, in many applied situations, the ability to obtain preliminary results as the data are collected and entered can be extremely valuable. Immediate and evolving data visualization can be particularly useful when analyzing issues that are time sensitive and related to safety, security, and health. Even if you cannot take the GIS into the field, data can be entered each time you return to the office or another place where the GIS is accessible, to begin the process of data entry coding and preliminary visualization. With access to a cellular network or WiFi, data can even be transmitted back to the office in real time, for near-instantaneous analysis.

Data collection considerations

Optimally, when you collect your own data, they can be linked to an existing **basemap**. A basemap is a map layer that has relevant geographic information appropriate to your study, including a coordinate system, projection, and scale. Some common feature types present on many basemaps include political boundaries, census tracts, ZIP Code or area code boundaries, or major landscape features such as roads and water bodies. Dependent on the level of detail required for your study, some common basemaps might include national or state data containing several of these key features. For more detailed studies, basemaps might include the address ranges along streets by block, census block groups, or other detailed location information. Excellent sources for the thematic layers on which to overlay or with which to link your project-specific data include government agency websites for the jurisdiction of your study (city, county, state, or federal) or data available through sites such as ArcGIS Online, which provides a variety of ready-to-use base layers to serve as a starting point for your GIS analysis.

When your study can be tied directly to an existing basemap, the process of getting your own data into the GIS will be greatly simplified. It will not be necessary to collect detailed spatial information for your sample units; instead, you can simply record the common spatial identifier along with the other information required for your study. For example, you might use a phone company area code map to link respondents in a phone survey to a geographic location based on their phone numbers. In this case, simply recording the area code or prefix dialed would be sufficient to map a respondent to a location on the map, while retaining an appropriate level of aggregation necessary to ensure privacy. Of course, with the dominance of cell phones, area codes and telephone prefixes are less connected to the geography of respondents and more to the origin of the phone contract, which may provide a somewhat different, albeit potentially interesting, perspective on the responses.

Regardless of the specific method, so long as you are able to directly tie study data to an existing basemap, the process of linking your data to the GIS requires that you know the coding scheme used on the basemap. In many cases, the codes used for units of analysis will have one-to-one correlation to the map units (e.g., area codes on the GIS map would be the actual area codes). However, in some datasets, the coding scheme used for the GIS map may be less apparent (e.g., county names might be coded with numbers or abbreviations in the GIS).

When no direct or obvious coding scheme exists, you have two options. Optimally, you will be able to determine the coding scheme that was used on your basemaps in advance of your data collection and use those same codes as you assemble your new data. In some cases these codes may already be incorporated into an existing GIS data layer, and in others there may be a separate metadata document, or data dictionary, that defines codes used by the organization or agency that authored the data layer.

If the codes used on the basemap cannot be obtained in advance, or are not readily available, you may still go forward with data collection but will need to plan on determining and adding the necessary coding information before linking your data to the GIS database.

Address matching

Address matching is another popular method for linking data to a geographic location on the map. This process uses a map layer with street names and addresses coded into the data.

When you provide an address, the GIS will attempt to match the street address to its proper location on the map. This is accomplished in much the same manner as you might locate an address driving around your neighborhood. Street addresses can be located by determining what address ranges fall in a given block and on which side of the street the odd- or even-numbered addresses fall. This process works well in a GIS so long as the underlying street and address information is properly stored within the database (some problems in this regard are discussed in the section on database concepts later in this chapter). The benefit of address matching is that data are mapped to a relatively specific location without the need to collect ground coordinates in the field with a GPS receiver.

Using the Global Positioning System

GPS is a readily available and inexpensive technology that complements GIS. Most smartphones contain GPS capability, as do many newer vehicles. Accessing the coordinates in a manner that is GIS-friendly is not necessarily straightforward with such devices, and it may require some effort to locate the needed information in the documentation for the device. Optimally, you would want to capture x,y coordinate positions as latitude and longitude with a known map datum. Most such systems will capture latitude, longitude with a WGS84 datum, so this is a reasonable assumption if you cannot confirm this in the technical documentation.

Alternatively, a significant number of inexpensive, handheld, consumer-grade GPS units are available at outdoor stores or through consumer electronics retailers. For many research applications, such units are sufficient. The advantage to stand-alone recreational GPS receivers is that they are typically more configurable than those intended for vehicle navigation or embedded in a smartphone.

With any GPS unit, determining your location on the globe takes only a few seconds. The GPS receiver obtains a signal from a set of satellites in orbit high above the earth. These satellites constantly transmit the information necessary for the calculation of position in a variety of coordinate systems and datums. With the touch of a button, the GPS unit can store these locations in memory for later download to your computer for mapping in GIS along with whatever additional data you collected at that location.

High-end GPS receivers provide additional capabilities, including fully functional onboard computers, which may include configurable database applications or even GIS software, such as ArcPad, for direct data capture and display. Some units incorporate additional features, ranging from external antennas for improved positional accuracy to cell phone and wireless connectivity, digital cameras, and additional sensor packages and range finders to measure offsets for mapping additional nearby features from a single location.

In the context of data collection for a GIS-based study, simply taking along a GPS receiver to acquire location information at each sample site will provide the additional data required to link any of your other field-based data to the map. Of course, for some studies, there may be issues of privacy you will want to consider before collecting detailed GPS coordinates. For instance, a survey respondent may not take kindly to you recording a GPS location on her front porch while asking a series of questions. However, GPS data may be especially useful in getting location information for environmental data, for example, the locations of vacant lots, city parks, or illegal roadside dump locations. GPS can be especially useful in contexts where data may be difficult to obtain through formal channels.

The two most important concepts and considerations when using GPS for location information are that (1) GPS is not as accurate as it appears but is accurate enough to be useful and (2) GPS data, like any data source for your study, must be in the same coordinates and datum as your other data. Let's look at each of these concepts in turn.

The issue of accuracy is initially addressed by reviewing the specifications of the GPS receiver you will be using. The product manual will typically include a statement of spatial accuracy for both the horizontal (x,y) plane and the elevation (z axis). Also note that specifications will typically state that these levels of accuracy are achieved some percentage of the time, for example, 90 percent of the time, which allows room for larger error at other times (the remaining 10 percent). Most GPS units, even consumer units with limited capability, will report your location to within a fraction of a meter in all directions, even though it is unlikely you would get a location reading that is truly that accurate. Just because you are provided values that include several decimal places, you should not always take them as accurate. Fortunately, for most GIS analysis, being within a meter or even several meters of the true location will be more than sufficient to address most questions. In fact, for many studies involving potentially confidential data, we typically want to be less accurate, perhaps to within just a few thousand meters.

Most current consumer-grade GPS units, including those in smartphones and vehicle navigation systems, or those created for recreational use provide accuracy on the order of plus or minus five or ten meters on the (x,y) plane. This should be more than sufficient to locate your data in a particular neighborhood or street but might present some difficulty if you need to locate which side of the street you were on. If being off five or ten meters is problematic, much more accurate (and expensive) GPS systems and techniques are available to achieve accuracy down to centimeters.

Although there is a great deal of math and physics behind the function of the GPS system, the end user can learn most of the necessary methods simply by reviewing the manual that comes with the particular unit and with a day or two of practice. Getting the location information out of the GPS receiver and into the dataset in the GIS is the final step. At a minimum, all GPS units will record a location ID (typically starting with "1" and counting up, although the user can also assign IDs) along with x,y,z coordinates and the date and time of collection. In conjunction with a survey or data collection sheet, all information collected in the field can be easily linked to its proper location on the map.

Data stored in the GPS receiver's memory can be downloaded to your computer. Some GPS units include software tools that may allow you to link to a laptop to transfer all of the information directly into a database or other software program. Other GPS software may require that you export the data and then bring them into the GIS as a separate step. If you will be collecting and integrating GPS data on a regular basis, it will be worthwhile to identify software tools to facilitate the direct translation of the GPS data into a format that is readily compatible with your GIS software. One such tool that works well with many common consumer GPS units is GPS Utility (http://www.gpsu.co.uk/). The GPS Utility software is inexpensive and will save data directly into Esri shapefiles, making your GPS data readily importable into your desktop GIS for additional analysis. Most high-end survey-grade GPS units also provide software that is directly compatible with common GIS data formats.

Creating basemaps from scratch

Perhaps the least desirable option for generating data necessary for a study is doing so from scratch, because it is tedious and time consuming. However for some studies, especially those that have unique study locations or that require substantial historic data, this may be your only option. Data created from scratch are most commonly based on either maps or aerial images (photo or satellite). However, sometimes new basemaps will be generated from data collected in the field. The actual map or image may be used in its printed physical form or scanned into the computer as a digital image.

Regardless of the format, the basic process of creating the basemap is the same. The map or image is interpreted by the analyst, and individual features of interest are traced one by one into the computer through a process referred to as digitizing. For each object digitized in this manner, descriptive information (attributes) must be entered into the computer. For example, if you were to digitize a map of county roads from 1950 for comparison to roads in 1990, each road name from the 1950 map would need to be individually entered into the computer.

Of course, maps show only the information deemed relevant by the map maker, so to get an unbiased picture of the past, it is often preferable to work from aerial photographs (figure 7.3). This is because an aerial photograph provides a complete and unbiased view of the data as they were at the time the photo was acquired. By contrast, when a map is drawn, much of the detail is intentionally left out to avoid clutter on the printed map.

Figure 7.3 A color infrared vertical aerial photograph obtained near Denver, Colorado. This photo was obtained as part of the Federal National Aerial Photography Program, coordinated by the US Geological Survey, and was collected on September 5, 1988. Photograph courtesy of the US Geological Survey.

Quality aerial photographs are collected in many parts of the United States on a regular basis, and in some parts of the country, they go back to the 1920s. The process of delineating, tracing, and naming individual features in the photograph is the same as described for a map, with one added complication: maps generally show some type of location information (e.g., latitude and longitude), whereas photographs often do not. Thus, to make data digitized from a photograph useful, the photograph must first be correlated to the actual locations on the ground. If you are fortunate, a sufficient number of fixed locations will be identifiable in the historic photograph, perhaps key road intersections, buildings, or other features known to be in the same location. With several such locations identified, you may then match these to a map with marked coordinates or perhaps physically visit those locations with a GPS receiver in hand.

Without a more in-depth knowledge of map projections and coordinates and datums, map and image analysis techniques, and related issues, GIS data creation such as described earlier may best be referred to someone with this specific expertise, especially if this is to become the basemap for your entire analysis. For completeness, it is important to realize that creating your own base data is an option, although it is not necessarily the quickest or easiest of the options available.

If you are fortunate to be working with more recent imaging technologies, particularly data acquired with digital, rather than film, cameras, the data may be georeferenced at the time of acquisition or during postprocessing. Even some older imagery, for example, digital orthophoto quadrangles (DOQs) produced by the US Geological Survey, have been georectified and referenced so that locations are mapped to their actual ground positions.

Unit of analysis

A statistical concept that is relevant to any analysis and potentially imposed by the data when conducting a spatial study is the unit of analysis. The **unit of analysis** refers to the sample unit being analyzed. For example, when surveying people about their smoking habits, you could select a variety of different units of analysis. You may want to examine each individual respondent, or if your work is focused on differences in smoking rates throughout an entire state, your unit of analysis might more appropriately be at a county level. Individual responses would be aggregated at the county level, simplifying the data substantially. Selecting and altering the unit of analysis may have important implications in your analysis, depending on the area of the spatial unit selected. This is discussed in the section about the **modifiable area unit problem** in chapter 5.

One important point to consider in any statistical analysis of data is **independence** of the observations. When looking for datasets or preparing to collect new data, it is important to understand if individual observations are independent, that is, the observations are unrelated to one another. Nonindependence in data has a variety of causes, which are discussed in most introductory texts in statistics. The unique consideration in spatial analysis is **spatial autocorrelation**. Spatial autocorrelation refers to a situation where observations taken close together in space are similar (positive spatial autocorrelation). For example, people who live in the same neighborhood may be more likely to be of a similar socioeconomic class than people from different neighborhoods. Sometimes spatial autocorrelation is useful in pointing you to the presence of an underlying factor that influences the observed characteristics. If the relationship is negative (observations near one another are different), this would indicate a greater diversity of data in space. The Spatial Statistics toolbox in ArcGIS provides tools to examine spatial autocorrelation. Two of the most popular statistics are Moran's I and Geary's C. It is beyond the scope of this text to provide the details of these statistics; however, numerous resources are available in both the statistical and the GIS literature or in software documentation.

The bottom line in considering the unit of analysis is to select an appropriate **level of analysis** considering both the variables to be measured and the level at which those variables actually operate. Data at the individual level may not be relevant if they actually function at a higher level. For example, in electing the president of the United States, individuals vote, but for most states, all of the electoral college votes are ultimately cast for the candidate receiving the majority of votes in that state. This is what makes it possible for one candidate to win the popular vote, while another may win the electoral college vote. Conducting a study using individuals as the unit of analysis is not particularly relevant given that the selection of the president operates at the state level.

Sometimes when seeking data for use in a study, you need to be especially careful. Just because a secondary dataset is available at a particular level of detail does not mean that level is the appropriate unit for your analysis. Consider your own question, the level at which the variables operate, and then (if appropriate) aggregate the data to an appropriate level. Of course, it is not possible to obtain additional detail in data that were previously collected and aggregated by someone else; for example, if you know which candidate won a given state, the dataset may not give the breakdown of individual votes. You can, however, aggregate detailed data to a larger unit of analysis, for example, you could group individual responses to the county or state level. There is always a risk that you may commit an ecological fallacy (chapter 4) if you do not conduct your analysis at the same level as the generalizations you intend to draw from the data.

Database concepts and GIS

Some basic database concepts are important to introduce at this point. In simple terms, a **database** is a collection of organized information. Of course, modern databases are computerized, but in the past, many databases existed in a physical form, residing in filing cabinets, library card catalogs, or bound volumes. Many of these older datasets are still of great value in GIS analyses, and when this is true, coding the relevant information into the computer is necessary. Databases that already exist in digital format, or those being newly collected, must follow some simple rules to ensure they blend well with GIS.

The software used to store database information is, at some level, irrelevant. Of course, using a formal database software program can streamline the process, especially for complex datasets; however, for simpler data, you may prefer readily available spreadsheet software such as Microsoft Excel. In fact, even a simple word processor or text editor can be used as a database, so long as the basic rules are followed. In short, this means that for purposes of GIS analysis, you can consider a wide array of digital and nondigital data sources as potentially useful. Of course, ArcGIS incorporates its own database tools, allowing data to be coded or imported directly into the software. Many of the common formats mentioned (Excel or various text file formats) can be directly opened in ArcGIS as well.

Rules for GIS database development

What are these basic rules? First, let's consider the layout of the information on the computer screen. A database is typically organized in rows and columns. The rows represent individual **records** in the database; these might be individual survey respondents, field sites, or counties—in many cases, they directly represent your units of analysis or subcomponents of those units that may be aggregated during the analysis process. In figure 7.4, we see the individual respondents A, B, and C are represented in rows. In GIS jargon, we often refer to these as **entities** in the database.

Columns in the database represent the descriptive information collected about each record, for example, the answer given by the respondent for each question on the survey, field measurements, or other descriptive data. These are called the **attributes** in GIS. Figure 7.4 provides an example of a small database. There are three entities, A, B, and C—in this case, the respondents. For each entity in the database, we have recorded four attributes, in this example, their numeric answers to each of four questions, perhaps using a 5-point Likert scale. It is useful to introduce here the concept of a **cell**, one individual box in the table. In the example in figure 7.4, a cell might store information representing an entity (e.g., respondent B) or an attribute (e.g., 2 being respondent B's answer to the first question). The column heading, for example, "Question_1" in figure 7.4, is referred to as a **field** name.

	A	B	C	D	E
1	Respondent	Question_1	Question_2	Question_3	Question_4
2	A	4	3	5	5
3	B	2	4	3	3
4	C	5	4	4	5
5					
6					

Figure 7.4 An example of a simple database table with three entities, A, B, and C, each represented by a row in the database. Each entity's answers to four questions are recorded as attributes; each question is represented as its own column in the data table.

One item missing from the database table in figure 7.4 is geographic information that would allow you to easily link the data in the table to a GIS map—that is, a spatial component or identifier. If you would like to tie these data to a basemap, adding one additional column would easily solve the problem. Assume you are collecting data for a large urban area with the intent of analyzing results by ZIP Code. Adding a ZIP Code column to the table would facilitate linking your survey data to a US Postal Service ZIP Code map. The key to accomplishing this connection between a database table and a basemap is to ensure there is a common spatial attribute, in this example, the ZIP Code. Thus, a more effective database including a spatial identifier might appear as shown in figure 7.5. In ArcGIS, this is easily accomplished by using the **Joins and Relates** function to make the connection between a data table and a basemap.

	A	B	C	D	E	F
1	Zip_Code	Respondent	Question_1	Question_2	Question_3	Question_4
2	55113	A	4	3	5	5
3	55401	B	2	4	3	3
4	55112	C	5	4	4	5
5						

Figure 7.5 An example of a database table that incorporates a spatial identifier that can be associated with a map, in this case, a US Postal Service ZIP Code map.

Creating GIS-friendly data tables

So long as you follow the basic format described earlier, you will be well on your way to developing data that are readily compatible with a GIS. However, to facilitate an even smoother transition, a number of additional formatting considerations are relevant:

1. Consistent use of space and case
2. Format and coding of the data
3. Structure of the file saved by your software

Space and case

First let's look at some issues related to space and case. Although the Microsoft Windows operating system and many of the programs that run on it allow you to put spaces into file names, this is not generally good practice. This is because some computers and software, including ArcGIS, do not allow for spaces in file names. Another thing Windows ignores is the distinction between uppercase and lowercase characters. However, when working with databases in ArcGIS, variations in case can cause problems.

The reason these two issues cause problems relates to the fact that a significant amount of GIS data, and particularly older base datasets coming from large government agencies, were originally developed on Unix-based computers, where case matters. That is, a Unix-based computer considers "E" and "e" to be different letters, whereas a Windows-based computer views these as identical. A related concept is punctuation. For most computer databases, it is also good to avoid using punctuation, as these characters often have special meaning.

Although issues of space and case might not cause problems in every situation, given the realities of data sharing, downloading information from the Internet, and a variety of different software and hardware platforms, it is best to avoid using punctuation and spaces when naming database fields or data files. You may notice that in figures 7.4 and 7.5, in those places where a space might normally be expected, an underscore was used instead. Similarly, it is important to be consistent in the use of character case.

At this point, you might ask, "What about when I need to store somebody's first and last names?" Although it might seem appropriate to place the name in a single field, from a database perspective, it would be preferable to place the first and last names into their own, separate fields. Of course, in many social science research applications, you may need to protect your respondents' privacy by coding information as numbers instead of using names. There are situations where use of spaces or punctuation may be appropriate, for example, a city name, such as Los Angeles, can reasonably be coded in a single attribute column. However, it would be less appropriate to include the state name, separated with a comma, in the same cell, for example, "Los Angeles, California."

The reason for splitting city and state into separate columns is to facilitate sorting and analysis later in the process. This is no different than you might do when entering data into Excel or SPSS formats. In using separate columns for city and state, you could select only those cities in a particular state for an analysis comparing different states. Later, the same database might be used for a more detailed analysis comparing different cities done via use of the city name column. In short, the more you differentiate attribute components into separate columns, the more options you will have in the data analysis phase later in the process.

Data format and coding considerations

When preparing to code into the computer, be it your own, new data or a transcription of existing data, there are several important considerations. As mentioned in the previous section, computer programs take the data literally, so differences in spacing, case, or coding can cause significant problems when creating and combining datasets for an analysis. Humans have an uncanny ability to understand that all of the codes in figure 7.6 refer to the same real-world location, even though there are substantial variations in how the coding has been done (including the possibility of typographic errors).

	A
1	San Francisco
2	SAN FRANCISCO
3	S.F.
4	san francisco
5	s.f.
6	Sna Francisco
7	SanFrancisco
8	Frisco
9	The City by the Bay

Figure 7.6 Although most individuals would interpret any of the following codes as referring to the same location, a computer takes the data literally and would view some, or all, of these as unique values.

Unfortunately, a computer database would consider each of the codes for San Francisco as unique. This would result in an analysis that requested all data for San Francisco excluding any of the data records that are coded differently than the request made by the analyst. This is an important consideration when developing a coding scheme. Of course, if you are planning to link your data to an existing basemap, it is preferable to use the same codes used in the existing dataset. Similarly, when working with multiple datasets from multiple sources, it will be important to verify that coding is consistent. If coding is not consistent between datasets, one or more dataset may require editing or updating to facilitate interoperability.

It is not unusual for differences in coding to occur at political or jurisdictional boundaries between organizations, for example, two adjoining counties. For example, if you were attempting to conduct a regional planning exercise in a large metropolitan area consisting of several counties, it would not be unusual to find that each county in the region uses different zoning codes. Just as the variations in spelling or abbreviations in figure 7.6 would cause you problems, so, too, would several different counties each using their own zoning designation codes.

Of course, when there are opportunities to develop consistent coding schemes between organizations, often recorded in a **data dictionary,** many of these issues can be preempted prior to any individual or organization attempting to create databases of their information. Data dictionaries are used to set out the specific codes and definitions to be used when entering data into the database so that the codes are clearly understandable by anyone who uses the database.

One additional, computer-specific issue in coding is the difference between a **number field** and a **character field (string)**. Many software programs used for data entry, especially database, GIS, and even spreadsheet programs, differentiate data by the type of information being entered into the computer. Although it may seem obvious that numbers are numbers, a computer also can treat a number as a character. In ArcGIS, for example, you could store a value that appears to be a number a variety of different ways: a short or long integer value, a floating point number value, a double precision number value, a character, or a special case number—a date value. By contrast, letters and words are always treated as characters and thus cannot be inadvertently stored as numbers. Data tied to a GIS will ultimately reside in a database, and databases treat these data types differently when conducting analysis.

In short, what this means is that your computer will not view a number stored in a character field in the same way as a number stored in a number field, and furthermore, most mathematical operations will not work on character data in the same manner as they will on numeric data. Therefore, it is important to consider, in advance, the format your data should take before coding occurs. If your data values represent a real measurement (interval or ratio data) for a quantitative analysis, the numbers should be coded as number fields in the database. If you are using a qualitative approach (nominal or ordinal data), you may find that either numeric or character fields are appropriate.

One final comment on data entry and coding relates to the choice of software to use. Many of us are limited by two realities when it comes to software: (1) what we have available or can afford and (2) the software we know how to use or have time to learn. Of course, if you are reading this book, you are most likely contemplating the use of ArcGIS software even if you have not yet acquired it. ArcGIS and other similar software packages include a database component as part of their functionality; nonetheless, many people find it preferable to use an external software program for data entry and management of the nonspatial data. If you are already using another software package for data entry, you may decide to continue using it.

The one major advantage to using a true database program, be it the one built into ArcGIS or a stand-alone product such as Microsoft Access, is that this type of program allows for a high degree of control during data input. That is, when developing a database, each individual field or cell can be programmed in advance to accept only appropriate data. Using these techniques can significantly reduce data entry errors by allowing only acceptable data to be placed into the database. This may not be a major concern if you have a small amount of data, and in fact, programming a database may take more time than simply entering the data. However, in large projects involving multiple people in the data entry process, the ability to preprogram automatic checks to ensure proper coding can be very advantageous.

For example, if you are using a 5-point Likert scale, the database could be programmed to allow only the digits 1 through 5, rejecting any letter or number other than that specifically allowed under the predetermined coding scheme. Similarly, databases can offer lists of possible answers, which in the case of textual responses can reduce misspellings or other typographic errors.

Software output formats

Regardless of the software you use to enter your data into the computer, you will eventually have to find a way to make the data work with your GIS software. Of course, file compatibility can be problematic, and there are too many possible variations to discuss in detail here; however, we will discuss three of the more common formats. Typically, these are available in most of the major software programs, including ArcGIS, and are generally applicable to other GIS and database packages and datasets you may encounter.

Essentially, all of the mainstream GIS, database, and statistics software packages are able to directly read or import files in **delimited text formats**. Because much of the data you use will have been created or stored using these sorts of software, it is valuable to recognize that text formats, which are one of the simplest, are readily exported and imported from these software packages. However, it is also worthwhile to review the specific documentation to determine if your GIS can read formatted database files, thus circumventing this step. Historically, one of the most common database file formats was **dBASE**. These files are typically named using a .dbf file extension. This format originated in the early days of the PC and became quite popular as a database format in the 1980s and 1990s. Because of its widespread use in the database market for more than twenty years, compatibility with the DBF file format has been integrated into many other software packages, including programs like ArcGIS, SPSS, and Microsoft Excel, used to create and manage data. When you can use common formats such as dBASE to import and export between the software you use to create data and your GIS, you will inevitably avoid some of the more common errors associated with data sharing.

If using dBASE files is not appropriate for the particular software you plan to use, files saved in plain or ASCII delimited text are another excellent choice. Typically, you will find options to save your data to a text format within the Save As menu or an Export menu option in your software. The delimited descriptor indicates that the file will be saved with simple formatting that indicates where one data value ends and the next begins, most commonly using a comma or space, as shown in figure 7.7, and often named with a .csv file extension.

Notice that the comma delimited file simply puts the data from a given row all together using a comma to indicate the end of one cell and the beginning of the next. With this in mind, it should be easier to understand why using commas within the database (e.g., names entered as last, first) could be problematic when saving to a comma delimited format—the comma between the names would be confused with a comma used to indicate the cells of the original database table.

Zip_Code	Respondent	Question_1	Question_2	Question_3	Question_4
55113	A	4	3	5	5
55401	B	2	4	3	3
55112	C	5	4	4	5

Zip_Code, Respondent, Question_1, Question_2, Question_3, Question_4

55113, A, 4, 3, 5, 5

55401, B, 2, 4, 3, 3

55112, C, 5, 4, 4, 5

Figure 7.7 An example of a database table (top) exported into a comma delimited format. Notice that the columns are separated by commas (bottom). Had any commas been included as part of the data values, they would present problems. Column headings are retained in the first row, with each subsequent row in the list corresponding to an individual record in the database.

With the popularity of Microsoft Office, and in particular the Excel spreadsheet application, for many datasets, this has become a dominant data storage format. Although a spreadsheet is technically not a database, it can take on many of the same characteristics. The use of rows and columns as an organizing concept is especially compelling as a database-like structure. Given the frequency of Excel (.xls) files being used for data storage, ArcGIS added the capability to read these directly as data tables for use in GIS analysis simply by using the Add Data function. This saves the extra step of exporting data from Excel into a comma delimited format prior to importing into ArcGIS.

The Excel, dBASE, and comma-delimited text formats are some of the most common methods for moving data between software packages. If your preferred software does not offer an option to import and export in these formats, you may need to explore alternative approaches or use third-party data translation tools (although if your software cannot work with one of these formats, we recommend finding new software). In addition to the data conversion tools included with each individual software package, a variety of stand-alone data conversion tools are available both free and commercially.

Review questions

1. What is an example of a close-ended question? What are some of the advantages to using a close-ended question over an open-ended question?
2. What is an example of a situation where it might be more appropriate to use an interview over a survey?
3. Please explain what the term *unit of analysis* means. How is a unit of analysis different from a level of analysis?
4. A typical GIS database is organized into rows and columns. What kind of information usually gets recorded in the row? What kind of information usually gets recorded in the column?
5. What is a data dictionary? What role does it play?
6. Consider a GIS project that you would like to conduct. What are some of the steps that you would take to develop your project basemap? What kind of information do you want to include on it?

Additional readings and references

Arctur, D., and M. Zeiler. 2004. *Designing Geodatabases: Case Studies in GIS Data Modeling*. Redlands, CA: Esri Press.

Dillman, D. A., J. D. Smith, and L. M. Christian. 2009. *Internet, Mail and Mixed-Mode Surveys: The Tailored Design Method*. Hoboken, NJ: Wiley.

Fowler, F. 2013. *Survey Research Methods*. Thousand Oaks, CA: Sage.

Kaplan, E., and C. Hegarty, eds. 2006. *Introduction to GPS: The Global Positioning System*. 2nd ed. Norwood, MA: Artech House.

Kennedy, M. 2005. *The Global Positioning System and GIS*. 2nd ed. New York: Taylor and Francis.

Spencer, J., B. G. Frizzelle, P. H. Page, and J. B. Vogler. 2003. *Global Positioning System: A Field Guide for the Social Sciences*. Malden, MA: Wiley Blackwell.

Relevant websites

- **National Atlas of the United States (http://nationalatlas.gov/)**.
- **National Map of the United States (http://nationalmap.usgs.gov/)**.
- **National Aeronautics and Space Administration (http://www.nasa.gov)**: Provides a variety of imagery and monitoring data for the world or portions thereof.

- **US Census Bureau (http://www.census.gov)**: The official site for the US Census, demographic data, TIGER files, and other related data.
- **US Geological Survey (http://www.usgs.gov)**: Provides a variety of national datasets, including aerial images, geographic places, maps, and topographic information.
- **US Environmental Protection Agency (http://www.epa.gov)**: Provides a variety of national databases related to environmental quality.
- **Atlas of Canada (http://atlas.gc.ca/)**: An example of an interactive GIS website.
- **Natural Resources Canada (http://www.nrcan.gc.ca/home)**: Provides a variety of national datasets, including aerial images, geographic places, maps, and topographic information. (On the homepage, search for "GIS data.")
- **Atlas of Sweden (http://www.sna.se/)**: An example of an interactive GIS website.
- **Centers for Disease Control and Prevention (http://www.cdc.gov/)**: Provides an array of national databases related to public health, disease, births, and deaths.
- **Library of Congress Collections, Sustainability of Digital Formats (http://www.digitalpreservation.gov/formats/fdd/fdd000325.shtml)**: Provides a description of the .dbf file format and its relationship to other geospatial data formats.

Chapter 8

Public participation GIS

In this chapter, you will learn about various aspects of spatially based, public participation GIS (PPGIS) methods and **volunteered geographic information**. You will learn how to collect and organize PPGIS data in the field using simple, low-technology methods and computer-based approaches. You will learn about the collection of spatially based and spatially linked data through community engagement. You will also learn methods for integrating data from qualitative social contexts into measurable spatial variables.

Learning objectives

- Learn about public participation GIS (PPGIS) and how it is used
- Learn about volunteered GIS
- Understand the value of the inductive approach
- Learn the steps of an inductive research approach
- Learn how to conduct your own PPGIS session

Key concepts

inductive spatial research
multiple research
   methods
primary data

public participation
   geographic
   information system
   (PPGIS)

reputable data
secondary data
sociospatial theming
volunteered GIS

Public participation GIS and participatory GIS

At the outset of any research project, specific or detailed data will need to be collected to meet specific needs of the research being undertaken (primary data). Other relevant data may be available from an existing source (secondary data). When specific data for a community or region are required, it may be useful to incorporate local knowledge. Public participation GIS (PPGIS), or participatory GIS (PGIS), refers to methods through which primary data are collected in an organized manner from individuals with specific knowledge, based on such individuals' membership in a particular group or community related to the study site.

These methods are referred to as public participation or participatory GIS because the source of the data is engagement with the public. It is important to keep in mind that for each individual research project, the *public* could be defined in a number of ways, depending on the scope of the project and the stakeholders involved (we discuss this later in the chapter). Members of every community, region, or place have local knowledge, as do subgroups based on socioeconomic status, education, or vocation. PPGIS methods provide the researcher with the ability to document and collect specific knowledge in a systematic fashion.

The value in including locally based knowledge from people who are most familiar with a particular issue, geographic area, or situation is that it leverages the strengths of inductive research. As we discussed in chapter 3, an inductive approach to research begins with observation or data collection first and then works toward developing a theory or explanation for what the data show. Although many researchers value existing, or secondary, data, they often are not sufficient on their own. Existing data are helpful in establishing a basis for understanding a particular topic, but they aren't always enough. You may want to collect additional, primary data yourself. **Primary data** collected via PPGIS methods can be an important tool for capturing local people's spatial understandings and perceptions of their surrounding environment. Remember, primary data means that you collect the original data to use for your research project. When incorporated into your research alongside other methods of data collection, primary data provide a valuable means toward understanding a particular issue, problem, or geography.

PPGIS is an important research method—but it is actually more powerful when it is integrated with additional types of research methods. We refer to this as **multiple research methods**: when the researcher employs two or more research methods and sources of data in a research project. We want to emphasize the value of viewing a problem or issue under study from multiple angles. Implementing a spatially based, multiple methods approach provides the researcher with various perspectives and sources of data about the issue under study. Adopting a mono-focused spatial view of a problem or issue under study can tend to produce a limited or narrow view of the issue.

Even when you are using PPGIS (primary data), a good place to begin the research project is by finding and assessing secondary data. Why? It is important to begin a research project by examining all the current or existing data or information relating to your study site and research project. Each piece of secondary data or information can be integrated with the information you collect when using PPGIS. The main consideration for using secondary data (as discussed in chapter 6) is to make sure that the data you are using are reputable. **Reputable data** means that the data are coming from a valued or credible source. The primary data that you collect could come from field observation and PPGIS, which will highlight local knowledge and perceptions (see figure 8.1). After you have collected both primary and secondary data, you want to be able to integrate these and to map them spatially. Upon doing this, you will be able to draw better, more well-rounded spatial conclusions about your data.

Figure 8.1 A public participation GIS (PPGIS) session during which participants share, discuss, and document their local knowledge about pesticide drift and other community characteristics coded using colored markers. Photo by Steven Steinberg.

GIS as a tool for voice and empowerment

A GIS can be a tool for empowerment in multiple ways. First, a GIS facilitates the documentation of local knowledge and information in a spatial manner. Documenting and transferring the local knowledge and information that people have in their heads into quantifiable, visual graphics is powerful. Not only have you converted a group of people's knowledge and experience into real, hard data, but you have also provided these people with the means to interject their voices in defining their own situation or issue. This act of documenting knowledge in a digital format provides power to the people who created the knowledge because they can effectively share it with others.

It is a fact that people absorb, learn, and comprehend information differently, depending on their sociocultural, environmental, and economic backgrounds. With ArcGIS, data captured via PPGIS can be overlain with data from a variety of sources, including data from different groups, existing sources, or imagery, to provide a set of tools that visualize these data as a whole to provide one big picture. Locations where information is in agreement, where it is in conflict, or where there are unanswered questions can be easily identified and, as needed, quantified and visualized to guide further research or decision making. Because GIS caters to both the visual and nonvisual learner, it expands the range of people whom your data can reach as well as those who can contribute to the process, including those who may be illiterate or speak a different language. This is very powerful because it opens a door to community involvement to people who might otherwise be left out of important decisions or activities affecting their lives and communities.

PPGIS methods also allow you to share spatial files and information with others. For instance, using links in ArcGIS for Desktop software or an online mapping interface, you can further enhance the information associated with a map by linking sound files, including recordings of oral histories, or videos of customs or activities unique to people from a particular community. There is the old saying that a picture is worth a thousand words, and using GIS, one can expand which groups have a picture on display.

Using public participation GIS as part of mixed methods

As noted earlier, PPGIS is central to incorporating the perspective of the public, in other words, to getting people's voices and perspectives heard when, perhaps, in the past, they have not been. Several years ago, we conducted a case study in northern California focused on two different agricultural counties (Steinberg and Steinberg 2008). For this research project, we significantly incorporated a mixed-methods research approach that used PPGIS alongside other research methods and data. Our goal was to focus on communities and pesticide use. We wanted to actively incorporate local knowledge and experience with the topic into the study.

As we began the research project, we sought to answer a question that was important and of relevance to the community. Local community members indicated a concern about their children's health related to the pesticides being used in their local area. There were numerous anecdotal reports of children getting sick at the schools, many of which were located next to, or surrounded by, agricultural fields, but there was no real empirical evidence of these issues. So the communities naturally suspected a potential link between the spraying of pesticides near schools and their children's health (Steinberg and Steinberg 2008).

As in any research project, we sought a variety of data sources, including a number of secondary sources from local, state, and federal agencies (Steinberg and Steinberg 2008). These sources provided socioeconomic data from the US Census, data for various political jurisdictions, parcels and zoning, as well as information on infrastructure, such as schools and parks, and pesticide application information from the California Department of Pesticide Regulation (Steinberg and Steinberg 2008). This official statewide database includes detailed information on pesticide applications for individual active ingredients in pounds for each public lands survey section in the state. However, much of the information that we felt was important to gaining a broader picture of this issue was not contained in these secondary sources. Because a significant number of farm workers are migrants and, in some cases, may be undocumented, information about their experiences is not necessarily captured via formal channels because of their fear of engagement owing to immigration status. In this instance, a PPGIS process was determined to be an ideal approach to capturing information because we would have face-to-face time with the participants and could answer any questions and keep their participation anonymous in terms of data collection. We deemed the PPGIS a good method for hearing the voices and understanding the knowledge of people who had not yet shared their thoughts, feelings, and opinions.

We created basemaps in ArcGIS from existing data for use in the PPGIS sessions (Steinberg et al. 2008). These maps consisted of high-resolution imagery from the National Agriculture Imagery Program overlain with county parcels and the locations of local landmarks, streets, freeways, schools, and shopping centers to provide participants with landmarks, making the maps easily understandable to the local people involved in the

PPGIS sessions. We worked with a local community-based organization and held the sessions at their usual meeting locations and times.

Although it can sometimes be convenient to enter data directly into the computer, this is not always the best choice, because there can always be glitches with technology, and it is a lot easier for a group of people to gather around a giant map and to draw on that versus gathering around a laptop computer screen. In past studies where we have used PPGIS, we opted to use printed maps with markers for several reasons. Some participants may not speak English, so being able to point at the maps and discuss them with others can help to facilitate the process. Additionally, the printed maps permit us to focus attention on the objectives of the session rather than the "gee-whiz" aspect of the GIS technology, which may not be familiar to study participants and can overwhelm the data collection process.

Many groups are present as part of this network, and each one has different interests in the use of the land and local natural resources. Just to provide some more specific details from the *Esri Map Book*, "Amazonia covers approximately 7.8 million square kilometers shared by nine countries. There are 33 million inhabitants and 385 indigenous groups. The protection of socioenvironmental diversity has been consolidated through the recognition of the indigenous territories (IT) and the constitution of protected natural areas (PNA)" (Esri 2013, 28).

Figure 8.2 is interesting because it highlights the notion that research can be much stronger when the researcher incorporates multiple perspectives on the data. This map presents or shares multiple pieces of information important to issues of land use. First, it illustrates the land of traditional indigenous occupation and use to distinguish between the land that is "officially" recognized and the land that is being used but is "not officially" recognized. The merit in presenting these various types of lands on the same map is that it helps the reader or policy developer to get a better sense of the various needs, values, and actions that local people may expect given these various factors. The kind of information that is reflected here is based on local knowledge and experience regarding local boundaries and land use patterns and is very important to consider whenever decisions are being made about such resources and their use. In the next section, we explore the role of technology in the field.

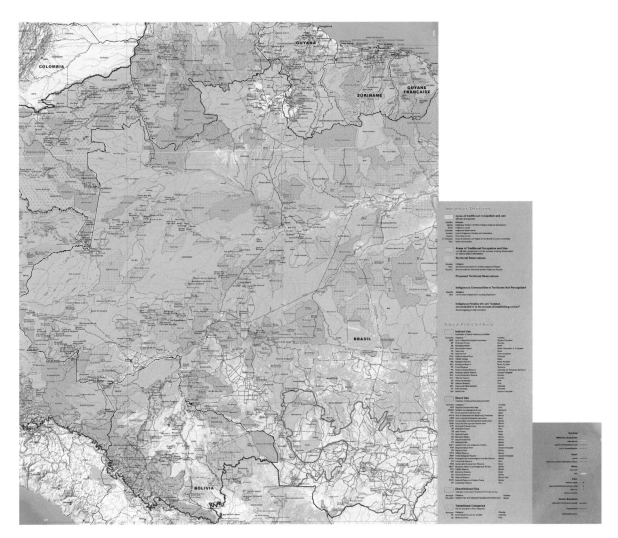

Figure 8.2   This map presents a collection of data and information integrated from a variety of different countries, organizations, and groups. The map is unique because of the holistic story that it tells. The map was created by a group called the Amazon Georeferenced Socio-Environmental Information Network (RAISG). The RAISG "is a space for the exchange and networking of GIS-based socioenvironmental information. That information supports processes that actively link collective rights to the promotion and sustainability of the social and environmental diversity of the Amazon region. The main objective of RAISG is to facilitate cooperation among institutions that already use socioenvironmental geographic information systems in the Amazon region through a methodology that coordinates collective efforts through an accumulative, decentralized, and public process of information sharing."
Map courtesy of the Amazon Georeferenced Socio-Environmental Information Network, 2012, and Gaia Amazonas Foundation. From *Esri Map Book*, vol. 28 (Redlands, CA: Esri Press, 2013), 23. Data from Bolivia (SERNAP, Viceministerios de Tierras); Brazil (IS); Colombia (IGCAC, UAESPNN, IDEAM); Ecuador (Ecociencia, Ecolex, MAE); Guyane Francaise (DEAL); Guyana (DCW); Peru (IIAP, MINAM, SICNA); Suriname (DCW, WDPA); and Venezuela (IVIC, Simon Bolivar Geographic).

Does using GIS mean I have to be "high-tech" in the field?

Linkages between qualitative research and GIS were relatively limited in early applications of GIS. Because GIS had its roots in applications of natural resource management and census demographics, the initial focus of GIS developed around quantitative data types such as counts of people or the areas and board-feet contained in a stand of timber. Qualitative researchers often pride themselves on being very low tech and having the ability to act as a voice for the people or topic under study. The reality is that to conduct a spatial analysis in ArcGIS, it is not necessary to take the technology to the field, as we demonstrated in the previous example working with farm workers in California. In many situations using participatory approaches in the field, low-tech data collection is often easier. It also avoids the perception that you value the technology above the knowledge and experience of the people from whom you are gathering data.

Ultimately, the high-tech tools will be used in conducting the data analysis. Although using technology in the field is sometimes appropriate, and may save steps in the data entry phase, the real power of GIS comes with its analytical capabilities. For example, determining spatial correspondences using overlay and spatial statistics tools in ArcGIS provides a means to both qualitatively and quantitatively assess data captured in the spatial format. Qualitative themes attached to place are readily transformed into measurable, quantitative statistics, effectively leveraging the power of both data types and analytical frameworks. As a researcher, you can maintain the utmost integrity of the value of human voices, perceptions, and oral histories in your project. You can still collect qualitative data in whatever format is most appropriate to your research question, be it oral, recorded, film or video, or handwritten. The key to using GIS for qualitative data is for you as the researcher to be clear in your mind about "what" qualifies as data. For a qualitative researcher, the notion of what qualifies as data is fairly broad.

Using a computer to record the data

There may be some instances where, as a researcher, you want to consider capturing data directly into your computer. For PPGIS, a computer may be appropriate if a capable technician is available to enter data as participants discuss the maps. If the size of the group is not too large, participants may annotate maps directly in ArcGIS using the drawing tools. In many situations, it may be preferable to gather data using other approaches (e.g., audio or video recorder, paper and pencil) and then transfer these data to the computer later. It all depends on the technological capability and understanding of the technology that your population has. If you planned to somehow record interview data, you would probably focus on transcribing interview data. It is especially useful to transcribe the interviews as soon as possible after completing them and while they are still fresh in the mind. Any of these alternative methods can always be incorporated into a GIS at a later time.

Also consider your participants' comfort level with technology. Will the computer distract your study participants? Would it help to have the ArcGIS available to provide interactivity to maps or simply to use printed maps with which respondents can interact and on which they can mark information for your data collection? You may also want to consider factors such as access to electricity to power your computer or recharge batteries as well as climate and conditions where you will be conducting your PPGIS session.

Also, it is important to consider how you will safely and reliably store and back up your data while in the field—computers should have removable media for backups. Will your computer be secure? Obviously it's not desirable to risk damage or loss of your computer, but even more important is ensuring the safety of the irreplaceable data stored on it. If you are using paper maps for PPGIS sessions, how will you maintain them? These, too, can suffer loss or damage. However, one of the values of hard-copy maps ("hard copy" meaning that the map exists in physical form) in collecting PPGIS data is that you actually have the data on paper, which serves as a relatively resilient archival material.

Internet-based methods

In some cases, it may be desirable to conduct a participatory mapping activity directly via the Internet. If you will be working in a location with a reliable connection to the Internet, you will be able to eliminate some of the concerns related to taking a computer and GIS to the field. For example, all of the data you collect are immediately stored on a server, making loss of or damage to your field computer less problematic. Additionally, in some facilities where you may opt to hold a session, the power, computer hardware, projection systems, and connectivity may be in place, eliminating the need to take your own technology to the field. A website designed around the ArcGIS Online service, or similar cloud or web-based technologies, can provide a consistent and reliable means of delivering the participant experience desired in the session, compared to taking a full-blown GIS to the field. Of course, it is important to confirm that available equipment and connectivity will meet your needs. A slow or unreliable connection to the Internet could lead to significant problems, and it is always a good idea to have a backup plan should something not function as anticipated. One additional benefit of a web-based approach is that it can give you the ability to readily share results with the community using an interface with which they have some familiarity and with data they contributed.

Volunteered geographic information

Data collected from the public in a slightly less structured manner than PPGIS provides an additional means for gathering data via the Internet. **Volunteered geographic information** (VGI) is differentiated from the participatory methods described earlier primarily in that data are compiled via web-based mapping tools and wikis from anyone who chooses to submit them. In this sense, participants are not targeted via a participatory session, as is typically the case when planning data collection via PPGIS; instead, a VGI approach involves setting up a website where anyone may submit data, often anonymously. Unlike PPGIS, you are not selecting the participants. The providers and background of the volunteered data might not be available. The fact that data come from whoever chooses to submit them does not necessarily decrease the data's value or accuracy; in fact, volunteered data are often of excellent quality. Many websites that aggregate volunteered data take advantage of their communities of participants to review and correct errors.

From a formal research perspective, data acquired via VGI will not offer you the same level of control over the sample population as can be obtained via more traditional sample design and data collection methods. If the ability to define and control the sample population from which your data will come is important to your research objectives, a VGI approach may not be appropriate. It is possible to obtain a variety of geospatial data types from VGI sites. For example, traditional basemaps, such as roads, have been developed by web communities such as OpenStreetMap (http://www.openstreetmap.org). Another example, WikiMapia (http://wikimapia.org), includes a variety of geographic features delineated as polygons with annotations indicating the locations of neighborhoods, parks, water bodies, schools, and shops, among other features its users contribute.

As with any other data collection, it may be safer to mark all of your data points or features on a hard-copy map. This information can then be transferred into ArcGIS at a later date. Computers occasionally have technical problems, get dropped, or run out of power. Paper and pencil do a much better job surviving a fall than the average laptop computer, not to mention that your pencil is unlikely to run out of batteries and need to be recharged. Sometimes a low-tech data collection methodology is better for fieldwork; that way, if some glitch happens with the technology, you will be safeguarded against losing data.

Maps of your research area

For successful PPGIS, it is essential to provide participants a relevant basemap from which to work. The basemap should provide sufficient detail to orient the participants to relevant locations and familiar landmarks. If you are using printed maps, consider the scale that will be appropriate for answering the questions you will be asking participants to address as they annotate maps. If you desire relatively precise locations, or if you will ask participants to delineate lines and polygons, you may require a large-scale map, or for larger areas of interest, perhaps a series of maps, especially when using printed maps. Consider how annotations will be made, for example, will you use color to differentiate data types? If so, it is important to test the colors you will use to be certain they will be visible when drawn overtop the printed map and that they will be stable over time so that you can correctly capture the data into ArcGIS later.

For researchers opting to work in a digital environment, making annotations using ArcGIS can provide additional benefits. For example, map layers or labels may be toggled on and off to help orient participants, and maps can be panned and zoomed. In addition, digital annotations can be saved in separate layers for each session or participant and coupled with metadata about the session, such as characteristics or demographic information about the participants. Digital annotations are also much easier to edit if an error is made during the session.

As with any mapping project, you will need to identify, or potentially create, the appropriate base layers for your study location. Figure 8.3 presents a good example of a spatial basemap. In many parts of the world, recent digital map data or imagery are available and can be used. Other regions may lack readily available data, or data may be out of date, be at an inappropriate scale, or contain an insufficient level of detail for your purpose. It is also possible that for some areas, the required base data will not be available in digital form, or if they are, the data may be proprietary, requiring purchase or licensing. You can work with a hard-copy map; in these situations, you may need to copy or scan the map data to incorporate them into your PPGIS process. So, given the vast array of digital and physical map data and imagery, chances are that an appropriate basemap of your research site will exist in some form you can begin with. If available data are not at a level of detail required for your study, we advise starting with the best map that you can find. You can add additional detail as you visit the field and gain more knowledge about the study region, so long as you are aware what information is missing and you are able to plan for and acquire it within the parameters of your study. Of course, keep in mind that the best source of both current and historic information about your site may well be the individuals participating in your PPGIS sessions. Don't overlook the value of their knowledge and experience about their community or region.

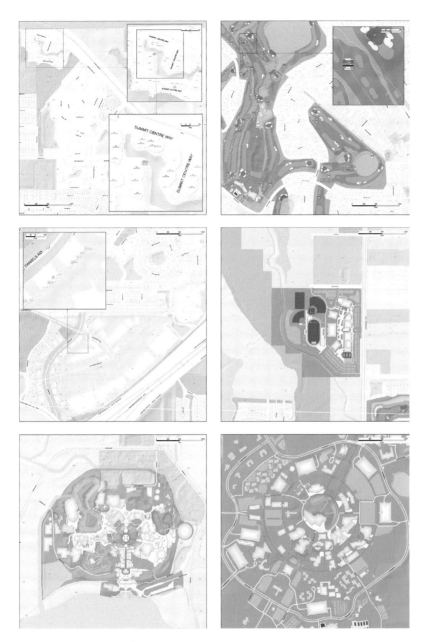

Figure 8.3 *Cartographic Excellence in Creating a Community Basemap* is an example of a spatial basemap. The basemap presented here was created by the Orange County Property Appraiser located in Orlando, Florida. The figure presents a series of six maps that were designed to serve as "a common basemap for all GIS applications" and that were "viewed by 5.2 million web users in 2012." The goal of these maps was to "enhance the cadastral map with value-added cartographic features, transforming the traditional tax map from a flat product to a rich cartographic community basemap." It is important to remember when you create a basemap to make it as real as possible and to have it contain cartographic features that the map users can recognize in understanding the map. Map by Manish Bhatt, Frank Yang, and Dan Duckworth, courtesy of Orange County Property Appraiser, Florida, which is also the source of data. From *Esri Map Book*, vol. 28 (Redlands, CA: Esri Press, 2013), 5. Data source: Orange County Property Appraiser.

Qualitative data and GIS files

A GIS can be used to portray both quantitative and qualitative data. To use qualitative data in a spatial format, one would first require identification of the type of data being used. Data files can exist as words, images, audio files, videos, or artwork, and any of these can be linked to one or more locations, such as the location where the data were collected or perhaps the locations to which the data refer, for example, the journal entries of a pioneer traveling westward. The integrating capabilities of GIS are perfect for generating themes that emerge from a variety of data sources.

Throughout this book, we present a variety of different ways you can collect, organize, analyze, and output interconnected data from multiple sources to provide an overarching understanding of your research topic, or in other words, to develop an understanding and compelling telling of the story. Whatever type of qualitative data you have, you will engage in an inductive coding process with these data to produce patterns of themes using the GIS spatial patterns that emerge from the themes associated with your data—a process we refer to as **sociospatial theming**. Key aspects in conducting qualitative research include the ability to identify themes as they emerge around specific topics. To be useful, this information must be coded to represent the "theme." The process for conducting content analysis and sociospatial theming is discussed further in chapter 12.

As coding of themes in qualitative data is carried out, the resulting codes can be added to tables associated with your spatial data in ArcGIS. A primary consideration in terms of populating your attribute table is to develop clear definitions for the pieces of information that you want to capture, along with clear coding systems to represent this information (as discussed in chapter 7). Content-based field notes can be linked to geographic locations on a map showing where and from what population the data were collected. Coding qualitative data in this manner facilitates geographic location becoming an active, and important, variable in qualitative data analysis.

Conducting a PPGIS data collection

As described in the previous sections, there are a wide array of options and considerations when preparing to collect spatial data in a participatory session. In this section, we provide a step-by-step example of the process, as we have implemented and refined this methodology in our own research projects over several years. Although no single approach will fit every situation, the steps discussed here can provide a starting point for your own research design and data collection.

Decide which type of PPGIS

We recommend using a hybrid approach that integrates both manual and computer-based approaches. How to conduct PPGIS for your project depends on the group of people you are trying to reach. You can choose to gather your information from a group of people face-to-face using paper maps, a digital mapping tool, or an online mapping tool. In some instances, given the particular audience, it might be better to use paper maps. This might be appropriate when you can pull together a population for a meeting where a PPGIS exercise will be conducted. Conversely, if your population is dispersed and has access to and experience with Internet technology, then you might prefer a web-based approach.

Starting from the bottom up: The value of inductive spatial research

A researcher can take two different approaches to collecting data: deductive or inductive. PPGIS is commonly used as part of an inductive approach to research because it begins with "observations" or collecting data from the field. When using a deductive approach, the researcher comes to the field with a previously established hypothesis that he or she plans to test. A deductive approach is appropriate when the researcher has already established a clear idea of the topic on which he or she wants to collect data.

Choose a general topic to investigate
How do you develop the topic that you want to investigate? Presumably, before you ever go to the field, you have some general idea of a topic that interests you. This interest may stem from your own area of expertise or derive from something that you care about and want to investigate further. Developing a salient topic is an iterative process that comes from your own knowledge and experience as a researcher, integrated with the issues that are of importance to local people. Examples of broad topics are people's health and the environment or the state of local economic development efforts.

Define a geographic area of interest
Anytime you are conducting a PPGIS research project, you need to clearly define the physical, geographic boundaries of your research area. This is a necessary step because the researcher wants to know how far to cast his or her net in terms of gathering data about the place in which the research is being conducted. The study site area can be as small as a neighborhood or as large as a city, county, or state. Whatever your study site, you want to develop a good sense of where you will delineate these boundaries, both to ensure you have the necessary

base data and to manage the overall scope of the research. In essence, you are carving out the area where you plan to do your research. Sometimes it is useful to use existing government-defined boundaries, such as a city or county boundary. In our environmental health example from earlier in the chapter, we limited our study site to two counties in California. We then narrowed it by choosing six specific communities on which to focus (three in each county). We selected these towns because they represented places where our partner organization had already done preliminary work.

Define the community of interest

Once you have defined your geographic area of interest, you naturally move toward defining the population of interest, in other words, how would you define or describe the people who are included in your study? If you are already familiar with your study area, then this should be an easy task. However, if you are unfamiliar with the area, you should do some background research to discover what kind of people live in the area and potentially obtain relevant demographic and community data that may be helpful in developing your sampling frame and subsequent analysis. The community of interest could include people from different ethnic backgrounds, with different incomes, and with various levels of education. In the environmental health example, our community of interest was the farm workers and others who lived in the towns surrounded by agricultural fields.

Who lives in the community of interest?

The best first step to take to learn about your community of interest is to first find existing data about your community. Within the United States, an easy first step is to visit the US Census Bureau website to gain an understanding of the demographics of the people in your study community. Another approach is to review existing newspapers or online print publications related to your community. Still another approach is to engage in conversation with community leaders. Each of these methods is described subsequently.

Data about your community can come in many forms that may be available and collected from a variety of federal, state, or local government agencies. One such commonly used source of data about communities in the United States is the US Census. The federal government collects census data every ten years, and the data are available for download from the Census Bureau website in ArcGIS-compatible format geography files that may be linked to a variety of attribute tables also available from the Census Bureau website. Interim updates are made to the Census numbers based on modeled estimates. A researcher can access information online and can usually download the data in a spatial format. Preprocessed data in ArcGIS format are also available directly from Esri, providing an immediately usable dataset that can save considerable time and effort over pulling the data from the Census yourself.

RESOURCE: US CENSUS DATA

Sociodemographic and economic data can be researched online through the US Census (http://www.census.gov/), American FactFinder (http://factfinder2.census.gov/faces/nav/jsf/pages/index.xhtml), Geo-Stat (https://www.census.gov/geo/), or FedStats (http://fedstats.sites.usa.gov/) websites. These sources can provide data about population statistics, industry employment trends, employment classes, income and poverty levels, and numbers of employees, as well as comparisons of these data in terms of race.

Finding regional economic data
The best way to understand or discover more about a study site is to find data on the economic status of the region. What are the region's main industries? Is there poverty? If so, where is it? What are the drivers of business and commerce in your geographic area? These are just some of the questions one could ask. This type of information may be available from the local county, township, or city website. Sometimes, economic development efforts focus on a region that may consist of multiple counties or smaller jurisdictions. Try searching the web for the name of a particular place coupled with the words "economic development data" and see what you find. You might also find these data by contacting economic development agencies directly. The Census Bureau makes a substantial variety of economic data available, including data from programs such as the American Community Survey (http://www.census.gov/acs/www/) and from federal agencies, including the Department of Labor, the Department of Housing and Urban Development, and the Social Security Administration. Some of these federal programs tend to focus on larger, urban areas and may not provide detailed information for smaller towns and rural regions of the country. Many of these data are provided in tabular format, requiring you to relate them to a spatial basemap before they can be directly usable in ArcGIS.

Locating historic data
Local libraries, museums, or historical societies are excellent resources for learning about a community's past. Sometimes you can find this type of information online, but unless you can access a university library, a specific community's historical records may be a bit more difficult to locate. These data can help you develop a detailed history and understanding of a place and can shed light on questions such as the following: Who once lived here? What were the major sources of income and commerce? What was life once like? Who were the groups who lived in this place? How did they use the local resources? and How did they interact with one another? Understanding the evolution of a community's history can

provide context to what is going on today and, most importantly, can help to explain the reasons behind why some groups work or don't work together, or why some groups are left out of decision making based on decisions and history.

Note that, increasingly, historians are beginning to use ArcGIS to take historical accounts and bring them to life by mapping them out, often through animated maps presented on the web. This enables historians to have a visual data source to accompany the historic record. We discuss this in more detail in chapter 9.

Understanding a community through newspapers

Local newspapers can shed light on what is going on in your geographic area and community of interest. What is the media saying about different groups that live in your community? What types of issues are being reported? Although newspapers may not be explicitly spatial in how they collect and report data, essentially every story has a byline with an indicated location, and the text of a given article may contain additional spatial information. Whenever we prepare to go into a community, we spend time reviewing the newspapers for the last several years to get a sense of the important local topics. How can this information help? While finding out what is going on in the local community, this research also aids in identifying important groups within the community of interest. Every community has groups that hold the power and make the decisions. This power is not always based on a numeric majority but may be based on other things such as a group's social connections or social status. There may be different groups of people who live in a place but have little influence on important decisions made in the community. Understanding a community's sociodemographic data helps one to know what ethnic and socioeconomic groups are important to the community beyond simply from a numbers standpoint.

Using spatial data to enhance the understanding of a community

Gaining an understanding of spatial issues in a community is also necessary to begin working with underrepresented populations such as certain ethnic groups or people from lower income levels. Some methods for accomplishing this include examining socioeconomic demographics surrounding business locations. This information will shed light on the types of people who live near different categories of businesses. These individuals may be patrons or employees, or may be perhaps part of a particular demographic. Depending on the scale of your analysis, you may need detailed spatial information. For example, at a local or community level, many businesses are close to one another, so obtaining census data at the most detailed level would be essential. Block-level data are the highest-resolution product that the US Census Bureau distributes.

In our work, demographic data, including population, race, gender, age, household, and housing information, were downloaded from the American FactFinder section of the US Census website. These data were joined to the TIGER/Line block-level boundary shapefiles for the study area, projected to UTM coordinates, and imported into a geodatabase with the ArcMap application of ArcGIS. County parcel and land use data can also help characterize the space around a business location. The most current parcel data were obtained from the county planning department. The parcel attributes characterized land use, parcel size (acres), land value, and improvement value, and listed owner information.

Once you have researched what is going on in a community, you can begin establishing trust with the population group with whom you are trying to work.

Talk with local leaders

By talking with local leaders, one can get a sense for the community and different groups who live there, where they are working, and how best to work with these different groups. This is an important step to take in addition to reviewing the newspaper data. Why? Because sometimes the news that appears in a local paper reflects the opinions, perspectives, and values of the newspaper owner, not necessarily the entire community. Talking to people from different parts of the community can provide an understanding of what is going on and people's perceptions of each other. Local leaders can be city council members, economic development professionals, prominent business owners, and directors of community organizations or nonprofits, among others.

As an illustration drawing on our own experience in the field, for an economic development project in a small rural California community, we met with county economic development professionals and the director of a local Latino organization. From these meetings, we learned that ethnic groups, such as Latino and Native American groups, are often left out of economic development efforts. We learned that Latino people would likely value our efforts as they seek to highlight Latino community strengths and contributions. Much of the media on Latinos portrayed them negatively and as a strain on local economies. These meetings also informed us about the prevalence of restaurant ownership among Latinos and about the intense competition among Latino entrepreneurs. Churches and soccer fields were identified as rally points for the Latino community.

Approach local groups to see about conducting PPGIS

Anytime one begins a project, it is important to meet with a variety of people from different groups and organizations in the community. The relevance of various groups and organizations will depend on the focus of your research project. We suggest holding a meeting with groups relevant to your study area early on to determine their interests.

In the study of the Latino community described earlier, we met with groups and organizations within the community who might be connected to this particular population. From conversations with these groups, we learned about the need for positive Latino role models for Latino youth. We also learned that the community would likely value our efforts as they sought to highlight Latino community strengths and contributions.

We learned the community had not been significantly involved with larger community and economic development efforts, which was surprising given that the local Latino community contributed largely to local economies through their entrepreneurial success and provided many services.

After working to establish trust with diverse groups or sectors of the community who have not been at the table, you can begin efforts aimed at hearing the voice of the community.

Earn the trust of the local community

Earning the trust of the local community will take time, so patience is crucial. This process should be collaborative. Let the individual with whom you are meeting or representatives from the group know clearly why you are contacting members of the community. An easy and clear way to say this would be, "We are interested in learning more about what is important to your group." In response, you may be asked, "Why do you want to know this?" Be prepared with an answer, such as, "By learning what is important to your group, we can work to make sure everybody's voice is heard by the decision makers." It will also be important to explain that you are giving back to this group through sharing your findings as you move through the information-gathering process. Knowledge is power, so giving these groups the opportunity to learn more about themselves will empower them.

Developing a good working relationship with a particular group
The key to establishing a good working relationship with a community group is to be thoughtful and to establish the relationship long before any deadlines. In other words, develop your relationship over time. Nothing is worse than a researcher rushing to establish a relationship because he or she is trying to meet an impending deadline for a grant or other funding. Relationships working with diverse groups in a community will only succeed if they are established with the right intent and over time. If an individual cares about the group he or she is representing, that individual will be very careful to evaluate the value of establishing a relationship to the community members themselves. It has to be a two-way street, where both groups can benefit. The gatekeeper of the community will want to see what you or your community economic development organization is about before allowing you to connect with others in the community. If you really want to work together, you will need to begin by first understanding what is important to that group, which comes through listening.

Identifying a trusted person in the group with which you are trying to work
Trusted individuals can further shed light on what is going on in your community and can give you advice on what the population with whom you are trying to work really needs or wants. As you move along in your project, trusted individuals are also a resource for double-checking, or ground truthing, what you find throughout your research. As an outsider,

double-checking (ground truthing) what you find with the people with whom you are working is a crucial step in ensuring accuracy.

Attempting to move a relationship forward too quickly, or just at the moment when the opportunity happens to arise, is not appropriate and does not build the kinds of long-lasting relationships that you want to establish with members of the community. It is essential to lay this social groundwork in advance of a final deadline. It is also important to realize that in many cultures, establishing social relationships first is primary, and then the work comes next. Once a relationship has been established, it can be built on for future efforts. People from different sectors of the community will be able to work together more effectively when there is trust. But trust is not something that can be developed overnight. It must be earned and developed in a thoughtful manner.

Research local groups or organizations that represent the population with whom you are seeking to work. A good place to start would be to first identify the names of any formal groups associated with the population in which you are interested. Many communities have organizations geared toward supporting different sectors of the population.

For our project, we contacted a Latino nonprofit that offers supportive services to the local Latino community. Latino people can go to this organization for tutoring services, child care, bilingual services, health care, job networking, and education, among other things. After communicating with leadership early on in the process, we decided to partner with the nonprofit in collecting data at its location, because the organization was a central part of the local Latino community. The organization was also interested in our study topic: local ethnic entrepreneurship and economic development.

Organizing a face-to-face meeting

Once you have identified organizations that represent the population with which you wish to work, you can initiate contact with individuals at these organizations. It is best to make this kind of connection and have this conversation in person. You may need to set up an appointment to do this, but meeting face-to-face goes a lot further toward establishing trust than conversations on the phone. In some cultures, this sort of relationship building may require additional time and effort. Meeting over a meal or sharing stories can be an important first step before getting down to business. A single meeting is often not adequate to build the necessary level of trust, so take your time.

Establish agreements about how you will share your findings and data with the local groups

As we have discussed so far, establishing trust is key to doing this kind of work. A major part of trust building and keeping that trust with the community of interest is to set up an upfront agreement regarding what will be produced and who will gain access to these different products. It is very important, first, to establish a clear understanding of what types of final products the course of a research project will produce. Examples of final research project products include final scientific reports, the actual data files that were part of the project, and both maps used to collect data (e.g., the maps that people annotate in a PPGIS project session) and the final results maps that are generated from the coded PPGIS data as part of the project.

It is important at the outset of every project to sit down with specific community groups, neighbors, the general public, and special interest groups to determine what their expectations are for benefitting from the project. You also need to discuss what final outcomes of the project are going to be shared with the public and what form these findings are going to take. Are community members going to receive copies of the project results, or some form of the project findings more geared toward the general public than toward scientists? (See chapter 13 for a more in-depth discussion of this.) It is always a good idea to share your results in a manner that makes sense to the local people. This may include the researchers holding a public meeting to share the results in a presentation or producing materials that can be given out to the local public. It could also include more scientific reports that could be shared with specific groups.

Preparing for your own PPGIS session

You will need to decide on a variety of issues and materials to conduct your own PPGIS session. The following is a short list, on which we elaborate.

The right location

Before you hold your PPGIS session, think carefully about where would be the best place to hold it. It should be a place where your participants will feel comfortable and somewhere with which they are familiar. For some groups, this might be some sort of community hall or meeting place. If you choose to do a PPGIS with an already existing group that holds regular meetings, you could try to talk to the leadership of the group to see if you could integrate the PPGIS session into a regular meeting time. Sometimes you may want to gather information at a public event or special meeting. For instance, if the group with which you are working holds some sort of special festival or annual gathering where many members of the group will be present, you could collect data at that event. The key issue is to make the

PPGIS session accessible and easy for your participants. If you have the resources within your budget to design your own special meeting, you could try that approach—but you will need something to draw in your participants, such as a nice dinner or some other incentive or prize. This is discussed further in a following section.

Time

The time that you spend conducting a PPGIS session will be determined by how much information you want to collect on your topic and the venue where you are collecting the information. Our own small-group, two-part PPGIS exercises typically take from one to two hours at minimum. Of course, if you want to collect a good deal of information, then you may want to budget more time.

Map of the study area

In terms of size, you probably want a map that a small group of people could sit around and mark up with colored markers. You have a choice of using a paper map or a digital map; which you use will depend on your purposes and the type of group and meeting setting. Special whiteboard technology allows the public to mark up a digital map and captures all this information through the whiteboard.

Where do I get a paper map?
Local resources for a hard-copy or physical map might be the local visitor's bureau, chamber of commerce, city hall, or visitor's center. If you have access to GIS expertise (often located at schools, government offices, or consulting firms), you can develop custom maps for your specific needs.

Where do I find a digital or online map?
There are many online sources for digital maps. For domestic research in the United States, check with the local county government offices. Many counties in the United States now provide a variety of maps and data related to their jurisdictions. If you are doing international research, begin by contacting local government agencies to see what kinds of maps and digital information are available. In many cases, you may find that the maps and data included with ArcGIS or available through ArcGIS Online are excellent sources

for your PPGIS sessions. Figure 8.4 presents an example of a PPGIS map. The great value of a PPGIS map is that it allows for important local knowledge and ideas to be recorded, aggregated, and analyzed in a spatial format.

Figure 8.4   A sample of a PPGIS map. Participants were provided a large-format paper map of the area of interest. On the basis of a series of questions, participants marked locations on the map where particular features or events had occurred. These data were later captured as digitized data layers for analysis in ArcGIS. Image courtesy of the authors. Data are from the US Department of Agriculture, Farm Service Agency. Participatory GIS data are courtesy of the authors.

Clear directions

When you hold a PPGIS session, you want to make sure that you have very clear directions for your participants. You want to keep these instructions simple and be available to assist people as they work through the activity. It is also important to provide a predetermined color-coded sheet if you intend to use different colors to indicate different types of data. Strive to have all of the groups use the same color-coding scheme for their answers. Also, if your project requires PPGIS participants to complete or sign a human subjects form, please provide adequate copies and explain what these forms mean (i.e., that they protect the rights of the human subject participants), and have an extra copy that each project participant can take home.

Group monitors

Your constituents or stakeholders should work on the GIS maps in small groups. We've found the most effective group size to range from six to eight people. A key to success in a PPGIS session is having a research assistant or someone who is already a part of your project guiding the process. We find that at the beginning of a PPGIS session using printed GIS maps as the base, many participants are fearful of drawing on the maps, having a sense that they are "ruining" what are seen as very nice maps. Sometimes the monitor can break that perception simply by writing down comments or annotations for the first few minutes and then putting markers into participants' hands after they get comfortable with the session.

Colored markers or pens

You can have the PPGIS participants mark whatever type of information you are looking for on the map. It helps if you can color code this information, preferably using bright and distinctly colored markers. For instance, if you wanted stakeholders to geographically locate different types of plants or animals, you would specify a color-coded scheme ahead of time to signify the different aspects. Provide several sets of the same colors of markers or pencils for each group.

When done effectively, we have found that the resulting maps can be digitally scanned, and then, using simple color identification tools in a graphics program such as Adobe Photoshop, the annotations can be extracted into their own separate layers for conversion into features in ArcGIS.

A flat surface for the exercise

If you plan to collect data using hard-copy maps, you will need to hold your PPGIS session in a meeting room or another place that allows you to spread out the maps. If you only have skinny tables, you might want to push them together to create a larger surface area that could accommodate a fairly good-sized map (i.e., two-and-a-half by three feet). If you are going to be collecting your PPGIS data outside at a public event, make sure that you are prepared for wind—and have some rocks or map weights that you can put at the edges of the paper map to keep it from blowing away.

Incentives or prizes

If you are doing a PPGIS with stakeholders who have a great interest in the topic or issue, they may have a natural desire to participate in the PPGIS session and to contribute their thoughts and ideas. However, many other people may need some encouragement to participate. This can come in the form of a prize or incentive. It is important to consider and budget money for these incentives in your grant. Incentives can come in the form of gas cards, gift cards, specialized baseball hats, or bags—it really depends on the type of group with which you are working. For instance, one year we worked on a project at a community health fair that offered such a great prize—stylized water bottles—that everyone wanted to participate in the study, even people we were not interested in sampling.

Refreshments

A good way to hold people's interest and to increase participation in your project is to hold a meeting where you provide food or snacks for the local population. The better the refreshments, the likelihood increases that people will want to stay and participate in your meeting. The amount and type of food that you provide depends on the time of the day of the meeting. For example, if you hold your meeting during the dinner hour, you should have more food available than if you hold a midmorning meeting. In any case, offering your project participants something to eat and drink is important.

Review questions

1. What does PPGIS mean?
2. What is the value of inductive spatial research?
3. How does PPGIS differ from volunteered GIS?
4. What would you need to consider if you were to design your own PPGIS session?
5. Do you have to have high-tech tools to conduct PPGIS? Please explain why or why not.
6. Let's say that you want to begin research work in a community where you've never worked before. What would be your first step? How would you create trust with the local people?
7. What is the difference between primary data and secondary data?
8. Why would one say that PPGIS can be used as a tool of empowerment for local communities?

Additional readings and references

Esri. 2013. *Esri Map Book*. Volume 28. Redlands, CA: Esri Press.

Ghose, R., and S. Elwood. 2003. "Public Participation GIS and Local Political Context: Propositions and Research Directions." *URISA Journal* 15: 17–24.

Meng, Y., and J. Malczewski. 2010. "Web-PPGIS Usability and Public Engagement: A Case Study in Canmore, Alberta, Canada." *URISA Journal* 22: 55–64.

Ramirez-Gomez, S. O. I., and C. Martinez. 2013. "Participatory GIS: Indigenous Communities in Suriname Identify Key Local Sites." *ArcNews*. http://www.esri.com/esri-news/arcnews/spring13articles/participatory-gis.

Steinberg, S. J., and S. L. Steinberg. 2011. "Geospatial Analysis Technology and Social Science Research." In *Handbook of Emergent Technologies*, ed. S. Hesse-Biber, 563–91. Oxford: Oxford University Press.

Steinberg, S. J., S. L. Steinberg, and S. M. Keeble. 2010. "Using PPGIS to Study Rural Ethnic Entrepreneurship." *CSU Geospatial Review: Geographic Information Science in the California State University System* 8: 5.

Steinberg, S. L., and S. J. Steinberg. 2008. *People, Place, and Health: A Sociospatial Perspective of Agriculture Workers and Their Environment*. Arcata, CA: Humboldt State University. http://hdl.handle.net/2148/428.

Steinberg, S. L., S. J. Steinberg, J. L. Kauffman, and J. E. Eckert. 2008. "*Public Participation GIS Research and Agricultural Farmworkers in California*." Paper presented at the 28th annual Esri International User Conference, San Diego, CA.

Sui, D., S. Elwood, and M. Goodchild, eds. 2013. *Crowdsourcing Geographic Knowledge: Volunteered Geographic Information (VGI) in Theory and Practice*. Heidelberg, Germany: Springer.

Relevant websites

- **US Census Bureau, Geography: Maps and Data (https://www.census.gov/geo/maps-data/):** This website presents the geographic data associated with the US Census. The US federal government collects and analyzes sociodemographic data on the American population every ten years. The data files and results of these data collection efforts are posted on this website.
- **US Agency for International Development (http://www.giscorps.org/):** This website presents information about GIS Corps, which coordinates global volunteer opportunities for people to use their GIS skills to help communities that need assistance.

- **Esri Community Analyst (http://www.esri.com/software/arcgis/community-analyst):** This website presents information on the Esri-developed Community Analyst application. This is a very useful tool for people who are interested in accessing data and reporting for different communities.
- **PPGIS.net (http://www.ppgis.net/pgis.htm):** This website presents a wide variety of descriptions of and topics related to using PPGIS.

Chapter 9

Qualitative spatial ethnographic field research

In this chapter, you will learn the various aspects of **spatially based ethnographic research**. This means conducting ethnographic research that has a spatial component. You will learn how to collect spatial qualitative data in the field. We advise using a simple, low-technology process because it reduces the risks to data collection security and provides more data collection flexibility. In this chapter, you will learn about collecting primary, spatially based data by learning on-site ethnographic data collection methods, including case studies, oral histories, and participant observations. In essence, these are all forms of sociospatial documentation.

Learning objectives

- Understand how to integrate GIS into ethnographic field research
- Understand spatial ethnography
- Learn how to ground truth map data
- Learn what sorts of technology are best for the field
- Learn how to identify local sources of data
- Understand the role that cultural perceptions of technology play
- Understand the importance of public communication for results
- Understand how to integrate GIS into oral history research

Key concepts

basemap
case studies
case study research
ethnography
evaluation survey

ground truth
ground truthing
oral history
participant observation

qualitative spatial
  ethnographic research
sociospatial documentation
spatial evaluation interview

Sociospatial documentation

In many fields, it is now common to ask research questions that have a spatial aspect, such as, "Why do the people who live in one ZIP Code have poorer health compared to people who live in a different ZIP Code across town?" Increasingly, a person's health is being tied to the geographic location where that person lives. Other questions might involve ethnicity and the environment in which people reside; for example, one might ask, "Are more Superfund sites located in poorer communities?" One could explore this topic by overlaying locations of Superfund sites onto US Census poverty data. A researcher could incorporate another primary data type into this study by conducting actual field research, where the researcher visits some of the poorer communities to talk to the residents who live near the Superfund sites. In-person visits and interviews may reveal additional information and insights that would not be readily apparent from US Census demographics.

Sociospatial documentation is the process of noting where patterns occur spatially. Sociospatial documentation means that a person investigates the spatial patterns tied to a particular social group; it is a method that can provide a way toward enhanced description and understanding as part of telling a group's story. Figure 9.1 is a good example of sociospatial documentation.

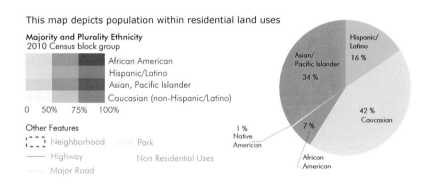

Figure 9.1 This map, based on 2010 US Census data, shows residential patterns of racial and ethnic diversity in San Francisco, California. This illustrates how a map can be used to portray qualitative data. The value of this map is that it clearly draws out the various different ethnic populations, which are color-coded and easy to identify by census blocks. The map is also a good example because it overlays the various neighborhood and community names with the census data. The map spatially portrays where the different ethnic groups live, as represented by US Census block group: African American, Hispanic/Latino, Asian/Pacific Islander, and Caucasian (non-Hispanic/Latino). "The Census Bureau collects this information by asking residents to choose the race(s) with which they most clearly identify and to indicate whether or not they are of Hispanic or Latino ethnicity. . . . The design of this map is intended to emphasize diversity by using colors with similar saturations. Neighborhood boundaries were overlaid for context." By visualizing these data, the patterns and clusterings of various groups of people are clearly visible throughout the city. By Alexandra Barnish and Larry Orman. Courtesy of GreenInfo Network, San Francisco, CA. From *Esri Map Book*, vol. 28 (Redlands, CA: Esri Press, 2013), 15. Data from US Census.

Integrating GIS into field research

This section provides examples of how you can incorporate GIS into various types of qualitative research. Some questions you should consider when using GIS as part of field work and the data collection process are as follows:

1. Will you bring a laptop with GIS software to the field location under study?
2. Do you have access to a basemap of your research area or other map data that represent one or more of the key variables? In what formats are the map(s) provided?
3. Have you verified or ground truthed the source map(s)?
4. What local sources of data can you access? Are there any written sources of local, traditional knowledge and information?
5. What are cultural perceptions of technology (including computers) among people at your study site?
6. What will happen to the results of your study once you are finished? Who will have access to this information?
7. What is the public communication plan for your findings?

In chapter 8, we provided some guidelines for doing your own public participation GIS (PPGIS). In this chapter, we provide a version of possible questions to guide your integration of GIS into ethnographic field research.

Technology in the field

The first thing you need to decide as a researcher is whether you are going to bring a laptop or other technology with you into the field. Many of the same considerations regarding technology in the field mentioned in chapter 8 also apply here. What sort of a climate are you entering? Are you going to a country or locale that has electricity? Is it a safe environment where you will be able to store your equipment without it being stolen? Whether to bring your laptop boils down to security and the characteristics of the field site you will be visiting.

Keeping things simple is often preferable, and although a laptop loaded with ArcGIS software might be capable of meeting your needs, it may be just as appropriate to use a smartphone with Global Positioning System (GPS) capabilities to record locations of interviews, create audio files of interviews, or capture text-based data entry forms. Smartphones, inexpensive tablet computing devices, and even traditional voice recorders are technologies that can serve well for data collection. If you do not require a full range of GIS capabilities in the field, it may be better to complete the mapping and georeferencing of your data when you return to the office.

Access to basemaps

Although digital maps for most parts of the world are readily available, these may not always be at the level of detail required for a community-based study. More detailed GIS data or basemaps may be available online from local county or city levels of jurisdiction. A **basemap** is a map that reflects your study area. It can exist in digital or hard-copy format. Having a basemap or creating one is essential for fieldwork. Some places where you may choose to do research will not have basemaps that are easily accessible online. A basemap is central to the research process because it lays out the field area where your study takes place and will be referred to continuously throughout the data-gathering process. Each country is different, and depending on the political, social, and economic orientations of a country, authorities may or may not choose to publicize geographic data and maps of areas located within their jurisdictions.

Although a detailed basemap can be essential for participatory approaches such as discussed in chapter 8, a basemap may not be necessary for collecting other data types, such as delineating geographic components, for which GPS coordinates or tracking can serve while in the field. A basic handheld GPS receiver can effectively capture geographic information on-site (figure 9.2).

Figure 9.2   A handheld GPS device that can be used to collect geographic data while doing field research.
Matt Cooper/Shutterstock.com.

Ground truth map data

In any study design, there is a risk of abstracting real-world concepts to an extent that they no longer represent reality, thus making the analysis and results of the study of little value. For this reason, it is important that you **ground truth** your data—that is, that you ensure that what is represented on your map or in your data file matches reality through on-the-ground field observation and/or assessment and comparison of your mapped data with other sources (e.g., other maps, aerial photos). If you are planning to conduct fieldwork, you may not always have the ability to ground truth some features of the map until you are physically on-site. You might note that the map you are able to acquire for an area is several years old, but what will be less apparent is how much things have, or have not, changed since the map was made. An old map is not necessarily bad; sometimes little has changed and the data are perfectly appropriate.

Regardless of the age of your map sources, it is a good idea to engage in some amount of verification of your data, or **ground truthing**. If you have access to the site, you might do spot checks of the map, especially if you have a sense of features that might be new or that may have changed since the map was originally made. If you are unable to visit the site in person, you can compare the map to an alternative source. For example, more recent aerial imagery collected from an airplane or satellite can serve as a useful source for comparison; using such an alternative source is sometimes referred to as surrogate ground truthing because the imagery acts as a surrogate for an actual field visit. Furthermore, imagery can provide additional detail necessary to locate sites in the field that are not depicted on standard maps. For example, a map might not show each individual house, barn, or shed, but these features may be clearly visible in an aerial image of the same location. With extensive global imagery now readily available online, it is possible to explore and capture relatively detailed and recent information for most parts of the world.

To use surrogate ground truthing, although a physical visit to the field is not necessary, you will need to get a sense of appropriate indicator variables that are visible in the alternative data sources. An indicator variable is something that can substitute for the real variable of interest. For example, using the imagery, you might be able to infer particular land uses observed in aerial images. In an American agricultural region, an observer might see expanses of crops, often organized into regular geometric patterns, with a few buildings and roads intermixed, showing a low population density. This would look very different than a rural region dominated by other industries, such as forestry or mining, and would certainly have very little in common with the land use patterns an observer would see in a large urban area.

Therefore, one could infer by observing aerial images of various regions which particular communities might be classified as agricultural based on the observable land use. Similarly, aerial images of an urban area with high population density may indicate the presence of densely packed housing, and the presence of industrial and manufacturing facilities would serve to classify an area as a very different urbanized community. In all three examples, the aerial images can assist in determining the classification of these communities based on land use. The indicator variable for these two conditions is land use. Although land use is not directly related to ground truthing per se, it is worthwhile to note that land use analysis might be useful in selecting sampling locations when designing a study.

If you wanted to ground truth a map that illustrates different population distributions by ethnicity, an image may not serve you well. You may be able to see houses, but you would not be able to see the ethnicity of the people living there. Instead, you might opt to ground truth the data against other sources of population data, such as a national census. The goal of ground truthing is to determine if the data you are using for your study are reasonably current, accurate, and appropriate to your goals.

It is important to keep in mind that you need not rely solely on existing data; you could gain invaluable ground truthing through eliciting the help of local experts: people who are familiar with the lay of the land and with local social and geographic features. This is part of the notion of PPGIS discussed in chapter 8. Of course, you should be clear about your plans to conduct research in the area prior to soliciting the help of local people. Clearly communicate your research objectives early on in the data-gathering process to ensure that locals are aware that the ultimate goal of your presence is to collect data about their community and that they know whom it is going to benefit. There might be some cases where the research method you are employing will preclude total openness about your purpose in being on-site (e.g., participant observation), but as in any research, it is important to follow protocols appropriate to the study.

Identify local sources of data

In the previous section, we began to explore the role of local community members. In this section, we further explore the role of community members as people who can not only assist with the process of ground truthing but also serve as actual sources of data. For the social researcher, local knowledge is always an important data source. Sometimes local groups have preserved their knowledge about features important to the group as oral histories. Other communities commit such information to written form. Local sources of data might appear in the form of stories, dances, rituals, and ceremonies, none of which may be officially recorded, except in the heads of community members, in some cases specific community members (community elders, healers or religious leaders, or others, depending on the culture). An exciting part of your research might be to try to record this information in a form that is accessible to researchers, other members of the community, or future generations, or simply to preserve the knowledge.

An advantage to finding locally stored data, especially data that are not accessibly recorded but rather exist in the form of stories, rituals, and the like, is that the data can then be incorporated into your study. For instance, if you are interested in documenting the geographic location and relevant social information concerning the sacred sites of an indigenous group of people in Latin America, the maps to which you have access may be rough but may give a good working picture of your study site. Probably missing from these maps would be the sacred sites important to the local indigenous people. When you are in the field collecting data, a GIS could be useful for matching up your field notes with the geographic locations of these sites. As mentioned earlier, your starting point would need to be some sort of basemap of your area.

Cultural perceptions of technology

Prior to collecting data of any sort using any technology (even a voice recorder), it would be a good idea to investigate how people living in the local area view the technology. Does it make them uncomfortable? Are they afraid of it? Do they embrace it? For example, if a researcher were to conduct a field study of the Amish, the researcher would need to be cognizant that the Amish religion forbids them from having their pictures taken, as photographs are considered graven images. Any researcher who goes into the area to study the Amish would need to be aware of this and use alternative methods of data collection. It is no wonder, with the significant presence of tourists and researchers, that Amish children are taught to run when they see a car slowing down near their fields. Similarly, if you will be conducting a study in a part of the world that is largely unfamiliar with modern technology, it may be better to avoid using such tools than to risk your study by assuming that those whom you are studying will embrace the unfamiliar technology.

Results of the study

Presumably, when conducting any research, you will have an intent to carry out an analysis and produce results in some form. Before you even begin your research project, as a researcher, you will want to consider the types of results your project will generate. Will your project generate results or findings that are geared toward a specific field of study? Depending on who or what organization funds your research, there could be limitations or requirements placed on who gets access to the study findings or project report. These are important considerations as you move forward with a project. This information should be clear to the research team, because people they encounter in the field will ask about it.

Public communication

If you have engaged with people as research subjects in the field to collect data, it is your job as a researcher to explain to them how and where people from their community can find information about your study. One way to convey information might be by creating a project website, where you can post results and findings in a manner that makes sense to the larger public. A researcher can share the same information or findings in different ways, depending on the audience. In other cases, where web access might be limited or unavailable, results can be returned to the community through other means, including printed maps or reports. In chapter 13, we discuss in depth strategies and considerations for sharing and visualizing spatial information.

Now that we have reviewed the various steps to integrating GIS into field research, we discuss various types of social science field research and how GIS could be used in the context of such methodologies.

Ethnography

An **ethnography** is a detailed qualitative description of a group located in a particular setting or environment. It is often based on a group's interpretation or understanding of their situation, and it is the researcher's job to capture this understanding. To produce an effective ethnography, one has to spend time with a particular group in their local setting. Earl Babbie (2013) notes that an important part of conducting an ethnography is telling people's stories the way the individuals perceive of them. This does not involve the researcher coming in and critiquing or changing what people tell him or her but rather recording what the people say and the exact way they have said it. For example, a very famous ethnography in the field of the social sciences was written by William Foote Whyte (1965) in the early 1930s and was titled *A Street Corner Society*. This study chronicled Italian immigrant life in the geographic locale of North Boston. In particular, it focused on the role of culture and the interaction of street gangs in the neighborhood. Geographic location plays a central role in the social organization of groups of people, especially in urban areas. An ethnography seeks to document this. Figure 9.3 presents an image of what it may look like to have people gathering on a street corner outside of a building. As this photograph illustrates, in many small towns, people in a community may choose to regularly gather in certain select places. When doing ethnographic field research, it may be important to map out the local gathering points in a community or town, especially if one is interested in tracing the patterns of social interaction or social ties in the community. Part of understanding a community's story is getting a sense of where local spatial patterns of interaction exist and when these types of interactions occur.

Figure 9.3   "City Hall with many people standing outside," Kent, Washington, 1935. Clark's Photo Studio. White River Valley Museum Photograph Collection 873a.

A spatial ethnography?

A GIS could be integrated into this type of research by having people contextualize or environmentally situate their stories for you over time. For instance, let's say that you are studying the homeless population of San Francisco and that you want to engage in an ethnographic approach. Part of your study might involve collecting the perceptions and stories of what it is like to be homeless, from homeless people. You may interview homeless people who have been on the street for several years and record their stories about being homeless over time. For example, you might map locations they describe as having been good for sleeping, getting meals, or panhandling at different times, in the past and at the present time. You could then examine key elements or variables that arise from these stories in the context of where current homeless shelters are located. Such a study might elicit support for the location of new services or for relocating existing services for the homeless to better meet their needs.

Case study research

Case studies are useful when you have an idea about a particular place or event that could potentially serve as a model for other, similar places. As the researcher, you can conduct a case study with the idea that a particular community is a model of a successful community because it has a thriving economy, local residents appear happy, and health is a major focus for residents. To support or disprove this hypothesis, you could carry out a case study. The information that you discover in the process of conducting your case study may or may not confirm your initial ideas.

In a case study, a researcher seeks to record in great detail a multitude of factors related to a specific geographic or social location. A sociological example of a case study could focus on a particular organizational situation and/or place. The researcher spends time in the community, gaining an understanding of the people, place, and interactions that occur there. A case study is an excellent method when using the grounded theory approach (see chapter 3).

The concept of a case study can be applied to either social or natural science research and their integration. You could conduct a case study of a community as your unit of analysis or conduct a case study of a particular habitat or industry that impacts a particular species of wildlife. For instance, a researcher interested in South African white shark cage diving conducted a case study on this phenomenon (Dobson 2005). Similarly, you might use a case study to show how a community or region has been affected by extreme weather. The value of a case study approach is that it provides an in-depth assessment of a particular issue or case. Case studies can often become more useful when you take a comparative case study approach, where you compare the findings from an in-depth examination of a single phenomenon in two similar or contrasting geographic locations (such as urban and rural).

What role would a GIS play?

Imagine that you are interested in conducting a case study, not of a particular place, but of a particular organization: a local senior citizen's center. The center has a good reputation for providing food to seniors who are shut-ins in a particular city. To conduct a case study of this organization, you would need to gather as much information as you could about its outreach programs. For example, where do the seniors served by the center live? How does the center organize its food distribution efforts? Does the senior center draw on donations of food from these older residents' home communities? Are there times of day when locations where the seniors live are clogged with traffic? How does the senior center work around this? Obviously, the senior center has been successful in keeping its constituents happy and has found a way to accomplish its goals on a limited budget.

Using GIS to help document and tell the success story of this organization within its particular spatial context could be very helpful to other organizations that have similar goals or to other types of social service outreach organizations. Case studies provide extensive information about successes (and possibly failures) to others so that they do not need to reinvent the wheel or attempt numerous different approaches before finding one that will work to meet their goals.

A GIS could be very useful to those who use grounded theory. Grounded theory is a research approach in which the researcher does not go into the field with a traditional hypothesis or idea about what he or she is going to find. Instead, the researcher allows the fieldwork to generate concepts or ideas. In the preceding example, a researcher would go into the investigation with no preconceived notions about what makes the senior center successful; rather, he or she would simply collect the data and see what patterns emerge.

Oral history interviews

Oral histories are an important way to collect data from people who do not necessarily conceptualize their lives as data. The stories that people tell about significant events in their lives can be very informative to a researcher who wishes to gain an understanding of a particular time and place. Oral histories can be collected in a written form, where the researcher conducts an interview and takes copious notes. They could also be recorded on electronic media, so long as the person being interviewed does not object. Using a combination of both written and recorded interviews offers an opportunity to capture the story as told by the respondent. Digital recordings can be stored on a computer as part of the GIS database linked to the location about which the respondent is interviewed. Written notes and transcriptions of the recording are also useful in conducting qualitative analysis and for linking the interview information in the GIS database using key words or concepts.

How to integrate GIS?

There are several ways to integrate GIS into oral history interviews. First, the GIS can be used as a data organization and visualization tool. Imagine that you are conducting oral histories about how people have used the Mississippi River over time. You plan to interview people who live at different locations along the river and whose families have had different degrees of interaction and experiences with the river over time. In your study, you want to investigate what sorts of factors, including spatial location, have affected people's experiences with the river over time. For instance, a person who lives close to a busy commercial port might have a different view of the river than someone who lives in a peaceful, remote location along the river. Similarly, a person who interacted with the river on a daily basis as part of her job might view the river as a positive source of economic livelihood or as an annoyance if weather or river levels prevented her ability to work.

As you collect your data, you could incorporate contextual factors about the environment, such as the number of people who live in the community where you surveyed the informant, the number and locations of ports or industries, or the presence or absence of oil spills in the region or of nature preserves along the river. You could then create files for the different geographic locations on the map and attach coded data regarding the environmental and social contexts that are important to your study. This would help you see patterns in potential factors affecting people's perceptions of the Mississippi River.

The second way that a GIS could become part of the data collection process for oral histories is by using maps portrayed with the GIS to visualize information for research subjects regarding a particular issue or problem. The oral history method is useful for studying the social and environmental contexts. For instance, you might be interested in researching the social and physical transformation of a particular neighborhood over time. You could use the GIS in the course of interviewing longtime residents of the community to interactively gather an environmental and social history of the neighborhood under study. You could tape record them sharing their historic memories and interactions. Then study participants could be shown various historical maps of the neighborhood, and they could point out relevant and important features or buildings (e.g., a local town square, parks, neighborhood gathering spots where people interacted or gossiped), which you could then mark on the GIS map with a symbol and the story recorded elsewhere. By taking this approach, you turn the GIS into an interactive data recorder as well as a technology to assist people in relating their oral histories, remembering stories and important events from a time gone by.

The GIS is perfect for both the portrayal and recording of historic information. Such information may exist in people's heads or on old maps and in historic photographs. It takes a skilled person or group of people to collect such information.

Review questions

1. How could you integrate GIS into ethnographic field research?
2. What role could GIS play in case study research?
3. What are the advantages and disadvantages to having technology in the field as part of your data collection process?
4. If you choose to take a high-technology approach to data collection, what is your number one consideration to make this a success?
5. What are the ethical considerations for any researcher who chooses participant observation?
6. Why is data cataloging such an important step in the research process?

Additional readings and references

Babbie, E. 2013. *The Practice of Social Research*. 13th ed. Belmont, CA: Wadsworth/Cengage Learning.

Bernard, H. R. 2000. *Social Research Methods: Qualitative and Quantitative Approaches*. Thousand Oaks, CA: Sage.

Cope, M., and S. Elwood. 2009. *Qualitative GIS: A Mixed Methods Approach*. Thousand Oaks, CA: Sage.

Dobson, J. 2005. "Exploitation or Conservation: Can Wildlife Tourism Help Conserve Vulnerable and Endangered Species? A Case Study of South African White Shark Cage-Diving Industry." *Interdisciplinary Environmental Review* 7, no. 2: 1–12.

Matthews, S. A., J. E. Detwiler, and L. M. Burton. 2005. "Geo-ethnography: Coupling Geographic Information Analysis Techniques with Ethnographic Methods in Urban Research." *Cartographica: The International Journal for Geographic Information and Geovisualization* 40, no. 4: 75–90.

Murchison, J. M. 2009. *Ethnography Essentials: Designing, Conducting and Presenting Your Research*. Somerset, NJ: Jossey-Bass.

Schensul, J. J., M. LeCompte, R. Trotter, and M. Singer. 1999. *Mapping Social Networks, Spatial Data, and Hidden Populations*. Ethnographer's Toolkit 4. Washington, DC: Alta Mira Press.

Schensul, S., J. J. Schensul, and M. D. LeCompte. 2012. *Initiating Ethnographic Research: A Mixed Methods Approach*. New York: Rowman and Littlefield.

Whyte, W. F. 1965. *Street Corner Society*. Chicago: University of Chicago Press.

Relevant websites

- **Ethnography Matters (http://www.ethnographymatters.net):** This website presents innovative and cutting-edge information on technology, research, and the future utilizing an ethnographic perspective. It focuses on the student experience with ethnography and the important role that context plays in research.
- **The British Museum (http://www.britishmuseum.org/research/research_projects/complete_projects/featured_project_inca_ushnus/studying_ushnus/ethnography.aspx):** This website focuses on the role of ethnography as a research collection tool; on various research projects and publications that relate to role of ethnography in the Peruvian Andes; and to the role that it plays as a research method in better understanding rural, remote ethnic populations.
- **Digital Ethnography Research Centre (http://www.digital-ethnography.net):** This website examines the role of digital media and the various symbolisms and meanings that occur through such communication. The website is sponsored by the Digital Ethnography Research Centre (DERC), which was created in 2012. According to the center's website, its mission is to "foster cross-cultural, interdisciplinary and multi-sited research around this important field in the Asia-Pacific region and beyond."

Chapter 10

Evaluation research from a spatial perspective

In this chapter, you will learn how to integrate spatial thinking and analysis into evaluation research. In recent years, evaluation research has become increasingly common in many disciplines. Studies taking on an evaluation research approach seek to assess how well staff, projects, programs, and organizations meet their goals.

Learning objectives

- Learn what is meant by evaluation
- Understand different methods used in evaluation
- Explore sociospatial evaluation research
- Learn the steps required to develop a sociospatial evaluation
- Understand the advantages and challenges to conducting spatial evaluation research

Key concepts

evaluation research
geographic target area
indicators
latent population
manifest population
multigeographies
sociospatial evaluation
sociospatial research
target population

What is evaluation research?

The goals of particular research projects can vary, from exploring a theoretical idea or hypothesis for academic purposes to examining targeted, real-world, tangible problems with the goal of making a decision or taking some specific action based on the outcome. Pure research occurs for the simple sake of interest in a topic by a researcher or organization and may not have a clear or immediate translation into the real world. Such research may, however, contribute to theory and empirical research related to a topic.

Evaluation research is a form of applied research that has direct, real-world application. It can be quite diverse, potentially involving a variety of methods of data acquisition, including surveys, interviews, experiments, focus groups, or observations. Evaluation research can be described as "research undertaken for the purpose of determining the impact of some social intervention, such as a program aimed at solving a social problem" (Babbie 2013, 359). It can also be defined as "the systematic assessment of the operation and/or the outcomes of a program or policy, compared to a set of explicit or implicit standards, as means of contributing to the improvement of the program or policy" (Weiss 1998, 4). The similarity between these two definitions is their focus on assessment.

Evaluation research provides a formal method with which to assess and identify the strengths and weaknesses of any particular project, program, or policy. Using data collected through evaluation research, a company or organization can assess where, when, and how to retool programs or policies. Why might an organization want to retool its practices? The organization may want to run more efficiently, for example, by improving the quality or delivery of services. For example, a social service organization may seek better approaches to deliver its programs or services to the intended target population. Although a number of factors may contribute to achieving this goal, there is a strong possibility some may be spatially defined and therefore benefit from a spatial perspective. For example, is the location of the office difficult to access? The great value of this type of research is that people can see its benefits immediately because of its applied nature. Virtually anything can be evaluated, and evaluation research brings a level of accountability to an organization.

Why do evaluation research?

A person, company, or organization may choose evaluation research because there is a need to answer questions about resources and success and be better informed in making future decisions. In many cases, a funder, corporation, nonprofit group, or university will seek to assess the effectiveness of its programs. Evaluation research can provide important insight into how well particular programs or policies are meeting stated goals. It can also assess how well certain units of an organization are functioning and/or meeting their stated goals or mandates. Evaluation research can also identify and document the existence of a situation or problem and is often the first step in developing creative solutions to solve problems or situations. If a problem or situation has never been documented, it can be more difficult to convince others to recognize and focus resources on the problem to implement necessary changes. For instance, consider a situation where shopkeepers along Main Street say that the youths of the town are loitering too much around their stores and scaring off potential customers. The first step would be to document that the youths are actually spending their time in the area of concern, which could be accomplished by conducting a simple observational study, noting the presence of teenagers in relevant locations, the length of time they spend in those areas, and some categorization of what they are doing. This could be supplemented by interviews with customers and shopkeepers. If this perception is a reality, then the social data will exist to document it; however, if the perception is based on assumptions or perspectives of a minority of shopkeepers, this, too, may come to light.

Evaluation research questions

Examples of broad questions that relate to evaluation research include the following:

- What are the program, policy, or intervention strengths?
- What are the program, policy, or intervention weaknesses?
- What is the target population for the project or program?
- Who or what is not being helped by the existing program or policies?
- How are the programs, policies, or interventions helping certain populations?
- Which programs, policies, or interventions are succeeding, and which are not?
- What can be done to improve the resource use of these programs, policies, or interventions?

Successful evaluation research employs well-defined research methods to assess how well the project, organization, policy, or program functions. It helps to break apart and assess the various pieces of a larger picture. The outcome of an evaluation research project will be a numeric or written analysis with a data-driven assessment of performance. This could be unit based (e.g., how well is a certain sector of the company performing?) or program based (how well is the new Building Healthy Communities program working?) or intervention based (how well is the no-texting-while-driving campaign working?). In a nutshell, information or data tell a story. The key to evaluation research is that you gather data that relate to the concepts that you want to measure.

Information collected as part of an evaluation can be used by the client or decision makers to identify where the strengths and weaknesses lie within the organization and its programs.

When companies, organizations, agencies, or nonprofits put money and effort into projects or programs, the organizations' decision makers usually have a desire to understand what they are getting for their money. Is this money being well spent? Could a company or organization be getting better results in return for the resources it is putting into a given project?

Let's explore each of these broader evaluation questions in some depth.

What are the program, policy, or intervention strengths?

Every program to be evaluated will have areas in which it is successful and effective—characteristics that should be identified and maintained. Taking a spatial perspective, core questions should address accessibility to necessary resources, clients or customers, and awareness (are the right people aware of the program or product?). Usually, the people served by a program or policy are going to have a solid understanding of how well the program or organization is actually meeting their needs. An assessment of program strengths can come through using a variety of methods, such as interviews, surveys, or observation. We discuss some of these approaches later in the chapter.

What are the program, policy, or intervention weaknesses?

Of course, a major portion of the evaluation process is to identify areas of program weakness and develop a plan to improve on these areas. Identifying weaknesses may be less popular than identifying strengths but is still an important step. This aspect of evaluation requires sensitivity and tact. Instead of using the term *weaknesses*, the aim of the evaluation could be reframed as targeting something less threatening to an organization, such as *challenges*. A spatial perspective may help highlight issues arising from the organization's location; for example, if it is a Head Start program, can parents get their kids there, given current public transit routes and schedules? Does the target population have awareness? (For example, are billboards or ads on buses in the places where they will be seen?) Or, if the organization is a business, is there adequate access to the needed labor, raw materials, and markets?

What is the target population for this project or program?

As part of the evaluation process, it is important to identify the primary beneficiaries of the program—in other words, the **target population**. To determine this, it is important to differentiate between the population that the program, policy, or intervention is targeted to assist and the actual group that is being helped. At the beginning of a project, you will want to talk with people who have solid working knowledge of the project or program. These people might be managers or project directors or on-the-ground staff. You should begin by talking with people in the company or organization who work on this project to get a sense of their programs and who is helped by these programs.

So far, we have been assuming a social science type of program designed to help some aspect of society. What about evaluating an environmental program where the focus is more on biological than on social aspects? If you conduct evaluation of an environmental resource–type project, physical, biological, or environmental features may be important units to examine. Perhaps a timber company wants to evaluate how well its timber-harvesting plans are meeting the needs of the community and workers and the biological needs of the surrounding forests. The company would have to employ scientists to come up with **indicators** or measurements of all of these different aspects to then produce an assessment or report card of how well it is doing. In many resource and manufacturing sectors, international certification standards subject to independent assessment may award a seal of approval for sustainability and environmental stewardship.

Assessing these characteristics may involve a number of spatially important criteria such as the distance to market, the effects on watershed health, or the presence of particular species of plants or animals. To receive such certification, a company must undergo a rigorous evaluation process. Once the company or organization receives this certification, it can market its products or goods with the special certification's associated label. For example, if a particular timber company's product receives a special certification, its wood products may be more desirable to customers who value the principles associated with sustainability. It is important to note here that the certification is only of value if the people or organizations that conduct the evaluation are reputable and maintain the standards of the evaluation process. If standards are not upheld, then the certification loses its meaning. Therefore, it is important to define the appropriate objectives and indicators in advance and to document them as part of the evaluation process.

What populations or entities are the programs helping, the target population or another population (latent population)?

Any project or program has intended goals and an intended target population it is designed to help. We call the intended target population the **manifest population**. There may also be populations, groups, or entities impacted by the program or policy that are *not* part of the intended manifest population. In other words, they are a **latent population**. For example, a small town may have devised a project to improve the educational capacity of its immigrant youth by providing students with free tutoring to help with homework and learning the English language. Sometimes the parents may bring their children to these sessions and also attend them with their children. They, too, become unintended beneficiaries of this English-language tutoring program and thus are a latent population that benefits even though it was not necessarily targeted by the program. It is valuable to track these latent populations, which may be impacted either positively or negatively. Some initial background investigation can help in identifying both the intended and unintended groups with the potential to be affected by the project.

Latent effects can extend to environmental change as well; an environmental project may focus on protection of an endangered species (the manifest entity), but likely other, latent biological or ecological entities will also benefit. For example, by defining, setting aside, and managing an area designated for the protection of an endangered species, other plants and animals sharing the same habitat will inevitably benefit. Of course, given the complexity of our world, the reality is that most situations have effects in both the social and natural realms, and trade-offs must be considered.

Good examples of trade-offs are the natural resource management choices made over the years concerning suppressing forest fires. For many years, there was an attitude that all fire was bad, and every effort was made to suppress wildfire. This ultimately benefited those species that prefer late successional–stage forests because the suppression of fire prevented the natural "resetting" of these areas and clearing of excess fuels. Eventually, fuels built up to such a degree that when a fire did occur, there was a higher likelihood of a more severe event. By contrast, as "let it burn" policies came into place, concern about damage to life and property became important, perhaps most pointedly during the 1988 fires at Yellowstone National Park, when the fires came close to the historic lodge and other facilities near Old Faithful (figure 10.1). As one of the largest fires in US history, there was significant debate about fire policy during and after the event.

Figure 10.1   Fire crews attempt to water down buildings as fire quickly approaches the Old Faithful complex during the 1988 Yellowstone fires; the fire is "crowning," racing along the tree tops and spreading rapidly. Photograph by Jeff Henry.

Which programs, policies, or interventions are succeeding, and which are not?

Answering the question of which programs are succeeding and which are not requires information. It also requires the use of benchmarks to assess how progress is being made toward program goals. To be successful in any type of evaluation, the first step is to define targets or mini-goals that can be evaluated across space and time. In figure 10.2, we see an example of a simple rubric that includes benchmarks. To assess progress toward reaching each benchmark, a numeric value can be assigned to various programs and policies that are being assessed.

Specifically, how are the programs, policies, or interventions helping certain populations?

A good place to start when evaluating how well a policy is aiding its target population is with a list of the programs or policies that you plan to evaluate. Through talking to people within the organization, you can gain a better understanding of the different project goals. From this you can create a rubric or list of possible results and then come up with a numeric or quantified system for ranking each of these possible results, ranging from 1 to 5. Following that, you could go through the process of actually finding data or information that indicates how these programs or policies are helping (or not). Such information might exist in the form of quantified data (e.g., number of people helped or assisted in some way) and more qualitative data (e.g., stories or vignettes regarding how people have been helped by various projects). Figure 10.2 provides a sample evaluation form that presents a rubric and scoring procedure for how you might assess a nonprofit organization's progress toward meeting stated goals that involve assisting the homeless.

We see that various activities on the rubric, such as providing food and shelter for homeless families, establishing safe environments, and helping people who are down on their luck to develop life skills and reintegrate into society, are viewed as important for the evaluation. Spatially defined criteria, such as those represented by the first three examples in the rubric, could be served by GIS in two ways: first to define the boundaries of the desired service area (what is the area intended to be served) and second to assess the actual service area represented by the clients availing themselves of these services. For example, to define the service area, GIS might be used to assess all areas that are within a specified distance or travel time from the office. The evaluation could then draw input from observations or interviews of a sample of individuals within the area as well as completing an overlay of those actually using the service, regardless of where they come from (figure 10.3).

In figure 10.2, we also see an overarching score assigned to the organization that assesses the organization's program performance on a number of items. The meaning of this score is then interpreted. After conducting the evaluation, the assessed nonprofit could be described as providing "some support" to the homeless community based on the Homeless Organization Evaluation Benchmarks Rubric in figure 10.2. If the organization wanted to improve or strengthen its existing ability to meet homeless people's needs in the future, the organization now has a report card of sorts that it can use to guide its improvement efforts. And by representing these data spatially, the organization can assess where gaps in service exist and develop strategies for reaching out to those areas.

Sample Homeless Evaluation Rubric	Strongly Agree (5)	Agree (4)	Neutral (3)	Disagree (2)	Strongly Disagree (1)
Provides food for homeless families within service area		4			
Provides shelter for homeless families within service area			3		
Provides a safe environment for kids within service area				2	
Assists homeless in developing life skills			3		
Helps homeless to re-integrate into society				2	
Program enables homeless to get back on their feet				2	
Total (Scores may range from 6 to 30)	16				
Meaning 6 to 10: Very weak support 11 to 15: Some support 16 to 20: Moderate support 21 to 25: Strong support 26 to 30: Excellent support					

Figure 10.2   Homeless Organization Evaluation Benchmarks Rubric. The first three items are both spatially and service defined, whereas the second set is socially defined. For an actual study, additional details (e.g., what defines the service area and potentially differential weighing of factors) should be explicitly defined to meet the desired purpose of the study.

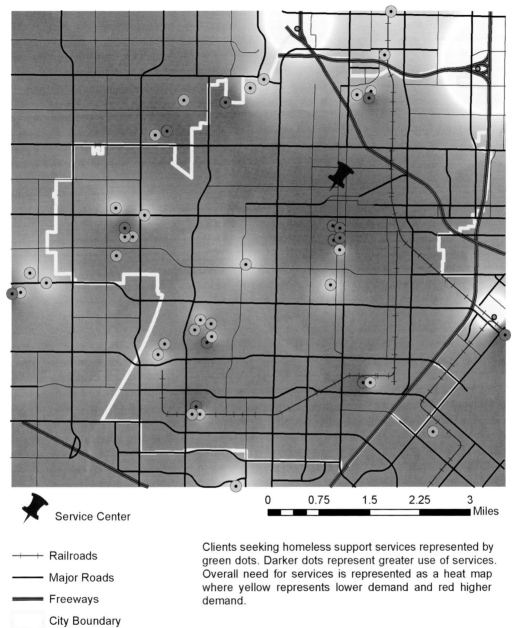

Figure 10.3 A hotspot map showing results of an analysis of demand for homeless services. Although the service center indicated on the map is situated near one area of demand, it does not address other hotspots in the region. Reviewing these results also indicates concentrations of demand coming from relatively distant locations, including **several at the periphery and outside of the city limits.** GIS and cartography by Steven Steinberg. Data for homeless population and service center locations represented on this map were artificially created by the authors for illustrative purposes. Transportation and city boundary layers, Orange County, CA.

What can be done to improve the resource use of these programs, policies, or interventions?

This is the type of question that comes at the end of the evaluation. Companies and organizations are always looking for ways to improve their efficiency and effectiveness. The evaluation research outcome will produce a more in-depth overview of existing programs and the resources used as a part of that process.

Sociospatial evaluation research

Now that we have introduced the topic of evaluation research, we further explore how this research approach can effectively incorporate a spatial perspective. An understanding of your organization's or company's performance from a spatial perspective can be very useful. **Sociospatial evaluation** means that you are actively incorporating spatial thinking and factors into your evaluation project. **Sociospatial research** focuses on the active consideration of space, place, and social indicators in a holistic fashion (Steinberg and Steinberg 2009). Because almost everything we do is associated with some location, it is not a great stretch to develop evaluation questions that explore these factors. In fact, many such evaluations have considered space for a very long time. What has changed is the availability of map-based tools and thinking. These spatial factors can consist of proximity (a contextual measure that accounts for who or what is near or far) and distance, buffers, or conceptual space.

In the late 1980s, well before GIS was in the mainstream of undergraduate education, I (Steve) was conducting a study of prenatal care at a clinic for underprovided women as part of a medical sociology course. There was concern at the clinic that many women failed to attend the full suite of prenatal appointments leading up to the birth of their children. The study was tedious and time consuming as it required going through each patient file by hand to create a list of appointments kept or missed, a total number of visits, and where each woman lived (recorded as a ZIP Code). In retrospect, the questions, and eventual results, of that study would have been ideal for a GIS analysis—alas, I had no idea at the time. Of course, the GIS software of the day was not particularly user-friendly or affordable for a college freshman either! Ultimately, the most common reasons given by women for missed appointments or ceasing to attend prenatal care were related to transportation. Interestingly enough, after lots of data compilation, analysis, and some rudimentary hand-drawn mapping, we discovered something quite surprising: the clinic was not close to any of the clients, nor to any public transportation. In fact, the only relevant individuals living close to the clinic were the doctors and staff, who lived in the surrounding, affluent neighborhoods.

While the concept of examining spatial relationships came to the fore, the same questions and data today would be relatively easy to examine in a spatial context in ArcGIS. Patient records might be electronic and supporting demographic data easily obtained in GIS format from the US Census Bureau. GIS layers of ZIP Codes, streets, bus routes, and other relevant base data would likely be available from the city or county. In fact, most of the time-consuming data compilation and analysis would be simple and straightforward by today's standards. The only aspect of the study that would not change significantly would be the interviews with staff and clients. Ultimately, an evaluation study of this sort conducted today could effectively highlight the demographic issues leading to the missed appointments and also help to identify potential solutions considering where clients come from, transportation options, or even locations for a new or relocated clinic that might better address patient needs.

Sociospatial evaluation research questions

A researcher should consider certain questions when conducting any type of spatial evaluation. The following questions apply to both natural and social science investigations:

1. What is the geographic target area for your project or program?
2. With regard to physical geography, what are the relevant natural and environmental features that affect the target population, for example, mountainous terrain, oceans, trees, expansive desert?
3. What human-constructed features of the environment affect the target population, for example, bridges, freeways, city boundaries?
4. Where are the programs, policies, or interventions both succeeding and falling short?
5. What are the geographical features that limit the success of various programs, policies, or interventions?
6. What are the spatial features of an environment that help to foster the success of various programs, policies, or interventions?

A more in-depth examination of these questions follows.

What is the geographic target area for your project or program?

Every project or program will have a geographic area that it covers. Defining the boundary of your **geographic target area** (your area of interest) is an important early step in the sociospatial evaluation process. How do you determine such a boundary area? First, you want to identify who the groups or entities are that your organization programs serve. Boundaries may be defined by jurisdiction, such as a city or county boundary (e.g., the organization serves teens in St. Louis, Missouri); a functional boundary, such as one defined by distance or time (e.g., our fire department will be capable of responding to any location in our community in five minutes or less); a naturally defined boundary (e.g., the Maumee River watershed); or even a conceptual boundary (e.g., individuals who participate in an online community for survivors of abuse). Your organization may serve different groups or objectives, and it is key to be aware of these. Once this is accomplished, you can work backward toward identifying the geographic boundaries of the entire project.

To get a sense for this information, we advise talking to people who have substantial familiarity with, and knowledge of, the project. Through conversation, you may create a list of the geographic areas—such as towns, cities, or neighborhoods—where the project or organization is operating. Another approach could involve showing the involved parties a map and asking them to draw a polygon around the boundary area. The important thing to remember is that sometimes the people who have the most knowledge about the projects are the ones with on-the-ground positions in the organization. This may not be the program manager or boss but rather lower-level employees—people whose opinions the organization does not often consider. The value in soliciting input from these individuals is that their feedback will allow you to gain the insight of people who have not had a chance to provide information in the past. They may bring novel perspectives and understanding to situations or problems.

What natural and environmental features affect the target population?

Another question to address is the geographic scope or breadth of your project. Does it cover **multigeographies** (which are multiple geographic locations)? You may have a project that affects large numbers of people in many different geographic locations, perhaps across a wide expanse or territory. These locations may not be proximate to (next to) one another, depending on the type and focus of your program. In such a case, your project or program consists of multiple geographies. To determine a geographic boundary for such an area means first looking at the diverse groups the program serves and then delineating the geographic boundary around this area. In some situations, multiple, disconnected geographic boundaries may be necessary, especially if the locations are spread out. In such cases, you may end up delimiting a broader boundary area around all of those locations, or you could draw individual regional boundaries around particular areas and examine each area as a separate unit of analysis.

Once you've narrowed your target geographic area, it may be important to understand and describe the physical environment where the target population lives. Will features related to this environment impact the way your population leads their lives? Similarly, will certain features affect the population's mobility patterns? For instance, is the area small and rural? Is it bounded by mountains on one side and the ocean on the other side—with nothing but small, winding roads connecting towns to one another (figure 10.4)? Or are there natural waterways in the area that facilitate mobility up and down rivers and/or canals? Do weather and time of year change the situation in significant ways? Developing a good sense for the physical features of your study region can be important, because they represent the natural environment in which your population is operating.

Figure 10.4 Residents of an isolated rural community, such as Mountain City, Nevada, operate in a very different context than those in a more suburban or urban community. Photo by Famartin.

What human-constructed features affect the target population?

Today, along with natural features, there are also physical features or structures that have been crafted by humans. Both natural physical features and human-designed aspects create the physical boundaries that compose an environment. Ultimately, the combination of these natural and manufactured features will affect people's mobility patterns. For example, consider the presence or absence of major freeways. When the freeways were built in the United States, many of them cut through neighborhoods and towns, often splitting towns in two. People had to adjust as the new motorways created new patterns of flow within communities. These instances of modern progress would make it difficult to get from

Figure 10.5 The Golden Gate Bridge with the city of San Francisco, California, in the background. An example of the built environment. Rich Niewiroski Jr./Shutterstock.com.

point A to point B without installing bridges or overpasses to cross the thoroughfares. Such newly created infrastructures definitely altered patterns of interaction and mobility.

Besides considering transportation routes, one should also consider the role of the town square. Town squares served a major role as the hub and heart of many communities and cities.

The town square, a traditional structure in many small towns throughout the United States, is an example of a space that facilitates local social interaction (figure 10.6). From a functional standpoint, the square creates a place for people to gather, share, and celebrate various activities together. From the past through the present day, in Old Town Orange, California, the Old Towne Plaza, or "the circle," as college students tend to call it, is the location of many community fairs and festivities throughout the year.

Figure 10.6 Aerial photo of Old Town Orange, California, including the circular plaza in the middle. From the past until today, the Old Towne Plaza has been used as a local community gathering spot. Image courtesy of US Department of Agriculture, Farm Service Agency.

As older small towns gave way to newer developments, suburbs created their own gathering places. In newer developments, planners have sought to replicate the community or small-town feeling through creating more densely packed housing that has a giant square of open space, park, and recreational areas for local social interaction in the middle (figure 10.7). These may consist of recreational soccer and baseball fields, swimming pools, playgrounds, and a school—all centrally located in the middle of the community. This style of development greatly facilitates outdoor recreation and activity, where people can and do interact with one another in a variety of ways, ranging from playing on sports teams to attending family parties and group dog playdates.

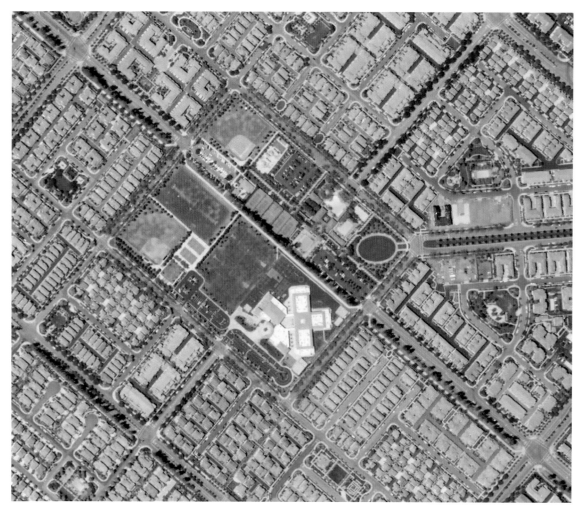

Figure 10.7   Aerial photo of Village of Woodbury neighborhood development, Irvine, California. The development is characterized by a large open space square in the middle and many smaller parks throughout the housing development. Image courtesy of US Department of Agriculture, Farm Service Agency.

A comparison of these two photos illustrates the differences in recreational space between an older, more established town, such as Orange, California (figure 10.6), and the newer development of Woodbury, located in Irvine, California (figure 10.7).

Where are the programs, policies, or interventions both succeeding and falling short?

In spatial evaluation, the goal is to assess where geographically or spatially efforts are succeeding and where they are falling short. The goal here is to connect each of the different projects and policies that are being evaluated to a particular geography and then to assess your strongly performing programs—in other words, you want to identify the projects or programs that are producing results! That should be easy to do because project managers love to highlight the areas where they've been successful and things are working well. It could be much more challenging to get the same program managers to highlight where the programs are not producing desired results.

Once you've determined your target population and its location, you can focus on conducting your spatial evaluation. It may be that some sectors within your geographic area are receiving successful services from your program, but others are not. Geographically, you could identify the differences between these two groups. As you conduct your spatial investigation, think critically about the spatial distribution of resources and the types of results that are experienced in different geographic locales. For instance, maybe your health outreach program to different ethnic groups in a particular area is working among blacks and whites, but this same program falls short in meeting the needs of the local Latino population. As a part of your sociospatial evaluation, you might want to assess some of the spatial evaluation questions we examine in this chapter. Effectively, you are looking to identify spatial patterns of weakness so that your organization can act on this information and do something to improve it. Being able to make good decisions about the future involves first developing a solid understanding of a situation. This comes from gathering good data.

What are the geographical features that limit the success of various programs, policies, or interventions?

In the previous section, we went through how to identify the natural geographic features that are a part of your geographic area. In this section, we build on that information. Every region or spatially bounded area is going to include certain geographic features that impact human interaction, mobility, and trade. For example, if your organization is trying to improve health access in a rural area that includes rugged topography with limited road transport and many rivers and streams, you must consider these factors as you develop the health outreach plan. Projects should be designed by carefully considering the people and their context. Geographic features that could potentially limit the success of various programs or policies, such as mountainous terrain that does not contain many roads, long stretches of desolate desert, or inhospitable rock-walled canyons that run through the middle of your region, should all be accounted for in the planning phase. Any or all of these physical features can limit people's mobility and their ability to communicate.

What are the spatial features of an environment that help to foster the success of various programs, policies, or interventions?

Examining the physical aspects of a place as you conduct an evaluation is essential. Drawing on the strengths of a place can be a natural part of the evaluation. If you catalog what you have, you can draw on this catalog in the future.

For instance, consider the notion of proximity. What is your boundary area near to? Is it located near a city or some important natural feature that can serve as a means of transportation, such as a river? In many of the southern states crossed by the Mississippi River, agricultural and cotton industries developed shipping lanes down the river. The same could be said for a town or community that is located along the shipping lanes of the ocean. We describe the features that can enhance the success of a various program or policy as positive spatial features.

Designing an evaluation research project

In planning for an evaluation project, time and money will impact the type of evaluation research that you choose to implement. The choice of what you do is going to be affected by how much time you have to complete the project and the number of resources you have to devote to the project. For instance, if your evaluation is being conducted over a longer period of time, you might choose a research approach that is more time intensive, such as a case study. If you have a shorter time frame, you might consider something like a short survey or a series of key-informant interviews.

In any case, we want to underscore the notion that a multiple methods evaluation is usually preferable to a single methods evaluation. Why? Anytime someone measures something using a variety of research methods, she will achieve a more well-rounded view of the topic. This is because different methods present slightly different perspectives on an issue. In essence, different research methods uncover various aspects of a project or program.

What are you evaluating?

It is important to determine what you are going to evaluate. Are you focusing on the functioning of specific groups within the organization or on the entire organization? Is the concentration on certain programs or policies of the organization? If so, what are they? You should specify the "unit" that you are going to evaluate. Examples of units might be departments or divisions within the organization or company. If you are focusing on subunits within the organization, such as departments, then these are your units of analysis.

Determining your goals using benchmarks

A major role of evaluation research is to develop the metrics or "benchmarks" used to gauge the success of a project, program, or policy. Benchmarks are directly tied to your project and what you hope to achieve as a result of it. To identify benchmarks, one must first be clear about the project goals. Examples of project goals might be meeting the needs of a certain population. For instance, assume you have a nonprofit designed to help the homeless in your town. More specific goals for the homeless project would be to identify the types of people served, how you benefit them, and most importantly from a sociospatial perspective, where you serve them. In this case, one is evaluating a specific program that was designed to serve the homeless.

Some of the benchmarks that might be involved in evaluating a homeless relief community program might be documenting the number of families that your homeless shelter serves. This could be broken down further to collect information of who benefited or received shelter from your center per week/month/year. Another potential benchmark might be to determine how many homeless people receive two meals a day from the center. Still another benchmark might be tracking the new or incoming and older or outgoing members of your homeless population. By tracking the homeless mobility you are determining where and how to best address their needs. One can also use this tracking program to see if homeless are moving on to a different geographic location. Perhaps they are re-integrating into society? To determine this one would need to do a little more investigation.

The benchmarks that are used as a part of evaluation research can also serve an important documenting feature for the nonprofit homeless organization. How does that work? Having data will allow the organization to portray who it is helping and how the organization actually makes a difference. In essence, the data from the evaluation research program can be used to portray the effectiveness of your organization, through using real, empirical data. Such information can be used to highlight your organizations impact on the population and community it is attempting to serve.

Most programs or projects have established and stated goals. In the homeless case study mentioned above, the overarching goal or motto of the nonprofit group administering the Bright Circle Homeless Services Center might be to meet the needs of the homeless in the community of Bridgeport.

Group versus individual focus

As you decide on the type of evaluation to conduct, it is very important to consider the area of focus. Are you going to focus on gathering your data from a group or on individuals?

If you choose a group focus, the will use methods that assess collective perceptions and attitudes. If you choose to focus on individuals, you will use methods that highlight the attitudes and perceptions of individual people. In a group setting, you will have more of an integrated discussion about ideas and issues. This can sometimes influence the thoughts and attitudes of each individual, depending on how willing they are to appear different from others in the group.

Sociospatial evaluative focus group

A focus group is a type of evaluation that you can use to gather people together to provide input on your project. Focus groups are sometimes more useful when you are clear about the topic on which you want to gather input and about the group of people from whom you want to collect input, such as stakeholders. In that sense, focus groups are a great strategy to use when you want to assess whether the needs of an entire group are being met.

As you talk with this group of people, you may want to have them indicate spatially, for example, by marking a map, where they feel the greatest needs are. In an entrepreneurship project that we worked on in northern California, we partnered with a local community assistance agency that had already established a good working relationship with the local Latino community. The main point is that they were trusted by the local Latino community because they assisted the community with various issues such as gaining housing, reading documents, and paying bills. This community group wanted to learn about local Latino people's degree of civic engagement. We assisted with the research design. First, we had a large group meeting where we discussed and defined what the concept community or civic engagement means to the various people. We broke down the discussion into various parts. Once that short discussion was complete, we broke the larger groups down into smaller groups of four to five people to have them discuss interaction and community gathering spots in the local community.

Because we could not find a list of locally owned Latino businesses, we asked people from the local Latino community who were our stakeholders at the meeting to indicate on a map where the local Latino-owned businesses existed. We also asked them questions to see if they could geographically mark on the map places where they went to get assistance with issues of health or business. We each assisted with running spatially based focus groups that were geared toward identifying where people in the local community gather and spend their time and seek advice or assistance for their businesses if they need it. We used differently colored markers for points of recreation versus points of socialization or sport (see figure 10.8).

Throughout the process, we guided the discussion of the various geographic points as well and helped to facilitate that the points identified actually ended up being marked on the map.

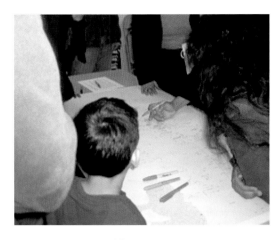

Figure 10.8 Spatial focus group mapping session, working to identify where people gather and where locals go to meet their various community needs. Photo by Steven Steinberg.

Evaluation interview

In an interview, the researcher sits with an individual in a one-on-one-type setting. An interview usually includes some close-ended questions that allow a respondent to choose from a list of answers, along with open-ended questions—these are questions where answer choices are not predetermined. The person answering the question is free to generate any sort of response. An example of an open-ended question might be, "What sorts of programs does your organization have to help the homeless in your community?" The survey question answers to an open-ended question may include information about overnight shelters, food distribution and sanitation. The advantage of the interview process is that you are interacting with the interviewee one-on-one and have the opportunity to ask additional questions and to seek clarification for answers through the interview.

An interview is a good tool to use when you are talking to members of a mobile population, such as the homeless, because it allows you to collect your data at one point in time and requires no further action once the data are collected.

In a **spatial evaluation interview**, you should ask questions that tie into the spatial aspect of your project. For instance, you might want to interview homeless clients in or near the places that they reside and leaders of homeless service centers to see the variation in how they each perceive the success of the project in and across different locations. Homeless people may visit different homeless shelters to meet their various needs. The daily mobility patterns of your population may indicate how to better meet the needs of a mobile homeless population. If so, you could craft questions related to where people who use the services of the homeless shelter sleep at night and where they spend their time during the course of the day. This information can lead to a more complete understanding and evaluation of where homeless people are and how best to meet their needs.

Evaluation survey

An **evaluation survey** consists of survey questions related to the topic that you are evaluating (see chapter 7 for more on spatial survey data). A survey tends to be more quantitative because the majority of questions will be close-ended questions, with perhaps a few open-ended questions included as well. A close-ended question provides answers or categories to choose from; an example might be, "Which services does the Bright Circle Homeless Shelter offer? Please check all that apply." The survey would provide a series of choices from which the respondent can choose:

a. shelter
b. food
c. job training
d. mail delivery

Once you have determined your topic, the next task is to design questions or indicators that measure your topic. This information can be captured using survey questions that have a spatial aspect. For example, homeless people may visit different shelters at different times to take advantage of various services offered throughout the day. You might ask a person seeking services at a homeless shelter to identify the general area where they choose to meet different needs. Here is an example of a close-ended question that might pertain to this topic:

1. I attend the following shelters to take a shower and clean up:
    a. 5th Street Center (195 5th Street)
    b. Concord Homeless Shelter (32 Pinehurst Lane)
    c. Interfaith Homeless Housing (125 East 7th Street)
    d. YMCA (8th and Broadway)
2. I attend the following shelters to eat lunch:
    a. 5th Street Center (195 5th Street)
    b. Concord Homeless Shelter (32 Pinehurst Lane)
    c. Interfaith Homeless Housing (125 East 7th Street)
    d. YMCA (8th and Broadway)

Evaluative spatial observations

Observations that are conducted by the evaluator could be of a qualitative or quantitative nature, depending on how they are employed. If you develop a rubric and make a quantitative assessment of the different goals or tasks you are evaluating, this would be quantitative observation. Instead, if you focus on writing up field notes and observations, this would be more of a qualitative evaluation.

When you conduct a spatially based evaluation, you are bringing in or noting the spatial locations of cases or whatever you are observing. Therefore you must somehow record this information. This can be done through using addresses or any other geographic feature, such as ZIP Codes or closest cross streets. The goal is to be able to tie the geographic information to the evaluation data you have collected. You can also take photographs to document the physical features of an environment. Nothing conveys a sense of place like photographs. These photos should also be consulted throughout an evaluation research project because they may yield important insights about the environment and its implications for your findings.

Evaluating existing data or documents

We have been operating under the assumption that one needs to collect new or primary data to effectively conduct an evaluation. When you assess existing data or documents, you pull together the various pieces of background data that can help to set the stage for your evaluation. Any company or corporation will probably have data on what types of programs it runs and information on the numbers of people these programs serve. If as an evaluator you find out that such data do not exist, you can then develop a process for collecting them. Sometimes an organization has been in the process of conducting its own documentation for a while to show the impact of its various programs. In this case, the researcher has a very positive thing, because she can use this information as part of the evaluation process. Of course, an evaluator wants to be cognizant of how the data were collected, so if you evaluate existing data or documents, be sure to consult the metadata, or information about the data, such as when the data were collected, who collected them, why the data were collected, and how. If metadata does not exist, ask questions such as, Who collected the data? Was any bias part of the data collection process?

Presenting the spatial evaluation

This section presents various components that we recommend a researcher makes a part of a final sociospatial evaluation report or presentation. To illustrate the various pieces, we have pulled in examples from a sample research report.

The spatial evaluation will be the end product for the company or organization. A final spatial evaluation should include many visuals and maps. The final spatial evaluation should comprise the following components.

Introduction to the evaluation

In this section, you are going to provide a brief overview of the evaluation and why it was conducted. What is the goal of doing this evaluation? How can the organization possibly benefit from doing the evaluation? Who is preparing the evaluation?

Project background

The project background should include the project purpose, the project need, and, finally, project objectives. Here you will describe the organization, company, or project being evaluated, including history about when the project or organization started, who funds it, and a discussion of the goals and purposes of the organization or project.

Methods

This section should describe what type of information you collected and how you went about collecting that information for your evaluation. It is important to remember that data can comprise any documents or material that inform the evaluation process, including any information you collected as part of the evaluation. If you used multiple methods, you should state that. Additionally, here you should discuss the various types of information that you collected. For example, did you collect your own data, or did you use existing data? Were the data that you used qualitative or quantitative?

Results

The sociospatial evaluation results section presents the various maps and descriptions of the spatial patterns observed, while assessing the data. You can highlight spatially and visually the areas that are succeeding or doing well and the areas that require further work. In presenting such an evaluation, you also want to make sure that you include any tables, charts, or graphs that might be relevant, such as shown in figures 10.3 and 10.9.

Conclusions and recommendations

In the conclusion, your goal is to make some overarching statements about patterns that you saw in the data. What do the themes and patterns in the data point to or indicate?

Additionally, based on your assessment of what was evaluated, you might want to provide recommendations about where and how to move forward.

Recommendations would include action steps to take based on the knowledge gained from the evaluation. By way of example, we list some policy recommendations that may have derived from an overall analysis of the data considered as part of the hypothetical homeless project we have discussed throughout the chapter:

1. Increase "mobile" health service to homeless population.
2. Improve homeless access to health insurance.
3. Increase transportation to overnight shelters for homeless.
4. Build community coalitions with members of the faith community to help aid the homeless.

We cover the connection between GIS and the development or design of policy in much more detail in chapter 14.

It is advisable to create both a physical report and a slide show that includes key points and visualizations. These two different forms of communication will enable you to share the study findings with others. Both should contain assessment materials from the project. A slide show presentation should focus on key findings, lasting perhaps fifteen to twenty minutes, to hit the highlights of what is found in the final report. The final report would contain all of the details about the data and patterns that were found in the data as a part of the evaluation. Chapter 13 provides much more detail on how to communicate the results from various research projects and findings.

An example final presentation is shown in figure 10.9. In many cases, one of the main issues facing lower income people is the issue of food availability and food access. Increasingly, small towns and urban centers have begun to focus on effectively meeting this challenge for lower income populations as well as homeless people. This has occurred through first assessing space, place, and location of existing different types of food sources and mapping out their geographic location.

The map in figure 10.9 geographically depicts areas around Tulsa, Oklahoma, with limited access to affordable, fresh food. The authors of this study used GIS to map locations of known fresh-food retailers, such as grocery stores, supercenters, and natural food stores. A point density analysis was then performed based on a two-mile radius around each establishment and was then symbolized to identify the number of fresh food retailers available to those residing within each area. Such analysis is very useful because it quickly provides an overview of and insight into food accessibility for a specific area, in this case, at the request of a member of the state legislature, who may not have had the time to obtain and digest this information effectively if it had been presented in a report or tabular format. Interestingly for this particular map, red was selected to highlight access to the greatest numbers of fresh food retailers within a two-mile radius; orange symbolizes access to two or more fresh food retailers; and green one fresh food retailer. Although the color choices may not be immediately intuitive (more typically, green is used to indicate a preferred situation and red a poor situation), the patterns of access are readily apparent on the resulting map.

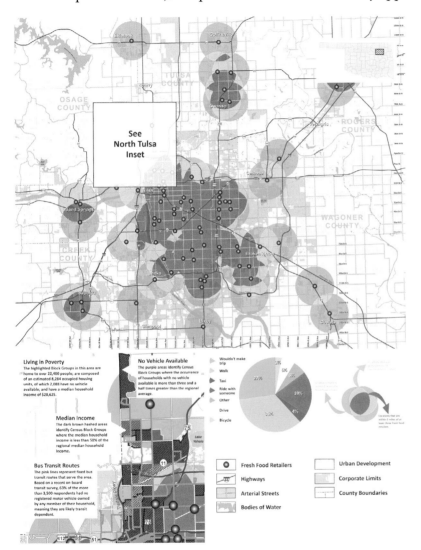

Figure 10.9   This map of the Tulsa, Oklahoma, area was developed to illustrate relative levels of access to fresh, healthy food. In this example, the colors are potentially misleading: the redder areas of the map are those with the highest levels of access (three or more stores within two miles), whereas those in green represent areas of lowest access (one store within two miles). The uncolored areas of the map are more than two miles from any store, representing the most poorly served neighborhoods. Map by Ty Simmons, courtesy of INCOG, Tulsa, OK. From *Esri Map Book*, vol. 27 (Redlands, CA: Esri Press, 2013), 55. Data from 2010 US Census, 2005–9 American Community Survey, 2010 Metropolitan Tulsa Transit Authority (MTTA) On-Board Transit Survey, National Center for Public Research (National Food Desert Awareness Month), Food Desert website, University of Oklahoma–Tulsa, School of Community Medicine.

The challenges and benefits of evaluation research

Conducting any type of evaluation research can present a challenge because this kind of research can create stress for people who feel that their performance, programs, or policies are being targeted. To effectively conduct evaluation research, researchers must be very thoughtful, careful, and sensitive in how they approach the topic. Evaluation researchers also want to be especially careful around issues related to confidentiality of information, especially spatial information. That being said, it is always prudent to have a clear understanding up front when you conduct an evaluation research project about which parties will see the final results—researchers must be careful about how spatial information is shared with the public.

Sometimes when a company or organization undergoes evaluation research, the employees can feel stressed, as if they are the ones being evaluated, instead of the project. This is because no one ever wants to be highlighted for not doing a good job. One way to mitigate employee nervousness is to hold a quick orientation meeting that explains what you are doing. Keep a positive, upbeat attitude and be very clear about when, where, and how the results are going to be used. Being open about the ideology of the study and the intended use of the evaluation data is central to establishing trust between the evaluators and the subjects (people being evaluated). Establishing a good, working relationship is an important part of great client/evaluator relationships, which ultimately are key to uncovering data and information as a part of the evaluation process.

As we have illustrated in this chapter, many advantages can emerge from conducting sociospatial evaluation research. The main advantage is that you are able to get a broader, more holistic and visual understanding of how a company's or organization's programs or policies are working and where they are not working to their full capacity. Having the ability to highlight performance hot spots or deficiencies will be very valuable to an organization or agency. Once the leadership knows the areas that need improvement, they can better target future resources to garner maximum efficiency. Organizations and companies are always looking for ways to be more efficient, and sociospatial evaluation provides that opportunity.

Review questions

1. What does the spatial perspective contribute to evaluation research?
2. Where do you begin a sociospatial evaluation research project?
3. What are some of the challenges to conducting spatial evaluation research?
4. What are some of the advantages to conducting spatial evaluation research?
5. What is a potential spatial evaluation research project that you are interested in?

Additional readings and references

Babbie, E. 2013. *The Practice of Social Research*. 13th ed. Belmont, CA: Wadsworth Cengage Learning.
Patton, M. Q. 2001. *Qualitative Research and Evaluation Methods*. 3rd ed. Thousand Oaks, CA: Sage.
Posavac, E. 2010. *Program Evaluation: Methods and Case Studies*. 8th ed. Boston: Prentice Hall.
Tripp, C. 2008. *Evaluation of a Public Participatory GIS Tool: A Public Planning Case Study*. Ann Arbor, MI: Proquest.
Weiss, C. H. 1998. *Evaluation*. 2nd ed. Upper Saddle River, NJ: Prentice Hall.

Relevant websites

- **American Evaluation Association (http://www.eval.org):** This website contains a variety of information on evaluation research and is the main website for the American Evaluation Society.
- **Policy Development and Evaluation, UNHCR, The UN Refugee Agency (http://www.unhcr.org/pages/4a1d28526.html):** This website shares various evaluation reports used to systematically examine the policies of various programs and practices sponsored by the UN Refugee Agency.
- **Natural Sciences and Engineering Research Council of Canada, Program Evaluations (http://www.nserc-crsng.gc.ca/NSERC-CRSNG/Reports-Rapports/evaluations-evaluations_eng.asp):** This website presents a collection of various program evaluation reports conducted by the Natural Sciences and Engineering Research Council of Canada.
- **Cornell Office for Research on Evaluation (https://core.human.cornell.edu/):** This website explains the value of evaluation research and how it works. IT provides news related to evaluation research in a variety of applied contexts and discuss current projects sponsored by the Cornell University Office for Research and Evaluation.
- **Centers for Disease Control and Prevention, Gateway to Health Communication and Social Marketing Practice, Research/Evaluation (http://www.cdc.gov/healthcommunication/research/):** This website explains the role that research and evaluation play in developing health communication and social marketing campaigns. It also provides a link to research summaries of relevant research projects related to health and communication.

Chapter 11

Conducting analysis with ArcGIS software

In this chapter, you will learn about the analysis tools available in ArcGIS, which can be used to analyze data prepared from both quantitative and qualitative sources. You will be introduced to various forms of analyses, including those you will likely want to apply as you begin to use GIS technology. You will learn about buffers, overlays, networks, map algebra, and raster analysis as well as interpolation, simulation, and modeling. You will also get an overview of analytical methods, extensions, and spatial statistics. The chapter addresses the means for linking external or discipline-specific quantitative and qualitative analysis with the spatial outcomes of ArcGIS. Finally, you will learn some of the common pitfalls to avoid when analyzing and interpreting results.

Learning objectives

- Learn a conceptual approach to data analysis
- Explore the most commonly used data analysis techniques for spatial data in GIS
- Learn the concepts of modeling and probability as additional techniques for data analysis
- Understand how to appropriately and carefully design and implement your research approach and report your results

Key concepts

buffer
cartographic categorization
compliment
cost surface
doughnut buffer
exclusive or
identity
intersection
least cost analysis
location error
map algebra
measurement error
modeling
Monte Carlo simulation
nearest neighbor analysis
network analysis
overlay
range
raster modeling
simulation
social networks
spatial interpolation
symbology
union

Approaching the analysis

As you prepare to conduct the analysis, you need to reconsider the original questions you had as a researcher. Data analysis is perhaps the most exciting part of the research process because it is where the researcher gets to delve into patterns in the data.

What questions did you ask?

Moving from a conceptual model to the logical model that guides your analysis requires that you have some sense of what it is you want to achieve through your analysis, even if you do not know the specific commands or procedures.

If you think back to the formulation of your original research question, you wanted to explore some key areas of focus as a part of your research project or investigation. Now would be a good time to list out and revisit those initial questions that you asked in chapter 3: questions about concept, questions about data, questions about location, and finally, questions for analysis. In the first stages of your project, you identified questions that you can now explore and answer by analyzing your data.

What tools and analyses are applicable?

To determine the tools and analyses that are applicable to your project, you must first answer the following questions: What type of data do I have? Are the data more quantitative (numeric) or qualitative (content based)? What sorts of relationships do you want to examine? Are you interested in comparing the results between two different groups? Are you looking to emerge with some sort of a predictive model for the future? You might begin also by thinking about how you want to use the findings from the data as a means to drive the types of analysis that you do as well.

Organizing the analysis

The best way to organize your analysis is to review your existing conceptual data model and logical data model (discussed in chapter 2). The factors mentioned as part of that model will help you to organize your analysis. A conceptual model highlights the flows and relationships between concepts in your study, whereas a logical data model spells out the types of data that you will need to examine to reflect the various concepts.

Analysis techniques

This chapter introduces several of the most common analysis techniques used in GIS. For each case, we present the logic and application behind the technique. Any of these may be used individually or in combination with others, and in some cases, more than one approach may lead you to a similar end result.

One of the important and powerful aspects of GIS is that there is not always a single correct approach to answering a question with GIS. In fact, when a classroom of GIS students is provided a set of data and an analytical question, each student will approach the analysis with a somewhat different tack, but all the students' approaches may be perfectly correct. This shouldn't be too surprising and can be demonstrated using a simple example.

Consider the following basic arithmetic problem. Pay attention to the process you use to solve the problem:

$$5 + 3 - 12 + 8 + 4 + 7 - 6 - 2 = ??$$

Individual approaches will vary. Some people will simply run through the list in the order it is presented. Others may do all of the addition first and then do the subtraction, perhaps to avoid dealing with negative numbers. Still others may work the solution by canceling out positives and negatives, for example, the + 8 cancels the − 6 − 2, whereas 5 + 3 + 4 cancels with the − 12, leaving the + 7. However you approach the solution, so long as you use valid logic, it will result in the same answer. Similarly, in GIS, a variety of approaches might be used in solving a problem, some with fewer steps and others with more. So long as the logic is correct, the specific approach may not be as important. Of course, if you are going to be running an analysis often, taking extra time up front to develop a streamlined approach to doing the analysis more efficiently may be worthwhile.

With this backdrop, we examine some specific techniques for your consideration. Each of these techniques should be available in most GIS packages you encounter, although the exact names of the tools may vary. If a particular approach is not available but is necessary for your analysis, this is a good indication that you will need to look for add-on tools or possibly even a different software package that provides the tools you require to complete your analysis.

Although finding the proper add-on tools may sound intimidating, it is actually quite easy. Websites and even user groups for the major software packages will point you to both free and commercial add-ons, and also to where fellow users will answer questions. In most cases, an add-on component or user can assist you in accomplishing almost anything you can imagine needing in an analysis.

Cartographic classification

A simple yet quite powerful analytical tool is based on data classification. **Cartographic classification** is a scheme used to classify different types of features or data. When mapping data, the categories used can make a huge difference in how the output appears, both visually and statistically. The same dataset can communicate very different information depending on how it is categorized. For example, if you consider an analysis of the retired population in a particular region of the country, you will obtain very different results based on how you choose to classify retirement age. If you use the US Social Security Administration's definition of age sixty-five, there will be a significantly smaller number of people in the category when compared to the same data using the AARP's defined age of fifty. Of course, there are benefits to using a higher age if you are the Social Security office; fewer people who receive benefits means the money will last longer. This also explains the logic of increasing the age to sixty-seven in the future. The AARP, conversely, garners much more political clout by including a large number of people in lower age groups.

If we were to display this population data on a map of the United States (figure 11.1), we would expect to see a visibly denser map of retirement-aged persons when using the AARP definition and might feel a greater sense of urgency about issues related to senior citizens. If we were to combine these maps with the population under the retirement age (those paying into the system), the picture might be even more dramatic. Such combinations might be accomplished via a variety of categorization tools offered in ArcGIS. These adjustments in categorization, classification, and visualization of your data in ArcGIS are made using the symbology settings for a given data layer's properties (figure 11.2). **Symbology** means what graphic icons (such as a tent for campgrounds or a dot for cities) are going to be used to illustrate certain aspects of the data layer.

Number of Senior Citizens in Lower 48 States by County

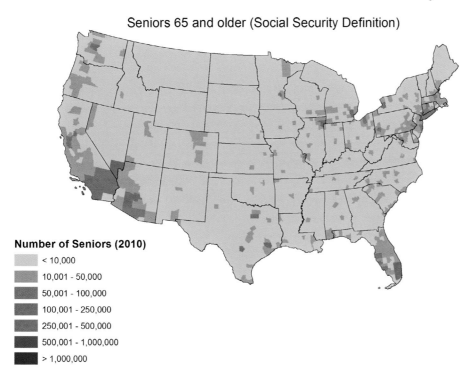

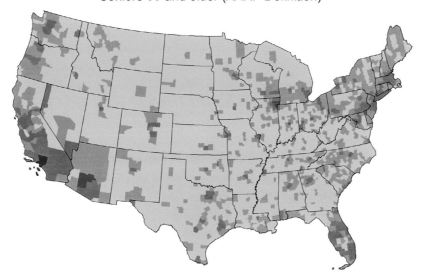

Figure 11.1 A comparison of maps showing the number of senior citizens in the lower forty-eight United States according to (top) the Social Security Administration's definition and (bottom) the AARP's definition. These maps are based on the raw number of individuals by county according to 2010 Census data. If one wanted to emphasize the number of seniors in the United States, clearly the AARP definition would make it appear that many more seniors are present. GIS and cartography by Steven Steinberg. Data from 2010 US Census.

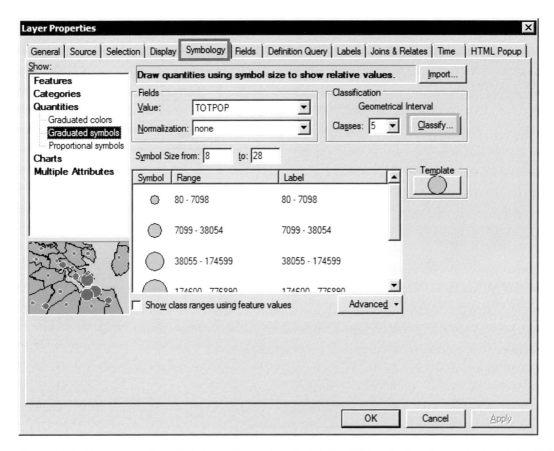

Figure 11.2 Users can assign attributes colors and symbols in ArcGIS by electing the Symbology tab within the Layer Properties (indicated by the red box). Choices made here can significantly influence the message your map communicates to the reader, as discussed in chapter 2. The representation of data into different categories (qualitative, quantitative, or descriptive data types) and ranges (for quantitative data types) is controlled here. Clicking the Classify button (indicated by the blue box) opens another dialog box that contains options to specify the statistical method for defining breaks between classes. Esri.

These examples have little to do with true analysis; they simply require a bit of manipulation of the category definitions or statistics underlying the map. However, many data collection processes impose classification categories up front when the data are initially collected. For example, many surveys will provide some level of confidentiality to respondents by obtaining responses falling into broad ranges. A range can be described as the possible scope of quantitative answers. There would be a high and a low quantitative point to this range. For example, income may be collected in fairly broad categories:

Less than $20,000 $100,000–149,999
$20,000–49,999 $150,000–199,999
$50,000–99,999   More than $200,000

When the data are precategorized in this way at the time of collection, the opportunity to reassign or break down the categories in alternative ways is lost. For example, using the preceding income categories, you would be unable to determine how many families have incomes between $30,000 and $60,000. While the opportunity to collect more specific data may not be feasible in all cases, when you have an option to collect and retain actual values for each respondent, ArcGIS will allow you to categorize and analyze the data at the time of analysis, thus providing far greater flexibility. Such data may also have greater utility in the long term because they will not be limited to the specific requirements of the initial study. In this sense, there are clear analytical advantages to collecting more detailed data, such as the ability to redefine breaks in categories as the specific requirements of the analysis change.

Buffer and overlay

Two of the most commonly used tools in GIS fall into the categories of buffer and overlay. Although it would be impossible to assign a precise value, it would be reasonable to estimate that 50 to 75 percent of GIS analyses are achieved using variations of these two concepts alone. In ArcGIS, a number of these tools fall in the Analysis Tools toolbox and provide easy access to several variations on these two key analysis concepts (figure 11.3).

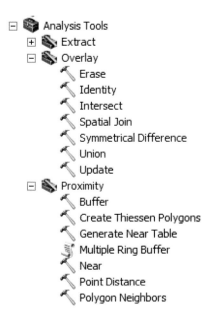

Figure 11.3 ArcGIS Analysis Tools toolbox, which provides capabilities relating to the basic concept of buffer (proximity tools) and overlay operations, among others. Esri.

Buffer

Buffering refers to a process of delineating an area around an object most often defined as a simple Euclidian distance (a straight-line distance "as the crow flies"). Essentially all mapped features in both the vector and raster GIS environments may be buffered, as shown in figure 11.4.

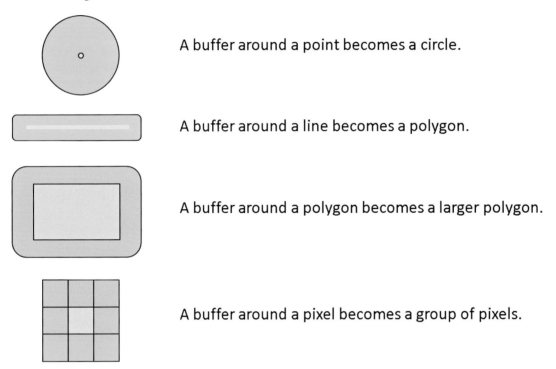

Figure 11.4   Graphic examples of a buffer drawn around each of the three basic elements of a vector GIS (point, line, and polygon) and a raster GIS (around a pixel). The initial feature is represented in yellow, and the resulting buffer is represented by the green zone, an area defined around the original feature. Buffers are useful in a variety of distance-dependent analysis questions.

Why are buffers popular as an analytical tool?

Buffers provide a simple yet effective method for analyzing a variety of spatial questions. Many GIS questions involve spatial concepts of proximity. Common phrases you might find yourself using as you conceptualize your own research questions include the following:

- Close by
- Within a distance of

- Next to
- Less than —— distance away
- Influenced by
- At least —— distance away
- Farther than

If you find yourself phrasing questions in this manner, there is a good chance that a buffer will be a useful component of your GIS analysis. For example, if you are a city planner interested in determining a good location for a new community park, you might want to be sure that the location selected is within reasonable walking or biking distance of currently underserved residential areas. This simple question involves two proximity questions:

1. Determination of desirable proximity (being close to underserved residential areas)
2. Determination of undesirable proximity (to determine underserved neighborhoods, you need to know which neighborhoods are currently too far from existing parks)

Using the concept of a buffer, such an analysis can be easily achieved.

Use of buffers does not necessarily require that the lines representing the buffer distance are actually generated on the map or computer screen. In many GIS software programs, these spatial calculations can be achieved in the computer's memory without you, the analyst, ever seeing the circle drawn around the point on the screen. This is important for two reasons. One, having all of your analytical buffers visible on the map or screen can lead to excess clutter, making it hard to visualize your results, so alternatively, some GIS software will simply highlight the features that fall inside the defined buffer. Second, if you need to know where the lines of the buffers fall, you may need to specify that you want the resulting data saved into a new data layer. This can be essential if your analysis requires that you use the buffers in a subsequent step of the analysis.

It is worthwhile to mention a few additional aspects of buffering in GIS. Buffers need not be a fixed distance. In some analysis situations, it may be appropriate to buffer only to one side of an object or to buffer at different distances around objects based on characteristics of the feature being analyzed (figure 11.5). For example, consider a buffer used to represent pollutants coming out of a smokestack. Although it might be a reasonable abstraction to assume that the pollutants will affect residents surrounding the facility, using a fixed (circular) buffer around the smokestack location is probably not quite realistic. Most could reasonably assume pollutants will be carried farther in the direction of the prevailing winds.

Another variation on a buffer is the **doughnut buffer**. These are buffers represented as concentric rings around the object and may be used to represent places that are considered desirable in the analysis when the goal is to be close, but not too close. For example, if you are locating a new factory to provide jobs in a community, there is a balance to be found: the factory should be close enough to be convenient for employees to reach but, at the same time, not so close as to be located in the middle of a residential area.

Finally, non-Euclidian buffers may use characteristics other than distance, such as time, as the metric for the buffer. If you are planning for the location of a new fire station, the goal may be to place it no more than five minutes' driving time from those homes and businesses that are in its designated service area. Of course, driving time is affected by numerous variables, perhaps the type of road (major artery or freeway versus neighborhood or city street), the number of intersections that must be crossed en route, and the *average traffic congestion* on the roads. In situations where buffer size varies based on other criteria, a look-up table of values for various criteria or a formula to calculate the buffer distance may be substituted in place of a fixed value.

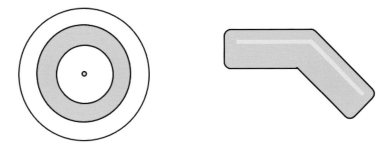

Figure 11.5 (Left) An example of a multiple-ring, or doughnut, buffer, in which the area of interest in the analysis is located no closer and no farther than a specific distance from the given location. The area meeting these criteria is represented by the green ring. Areas too close or too far are represented by the white rings. (Right) An example of a non-Euclidian, or variable, buffer in which the feature being buffered is a line. The buffer distance to the left (lower) side of the line is three times greater than the buffer distance defined for the right (upper) side of the line.

A second technique, overlay, must also be considered. Overlays are explored in the next section.

Overlay

An **overlay** is another powerful analysis tool that a researcher can use to address the concept of spatial correspondence, or things that occur together in space. Conversely, we can use overlays to look at negative occurrence or noncorrespondence. Overlays are most

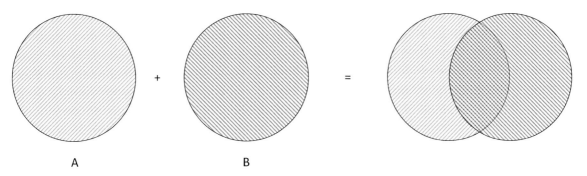

Figure 11.6   A Venn diagram showing two separate data layers (A) and (B) that are combined together while retaining their appropriate spatial position. In a GIS context, the spatial position would most commonly be the mapped location on the ground. Some portions of polygon A overlap with polygon B, while other portions do not. Similarly, some portions of polygon B overlap with polygon A, while other portions do not.

easily conceptualized by using common Venn diagrams of overlapping circles, as shown in figure 11.6.

If you have two sets of data represented as polygons A and B that are partially overlapping, as shown in figure 11.6, four basic analyses can be conducted to examine the spatial relationships between them. Each of these is presented visually in figure 11.7.

First, we can ask where **A and B** occur together, that is, where these datasets overlap in space (in GIS jargon, this is the **intersection** of the data). This might be useful in looking for sites that meet two or more key criteria. For example, I might be interested in looking for a home within a specific school district that is also on a south-facing slope to ensure good sunlight. An intersection of two polygon layers representing these boundaries of interest would enable me to narrow my search.

Second, we can ask where either **A or B** occurs in space (in GIS jargon, this is also called a **union**). Perhaps I'm interested in locating a home that is either on a south-facing slope or in an area of low canopy cover. A union of polygons representing these two options will result in all areas that meet either one (or both) of my specified criteria.

Third, we can ask what occurs within the area **contained by A**, including those portions of B that fall inside of A (in GIS jargon, this is sometimes called an **identity**). Perhaps not all areas meeting my desire for sun are available because some are city parks. An identity using a data layer representing private land (or private land currently for sale) would allow me to identify those areas that may be reasonably considered. Unlike the other three functions discussed here, this one will provide a different result if the order of the input layers is reversed.

Finally, we can ask where **A or B, but not both, occurs** (in GIS jargon, this is referred to as a **compliment** or an **exclusive or**). In the ArcGIS Toolbox application, this function is accomplished using the Symmetrical Difference tool. This operation provides an output consisting of those areas where there is no overlap between the layers. Continuing with our example of seeking a home with good sun, we may be interested in those properties that are located on south-facing slopes or with low canopy cover but do not wish to consider areas where both conditions coincide—perhaps out of concern that it would be too hot.

Examples of the resulting output for each of these overlay concepts (assuming polygon area–type data) are illustrated in figure 11.7.

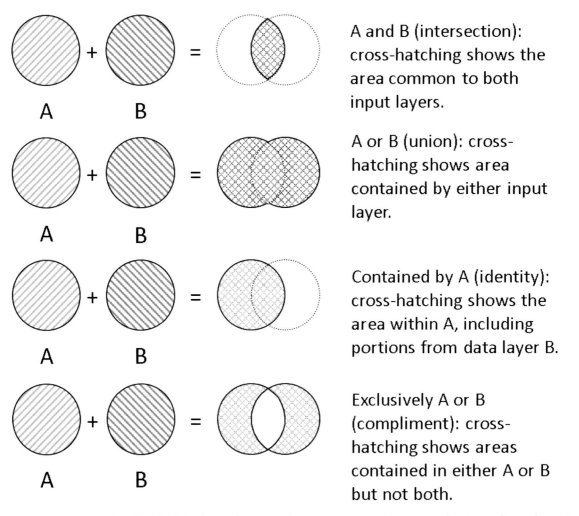

A and B (intersection): cross-hatching shows the area common to both input layers.

A or B (union): cross-hatching shows area contained by either input layer.

Contained by A (identity): cross-hatching shows the area within A, including portions from data layer B.

Exclusively A or B (compliment): cross-hatching shows areas contained in either A or B but not both.

Figure 11.7   Examples of each of the four polygon overlay types (intersection, union, identity, and compliment) and their associated results.

We can extend most of these overlay concepts to include the other vector data types, points, and lines. When considering points or lines, we typically will take one of two approaches. By creating a buffer around a point or line, you can generate a polygon for use in an overlay, as described earlier. Alternatively, you can examine interactions between other combinations of feature types, for example, a polygon and either a point or line or interactions between sets of lines or between points and lines. Such interactions may include those of containment (e.g., a point falling inside a polygon) or those of intersection (e.g., a road passing through a city or two roads crossing at an intersection). Examples of these and others are shown in figure 11.8.

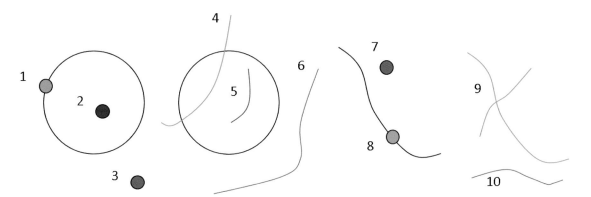

Figure 11.8 The results of overlays of vector feature types. Combining point locations with a polygon feature gives three possible results: (1) the point falls on the boundary, (2) the point falls inside the polygon (containment), or (3) the point falls outside the polygon. Combining lines and polygons provides possibilities, including (4) the line passes through (intersects) the polygon, (5) the line is entirely inside the polygon, or (6) the line is entirely outside the polygon. Combining lines and points, you may get results, including that (7) the point falls off the line and (8) the point falls on the line. Finally, (9) lines can intersect or cross or (10) they do not intersect or cross.

In the raster GIS realm, the overlay process works very differently, using a process commonly referred to as **map algebra**. We discuss raster approaches in detail in the modeling section later in this chapter.

Proximity polygons and nearest neighbors

When we ask questions about how close things are to one another, we can use two general approaches to assess this. The first of these, **proximity polygons**, looks at each location of interest and delineates polygons surrounding them that are closer to each individual point of interest than to any other. In ArcGIS, there is a tool in the Proximity toolbox called Create Thiessen Polygons (figure 11.3) that may be used to accomplish this task.

Typically, the locations of interest are represented as discrete points, as shown in figure 11.9. This might be useful in delineating service areas for a police or fire station, bus stops, community health clinics, or grocery stores. Though it is useful, this is a simplistic method that has some important limitations. Resulting service area polygons do not account for additional information such as locations and routes of roads or speed of travel. The underlying assumption is that all locations falling in the delineated area are closest to the central point and therefore allocated to that service area. Although one might presume that the fire station closest to an emergency should respond, the closest station in response time is not necessarily the geographically nearest location.

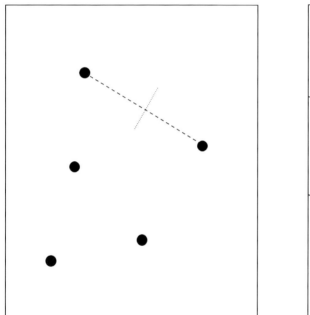 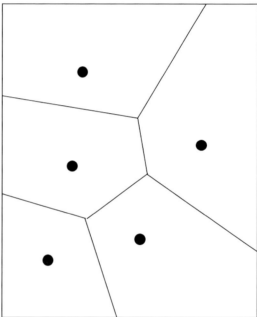

Figure 11.9   The construction of a Voronoi diagram. Assume that you are starting with the map on the left, showing five point locations (perhaps representing police stations). To delineate polygons that include the areas closest to each point, we use a simple fact. Any location just less than halfway between points must be closer to one particular point than to another. In geometry, this is defined by the perpendicular bisector of a line connecting any pair of points on the map (in the figure, the dashed connecting line is bisected at the midpoint by the dotted line). If you imagine doing this for every point pair, the bisectors will converge and create polygons around each point on the map, as shown on the right of the figure. These represent the proximity polygons for each point on the map.

A **nearest neighbor analysis**, also available in the Proximity toolbox as the Near tool (figure 11.3), does not delineate areas but rather determines the closest location to a point or points of interest. The result is added to the layer's attribute table, indicating the feature ID of the nearest point, and the distance in units is selected. A nearest neighbor analysis might be useful if you wanted to determine how far it is from your home to the closest grocery store, as shown in figure 11.10. Similarly, you can obtain nearest neighbor statistics for entire combinations of data layers, including the minimum, maximum, and average nearest neighbor distance between all homes and grocery stores in the data.

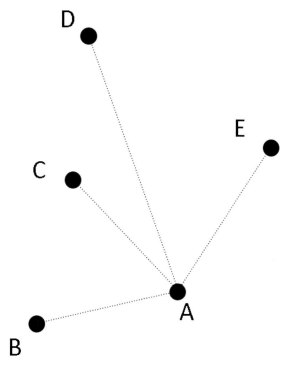

Figure 11.10   In a nearest neighbor analysis of points, the GIS would determine a distance measure between point pairs (shown for point A in the preceding example) and the nearest point. In this case, point B is closest to point A. If we had two sets of data, for example, a set of points representing neighborhoods and another representing playgrounds, we could take the nearest neighbor distance between each neighborhood and each park to get an average nearest neighbor distance for all combinations of the data rather than for a single point.

Social networks and network analysis

GIS can play an important role in analyzing concepts that are less evidently spatial. For example, consider the concept of social networks within a community. **Social networks** are important to a community because they represent the degree of social cohesion and communication within the community. Examining social networks therein could be useful in planning development. Understanding the link between individuals and groups in a community is useful when a researcher is seeking to organize local groups to accomplish community-oriented goals.

GIS can help in studying these social networks by mapping out geographic distances and locations between the members of a particular group or community. For instance, you might expect to find that social networks are related to the geographic proximity of individuals relative to the center of town. In figure 11.11, we present a simplified example of such a map.

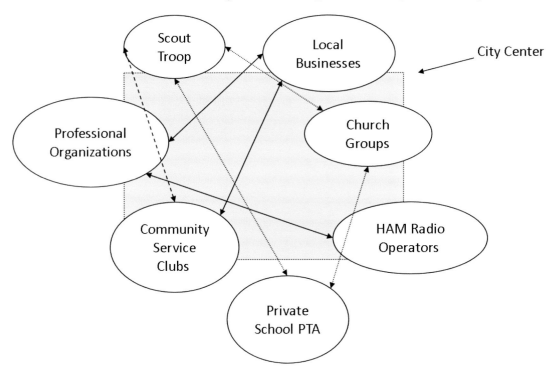

Figure 11.11   Social network connectivity of community groups and their geographic positions relative to the city center. The city center is represented by a tan, rectangular area near the center. Each group in and around the community is represented by a circle. These circles may represent an actual geographic position or perhaps a representative location based on the percentage of that group inside or outside the city center. Finally, the ties between and among these groups are represented by the arrows. The researcher would determine how the groups are defined (variables or entities) and the relevant attributes he or she collects about the groups, and then these data would be stored in a data table associated with the map shown in the figure. Any type of data could be collected on the groups.

A study that identifies different types of social networks, in essence, the connections between groups and individuals, can aid communication and response to any crisis situation. GIS allows for mapping differing degrees and levels of social cohesiveness and acquaintanceship networks, which could be key to responding to disaster. Using GIS to catalog and geographically locate where local skills reside and who possesses which skills and resources could be instrumental to improving emergency management and response in times of crisis, be it environmental or social. Additionally, an understanding of these social networks could promote community projects and increase community strengths. That way, when the community faces a need or an emergency, a plan could already be in place based on the GIS data of mapped skill sets and patterns of social interaction. Key information and local social mobilization could be quickly initiated using a GIS-based alert system. Similarly, a GIS could be very useful in visualization, planning, and disaster preparedness. It would allow for visualization of and response to disasters prior to their occurrence. Visualization techniques of this type are discussed further in chapter 14.

Network analysis

Network analysis is most appropriate when studying connections that follow specific pathways. In GIS, these are most commonly thought of as infrastructure or transportation related, for example, a pipeline or a subway. However, networks may be equally appropriate in social science applications for exploring the connections between people, social cohesion, strong and weak ties, community strength, and so on. To illustrate this concept, it will first help to explain the basic structure of a GIS network.

An important location in a network is represented as a node. The node acts as the individual sample unit or measured entity in the GIS database. Connections between nodes are represented by lines. The length or other attributes of a connecting line indicate characteristics of the connection. For example, in figure 11.12, the three nodes represent individuals (Jane, Bob, and Sue), and the connecting lines represent the ties between these individuals.

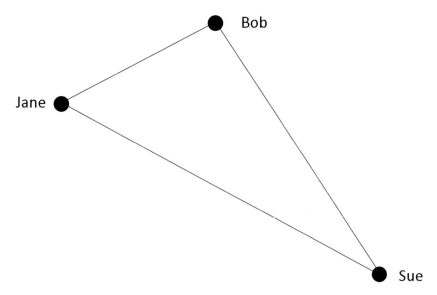

Figure 11.12   An example of a three-person social network. The black dots represent individuals (nodes) in the network, whereas the lines represent the connections between these individuals. The length of a line in this case represents the strength of the ties between each pair of individuals or how close the relationship is between any two of the individuals.

A quick glance at the network in figure 11.12 shows that Bob and Jane are closer friends than Bob and Sue or Jane and Sue. It is harder to tell if Jane or Bob is a better friend of Sue. However, one other important observation of this network is that it is fully connected, in other words, all three individuals know each other.

If we were to expand the network to include a fourth individual, George, perhaps this would no longer be the case. Perhaps both Bob and Jane are friends with George, but George and Sue have never met. Now we see that the network has a missing line (figure 11.13).

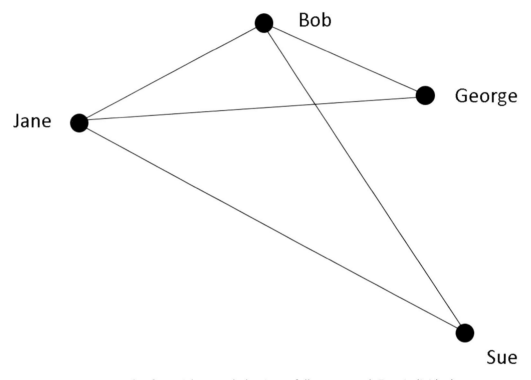

Figure 11.13   An example of a social network that is not fully connected. Four individuals are represented by nodes (black dots) in the network. One pair of individuals is not connected by a line (George and Sue), indicating that they do not know one another. However, an analysis of the network shows that George and Sue are indirectly connected via mutual relations, Jane and Bob.

You might be wondering if it is really possible to add a significant number of nodes and maintain the appropriate geometry, that is, to draw all of the needed connections and to place them all exactly the proper distance from all points with which they connect. The answer is yes; however, it may be difficult to visualize such complex networks graphically.

Although line lengths and even directions can reasonably represent a physical network such as a pipeline, they can become cumbersome surrogates for the links that must be incorporated into a large, interrelated social network. For this reason, you may need to use attributes in the database table to represent the connections and ignore the visual representation.

Fortunately, with GIS, the "geography" we represent does not have to exist in real space. In the case of social networks, it is not the physical location or distance between individuals that we are interested in; rather, we are interested in who knows whom and how well they know each other. It is entirely possible that "close friends" live thousands of miles apart in real space. Network connections can be based on a variety of attributes, even attributes that may not typically be mapped in geographic space. Connections as represented in the networks in figures 11.12 and 11.13 are closeness based on some defined measure of friendship.

Although a number of tools allow for visualizing conceptual networks of this sort, many of the tools are not easily integrated with a GIS analysis. Within a GIS context, a good option is to use a conceptual network mapping tool such as the ArcGIS Schematics application. This ArcGIS extension provides specific capabilities to emphasize relative relationships in lieu of geographic ones.

Conceptual networks, much like their real-world counterparts, are not always simple. Connections may incorporate additional complexity; for example, they may not always be two way or direct. Consider an effort to map relationships between these individuals based on economic rather than personal relationships—the network might change dramatically.

Using the example presented previously, but modified as shown in figure 11.14, let's assume that Bob owns a small accounting firm and employs both Jane and Sue. Economic flow is one way from Bob to both Jane and Sue (represented by arrows pointing from Bob to his employees). George runs a small business and contracts his bookkeeping services with Bob's company, so the flow is from George to Bob, and the thicker arrow may represent that there is a larger sum transferred between them. Last, we discover that Jane is married to George; thus, their finances are intermingled and money moves in both directions between them (represented by a two-way arrow).

Of course, seeing these networks on the computer screen is not really the point, and for large, complex networks, they may be difficult or impossible to visualize. However, by building these types of networks within a GIS, you can analyze the connectivity and flow within the network. Such analysis can be particularly useful when exploring how and where additional connections between individuals or organizations might positively enhance the flow of information or the effectiveness of communication and interaction between community organizations. It would also be useful in determining which individuals or organizations are the best connected, have the most influence, or other related characteristics.

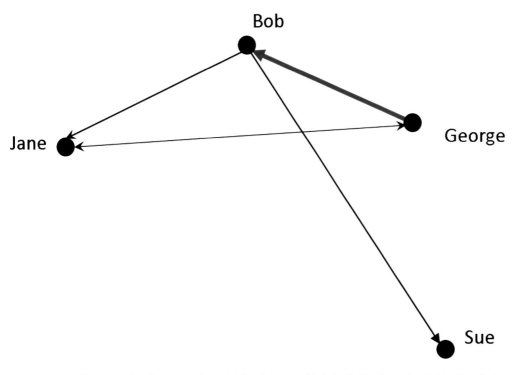

Figure 11.14   An example of a network model involving multiple individuals and variable flow between individuals. Each person in the network is represented by a node (black dots). The flow of money is represented by the arrows. The amount of money and direction of movement are represented by line thickness and directional arrows, respectively.

As you can probably imagine, such networks may become quite complicated; however, so long as the attribute tables behind the network represent the magnitude and direction of the flow or connection, the GIS can be used to analyze these connections, the number of steps between individual nodes, and the entity moving from node to node, be it real (e.g., dollars) or conceptual (e.g., goodwill). Again, the creative use of tools included in ArcGIS Schematics or Network Analyst can assist you in accomplishing this type of analysis task.

One final example to introduce in the network setting is the fact that geographic space and social network space may be very different when related to migrant laborers. Immigrants, from any ethnic group, may be more strongly connected to their immediate community of migrants, through commonalities including language, culture, and so on, than they are to the communities in which they find employment. Furthermore, their social ties to family members in other, faraway countries may be much stronger than to individuals living in the same town. Such a situation could result in community networks that are close, or even overlapping, in geographic space but that do not interconnect, as shown in figure 11.15.

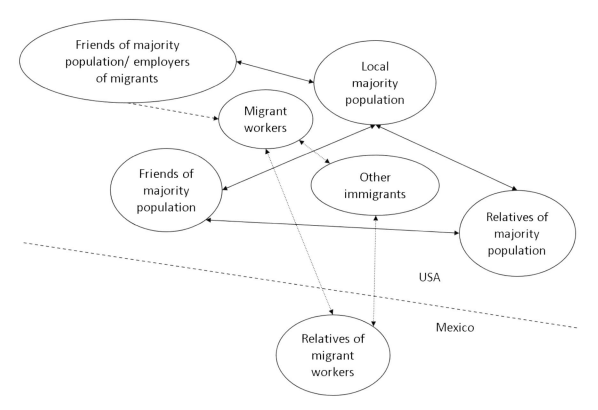

Figure 11.15 An example of mapping a social network for migrant workers in and among their work and community environments. For simplicity, in this figure, individual nodes are clustered within the ovals, each of which represents dozens or hundreds of individual respondents. Major connections between groups are depicted by connecting arrows.

Once again, the situation can become complex in a hurry, but within the GIS environment, such interactions in a social network are readily analyzed for connectivity and interaction.

Least cost analysis

Least cost analysis is a similar concept to the network approach described earlier and can be analyzed using Network Analyst once costs have been assigned to various road segments or pathways of travel. The idea of a least cost path is to find the most efficient (or least expensive) path through the data space. The routing methods incorporated into onboard vehicle navigation systems or even those included on most modern smartphones to provide directions to a specified location are based on these concepts.

The concept of cost might be operationalized in a variety of ways in this type of analysis. For example, in a network of individuals such as described earlier, we might be interested in the shortest social pathway between two individuals. Or in the case of community planning, what is the shortest physical pathway along the road network between the fire station and a burning building? How we choose to define the cost (which could perhaps be better described as a relative measure of a particular variable) is up to us. The GIS simply needs to have those values applied to the connecting pathways in the network, as shown in figure 11.16.

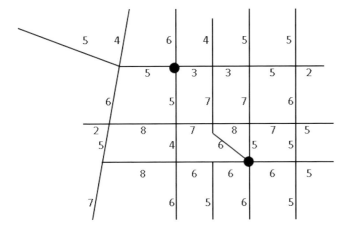

Figure 11.16   In the earlier example, each segment of the network is coded with a value representing the variable of interest, perhaps minutes of travel time between intersections of a road network. To determine the best route between the two points represented by black dots, the GIS would total the values along all possible routes to determine the most efficient pathway. More complex network models may incorporate values that change based on time of day, direction of travel, or any number of additional criteria.

For some analysis situations, the pathways may be unrestricted; that is, you may allow for connections to occur anywhere in space. In these situations, we look for pathways through the data in an unrestricted setting, making these approaches different than network models of lines that interconnect nodes. Such situations are best implemented in a raster GIS environment. In raster GIS data, the entire study space is filled with raster cells, thus allowing for all locations to be assigned a value. These values are incorporated into the analysis much like the lines in the network model.

Let's take the example of walking to work, as shown in figure 11.17. In the simplest scenario, you would move from cell to cell in any direction (including diagonally) and attempt to minimize the total number of cells you must travel through (the least cost). This analysis assumes that the cost, or effort, necessary to travel across all cells is identical, and if so, the least cost path is the same as the shortest path. Try to determine which sequence of cells is best using the assumption that all cells cost one move; odds are you will find multiple pathways that get you from home to work in the same, minimum number of moves. Make a note of your result.

Of course, this example is not entirely realistic. Some cells may be simple to traverse, perhaps flat, open spaces with nice sidewalks; other cells may be extremely difficult to cross, for example, they could be occupied by buildings, water bodies, or other obstructions. Cells occupied by such obstructions might be assigned a higher cost. A partial obstruction, such as a hill or a pond, may have only slightly elevated costs. Other obstructions may be total (e.g., there is an impossible obstruction that entirely prevents movement across a particular cell). If a cell is impassible, the cost assigned should be exceptionally high (e.g., equal to the total number of cells on the map) to ensure that no total cost calculation including it would ever be selected.

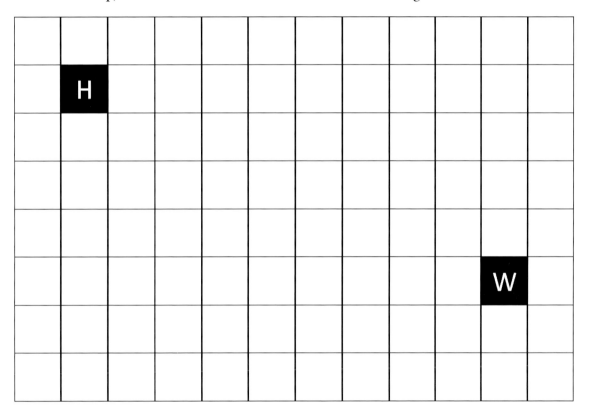

Figure 11.17   In a raster-based least cost analysis, the area under study is divided into cells. Each cell is assigned a value representing the "cost" associated with crossing through it. If all cells "cost" one move, the least cost path would be the same as the shortest physical path between the cells representing the locations for home (H) and work (W). However, if some cells cost more, the least cost path could route across a longer physical distance, while still resulting in a lower overall cost.

In a least cost analysis, each cell is assigned a value based on predetermined criteria. The criteria may come from actual fieldwork or from assigning costs to an existing map based on criteria operationalized for your variables under study. For example, you might assign a value for perceived safety obtained in a survey of residents of a particular metropolitan area.

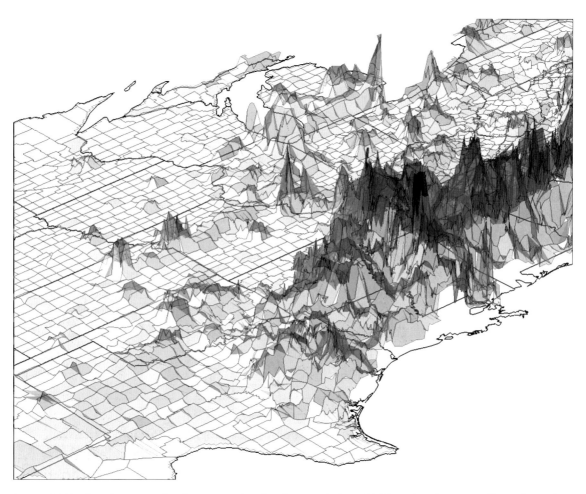

Figure 11.19 A surface map showing census data for the population of African Americans as a 3D surface. This perspective is to the northeast across the states of Texas, Louisiana, and Alabama, toward the Great Lakes. Darker colors and higher peaks indicate larger populations of African Americans. A strong population band crosses the South, whereas in the more northern states, populations concentrate around major urban centers. Peaks are visible along the lakes in areas near the cities of Milwaukee, Wisconsin; Chicago, Illinois; Detroit, Michigan; and Cleveland, Ohio. Although such 3D visualizations can be effective, it is also sometimes difficult to develop a perspective in which the peaks do not obscure important information behind them. GIS and cartography by Steven Steinberg. Data from the 2000 US Census.

Spatial interpolation and simulation

Spatial interpolation is another common raster analysis technique. Interpolation is a technique used to estimate values between sampled locations. A good analogy is regression in traditional statistics. A variety of linear and nonlinear regression techniques can be used to fit a line to an x,y plot of data points. Multivariate regression techniques are used in a similar fashion to estimate data for additional data dimensions. The primary difference between traditional regression and spatial interpolation is that rather than fitting data in hypothetical data space, as in traditional regression, the estimates in spatial interpolation are linked to real geographic space.

Thus, x,y values used in spatial interpolation are typically the true x,y positions in the coordinate system of the analysis. The estimated data value is referred to as the z value and may represent any data variable desired. Traditionally, z is used to represent elevation; when working in a spatial analysis context, we can interpolate for any variable in conjunction with the spatial positions of x,y,z. Interpolation is used in any situation where a complete census is not possible and you want to estimate data values in locations between sample points. Two underlying assumptions of a spatial interpolation are as follows:

1. Unsampled places nearby sampled locations will be more similar to the sampled location than locations farther away (distance decay).
2. Places between sampled locations are likely to have attribute values that are in between those of the sampled locations that surround it.

For example, imagine you randomly collect and map by neighborhood square footage of one thousand households in a large metropolitan area. It might be reasonable to conduct a spatial interpolation of these data to estimate home sizes in areas that your sample did not cover. The assumption here, much as in a traditional regression, is that values between known, sampled points are likely to have a relationship with and fall between the sampled values. Thus, if in one part of town, I measure houses at twelve hundred square feet, and a few blocks away, I measure houses of eighteen hundred square feet, it might be reasonable to estimate that houses adjacent to these known values would be of similar size and that those located roughly in between these sampled households might be somewhere in the middle, perhaps fifteen hundred square feet. Granted this may not be a valid assumption in every case or for every attribute type, but spatial interpolation is an excellent approach for many variables when you need an estimate of data values in locations where you are unable to collect them.

A CAUTIONARY TALE: DATA NONINDEPENDENCE CONSIDERATIONS

Some statisticians believe that spatial statistics are invalid because of the inherent nonindependence of spatial data. This stems from inherent autocorrelation in spatial samples. We would naturally expect data values geographically near a given sample location to be similar (i.e., if I sample air quality at a particular location in a city, I would expect the air quality one meter away to be more similar than the air quality sampled one mile away). There are additional issues near the boundaries of the study area and in the size selected for the study units. Spatial statistics are widely available and commonly used in GIS analysis. For better or worse, these tools are readily available in most GIS software packages, regardless of the end user's statistical sophistication and understanding of their appropriate use. Therefore, we will leave it to the individual reader to determine if these issues are concerns in his or her own analysis situation.

The idea of spatial interpolation, simply enough, is to get more for less. We can reduce the time, effort, and cost of doing a complete census by selecting a sample of data collection sites and interpolating values for areas in between. Interpolation can also be useful to estimate values resulting from incomplete or missing data. How to best select sample locations is up to the researcher. Although it is often preferable to do random samples, sometimes it is more effective and efficient to use other sampling frames, as discussed in chapter 5.

Spatial interpolation is done in raster GIS because the cells of the raster provide a framework to fill with interpolated values. As in any raster-based analysis, the cell size will influence the detail of the result. In figure 11.20, we see a map of the study area with sampled locations represented by the black cells. The value measured at each location is labeled. Conducting a spatial interpolation in GIS, we fill in estimates of values in each of the empty, white cells. The result of an interpolation of these sampled points is shown in figure 11.21.

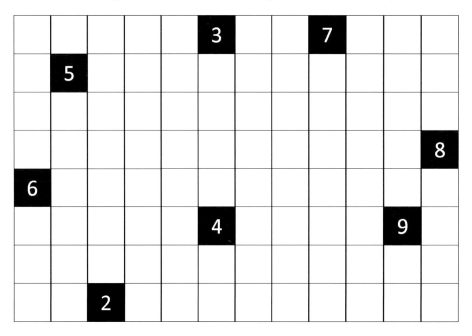

Figure 11.20   A raster map representing a study area shows values measured at randomly sampled locations (black squares). By using a spatial interpolation, estimates of data values for all unmeasured cells within the study area can be computed.

Note that this is only one example; variations on the interpolated map would be obtained by using different interpolation techniques. There is a substantial literature on spatial interpolators and their variations. If you find interpolation to be a useful technique in your work, it would be worthwhile to further study the topic. For most day-to-day GIS analyses, a simple interpolation such as inverse distance weighting (IDW), as shown in figure 11.21, is probably sufficient. However, keep in mind that this is roughly analogous to a linear regression in standard statistics. Just as linear regression makes a number of assumptions about the data, so does IDW. If you have cause to believe that your data are not linear, there are additional covariates, and a more complex interpolation may be appropriate.

Of course, any given interpolation is only as good as the data on which it is based. Therefore, in the preceding example, we have two potential errors. There is a possibility of an error in the actual measurement made at the sample point, due either to a misreading

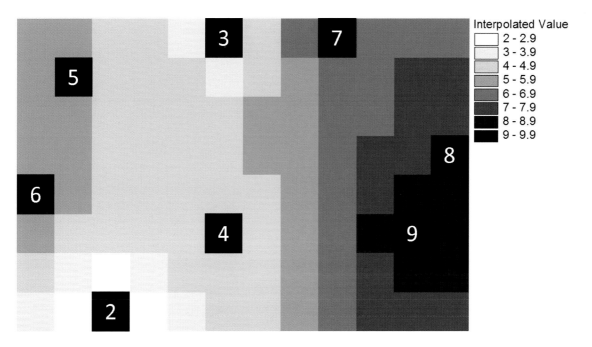

Figure 11.21 The result of a spatial interpolation of the map shown in figure 11.20. Cell values based on an interpolation of the values at the originally sampled locations (labeled cells) are represented in various shades of gray. In this example, the interpolation technique used was inverse distance weighting. This process looks at nearby cell values and applies weights based on their distance from the cell being computed. The primary assumption of this process is that nearby cells are more similar than cells that are farther away. Available interpolators will vary by the GIS software you use. Selection of an appropriate interpolator is best based on the nature of the data and phenomena being studied.

of the measurement device or to an error in entering these data into the computer (**measurement error**). The amount of error depends on the type of data and the techniques used to collect them. The other two errors are related to the (x,y) position of the data. If you mistakenly shift data in the x and/or y direction (**location error**), the results of the interpolation will be erroneous simply because the location and thus the influence of the measured point are incorrect. We can, to a reasonable degree, address both of these errors with a little more analysis.

Measurement error refers to the accuracy of the original data collection. However, in many situations, we can determine the magnitude of these errors through a little checking. For example, if I were to tell you that the measurements in figure 11.20 were made within ±0.05 feet, what this would mean is that a measurement of 3 could actually be anywhere between 2.95 and 3.05. Furthermore, most errors come with a confidence (e.g., 95 percent) indicating that even the specified range is not 100 percent correct; in fact, 5 percent of the measurements could have errors even greater. What does all of this mean to you? Returning to our example in figure 11.20, it is worthwhile to point out that many different **realizations** of this data could result. Each of the measurements on the map has a **range** of values that are reasonable to expect in addition to the values reported in the data. Each of these possible variations would result in a different realization of, or variation on, the results.

A variety of interpolation techniques are available, with some variation by the specific GIS or spatial statistical software used. In general terms, these fall into the categories of **exact** and **inexact interpolators**. Exact interpolators will always return the input value at the sample location; that is, if you provide a value of 5 at a sample site, the result will return a value of exactly 5 at that site. Using an exact interpolator is appropriate for data that can be measured reliably; if you are 100 percent certain that you saw five people living in a particular household, there would be no reason to allow an interpolation to estimate a number other than 5. By contrast, an inexact interpolator allows the value at the sample location to vary to obtain a better overall fit of the interpolated result. This may be appropriate when you have measurement error and you are less certain that the measured value is correct. In this case, allowing a measured value of 5 to be returned as 4.92 in the interpolation process might be just as valid.

If you consider these two options in a regression context, linear regression is simpler, but the line that is fit to the data in most cases does not pass through every sample value in the original scatter plot. By contrast, nonlinear regression may allow for the resulting line to pass through all of the data points, but at a cost. The resulting regression line is generally much more complex, and the equations necessary to describe such lines are similarly complex. Sometimes this complexity is desirable, but in many cases, simplicity is preferred, so long as the answers that result are sufficiently accurate to meet the needs of the analysis.

When we consider the issue of locational error, the same type of approach might be used, but rather than varying the measured values, you might vary the coordinate location values. A Global Positioning System receiver is not exact in locating where you are on the ground, and therefore, the mapped location of a sample could be off in any direction by plus or minus some amount. The approach is the same as already described. Of course, there is an added complexity that you might want to vary both the measurement and location values within their specified ranges to get an even more complete picture of the data.

So when should you use interpolation or simulation? Interpolation is appropriate when data are sampled across space and you have reason to believe that unsampled locations nearby are somewhat similar. This may not be the case with all data, so only interpolate if this assumption is valid for your data. **Simulation** is a powerful tool when you have some known variation in your data for which you want to account. Keep in mind also that simulation requires much more computer power because you will be running your analysis hundreds or even thousands of times.

Probability maps can be useful when the end user has an understanding of their meaning but can be confusing to a general audience. For example, it is likely most people will take an umbrella if the forecast is for a 90 percent chance of rain (a 10 percent chance of dry). Strangely, these same individuals will buy tickets for the Powerball with a mere 0.00000125 percent chance of winning. Many people do not understand probability values in a rational or logical sense. If they did, all those Powerball players would feel completely content walking around with no umbrella even when the forecast called for a 90 percent chance of rain (a 1 in 10 chance of staying dry versus a 1 in 80,089,128 chance for the Powerball win).

Modeling

Earlier in this chapter, we discussed the four levels of abstraction as we prepared to model some portion of reality in a GIS context. In general use, when we talk about **modeling** in GIS, we are referring to developing a series of analysis steps that are designed for a particular analytical situation. Models are based on some combination of theory, research, expert knowledge, and individual creativity from relevant disciplines. In most cases, you will develop and implement your own models in GIS. In analysis situations that are more common, there may be existing models built by a third party with a user-friendly interface within the GIS. The saying "why reinvent the wheel?" comes to mind here. If there is already a GIS approach that works for your situation, be it one that is freely available or one you can purchase, it may be worthwhile to stick to these known and tested solutions, especially when your application is oriented to problem solving as opposed to basic research.

The advantage of using models developed by others comes in the development and testing based on research and knowledge that may be out of reach of the average GIS end user. Preexisting models are readily available for a variety of natural resource applications and are starting to appear for social science applications such as crime analysis, community planning, and public health. It is important to note that many of these GIS models are simply presented in research journals or conference proceedings and are not available as ready-to-install software for your computer. Instead, such models provide you with a framework (a logical or physical model) that you can implement in your own GIS environment.

Some models are available as installable software that stands alone or in conjunction with a GIS software package. A caution when using such preprogrammed models is to be aware of the assumptions they use. Not all models perform equally well in all situations; thus, it is essential that, as the end user, you understand these assumptions and, when appropriate, how to modify them for your particular situation.

Raster modeling

Of course, it is not unusual to find that the model you need does not exist in an easy-to-use, canned format. This may be hard to believe given that web searches on "GIS modeling" turn up more than 1 million hits. But assuming that you cannot find the perfect model, you may need to develop your own. Earlier we suggested that modeling in raster using map algebra is a powerful approach for developing your own models. Put simply, map algebra is the mathematical combination of data layers in GIS. **Raster modeling** means that one uses map algebra as part of modeling and analysis.

In figure 11.22, we see a simple addition of two map layers using map algebra. In your GIS software, this is achieved by using a tool to organize the mathematical combination or by simply writing the formula directly using the names of the map layers involved. The actual software processes are relatively simple; the more difficult question is what mathematical combination of layers you should use. Two options come to mind in answer to that question.

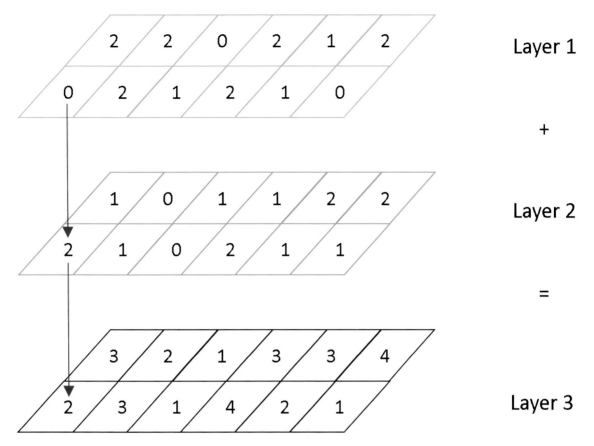

Figure 11.22 An example of map algebra used to add two map datasets (layer 1 + layer 2), resulting in a new, third dataset (layer 3). Cells are added based on their spatial overlap; thus, the cell in the lower left corner of layer 1 (0) added to the cell in the same position in layer 2 (2) gives a result of 2 in the resulting output (layer 3). Map algebra can be based on any mathematical function: addition, subtraction, multiplication, division, logs, powers, and so forth.

First, you could go through the process of abstraction described earlier in the chapter. In doing so, you would determine what data you want to include, how the data will be collected and coded and at what spatial scale (cell size), and how those data will ultimately be combined. Such a process necessitates that you determine how to combine the layers using a meaningful mathematical approach. For example, if you were conducting a study of criminal activity in a particular city, you could assign a value to each cell location representing the number of crimes at that location. Individual data layers might be used to represent individual months. By simply adding together various groupings, you could accomplish several outcomes. Adding all twelve months of data would aggregate crime rates on an annual basis. However, if we wanted to determine if certain seasons show higher crime rates than others, we could add together only the layers representing the warm months, the holidays, or any other combinations of interest. Similarly, if we had crimes broken out by type (burglary, murder, rape, etc.), we might find that certain types of crimes occur more frequently at certain times of year. Of course, this is a simplistic example and does not incorporate socioeconomic or environmental factors that would also be appropriate to consider.

Alternatively, there might be a **mathematical model** that has already been developed in answer to your question or that can be modified slightly from a similar question. Mathematical models are common in the social sciences, and in many cases, these models can be extended into a spatial context by implementing the equation in your GIS software. For example, a multivariate regression model could be implemented by developing a raster data layer for each variable in the regression equation. If the necessary data for the selected model can be found or collected in a spatial context, map algebra can be used to combine those data in geographic space, thus generating a spatial dataset as the result of carrying out the regression for each cell position in the input data layers.

Keep in mind that to accomplish this, it is not unusual to collect samples across the study area for one or more of the necessary layers and then to interpolate a complete data layer from the samples. In doing this, the interpolation process introduces an added level of uncertainty not typically accounted for in the original mathematical model unless it was designed explicitly as a spatial model.

Although it can be perfectly reasonable to extend nonspatial models into the spatial realm, it is worthwhile to do some pilot testing and assessment of the results before moving forward with a large analysis. However, with care and a little critical thinking, it is reasonable to implement mathematical models in GIS to determine how they vary in space. Ultimately, GIS modeling is a process of working through the steps of the abstraction process in sequence:

1. Examine the reality you are attempting to model.
2. Conceptualize it into terms that can be put into a GIS (data layers and relationships).
3. Work through the logical relationships of these component parts to develop a model that makes sense (and, if possible, is based on a previously developed model of the system in question).
4. Develop an approach to implement this logic in either a vector- or raster-based analytical process given the capabilities of available software and resources.

When to use GIS as a problem-solving tool

Ultimately, the utility of GIS and the particular tools and approaches selected depend on the analysis you want to make. GIS can be used as a tool to analyze almost any type of data and, in many cases, is very effective. However, this does not mean that GIS is a panacea, and it may not be the best solution to every problem.

So when is use of GIS appropriate? We reiterate a few indicators here for your consideration:

1. Your question has a clear spatial component, in which case GIS is excellent for creating, storing, updating, and analyzing your spatial data.
2. Your question will benefit from the spatial analysis capabilities of GIS (questions that include phrases like "near," "next to," "in conjunction with," "related to").
3. Your analysis is one that can benefit from existing base datasets or will build on them.
4. Your study will be revisited, in which case, the upfront investment in GIS and data development will pay off over the longer term.
5. The GIS visualization capabilities will assist in data exploration or mapping and in data output appropriate to your study, final reports, or other uses.

Potential pitfalls

The greatest risk in using GIS as an analysis tool comes through its unintentional misuse. To help you to avoid some common problems, we review them here. Several of these have already been discussed in greater detail elsewhere in the book and are presented again for the purpose of review; others are discussed in detail in the sections that follow.

Given the complexity of a typical GIS software program, there are numerous points in an analytical process at which you can make mistakes. Most of the errors related to the actual use of your GIS software can be eliminated by reviewing the function of each command or process in the software documentation and understanding how it works and what it does. Because there is no risk of permanent damage to your data so long as you keep a backup, the best way to understand what a particular command does is to try it.

Although this may seem obvious, a surprising number of individuals new to GIS are fearful of exploring a tool or command they have not used before. Experiment—just do so after you have made sure that your data are safely backed up so that if you are not happy with the results, you can go back to your original data to try something else. There is no harm in experimenting in GIS, and in many cases, exploring will be the best way to discover new and useful possibilities. No single book on the topic of GIS covers everything that is possible; in fact, much of what is possible in a GIS analysis has not yet been done because every new analysis or question presents new possibilities.

Most importantly, before you even begin an analysis in GIS, you should make sure that you have a clear understanding of your project goals. Knowing what it is you are studying, what data you will require, and how the data will be collected and measured are all essential issues that may be influenced by the GIS but should not be dictated by it. Remember, you should determine how you want to carry out the analysis as opposed to allowing the software to dictate what you do.

Other considerations in carrying out your analysis relate to the choice of scale, projection, coordinate system, and datum for each data layer, as discussed earlier in the book. A good rule of thumb is to minimize the number of different combinations used as much as possible. The more consistent your source data are, the less room for error there will be in any conversions or processing done within the GIS software. It is especially important to avoid using drastically different scales of data (e.g., orders of magnitude) because scale-related errors cannot be addressed effectively by the GIS software. Projection, coordinate, and datum conversions can, for the most part, be achieved with minimal error.

Of greater concern are the issues that fall outside the GIS software, those being the issues that you, the analyst, decide through the abstraction process. Looking at reality seems easy enough; the problem comes when you begin to simplify into terms that will work in the GIS. For physical features, the abstraction process is somewhat more straightforward: you can measure the location of a road, river, or town. However, much of the data relevant to the social sciences are qualitative or conceptual, making them harder to link to discrete mapped locations and categories. This is not to say that you cannot work with data of these types, but rather that you need to be conscientious about how you will make these links.

Revisiting the accessibility example

Let's refer back to the example of accessibility presented in chapter 2. In this example, we considered a variety of criteria that might help to define a woman's access to prenatal care at a public health clinic. Issues such as location along public transportation, available day care, ability to miss work, hours of clinic operation, and so forth, were suggested as variables that could be used to create an index of accessibility. The problem is that we may not understand and include all of the actual components that truly define the concept of accessibility. Thus, errors in our analysis may arise in the original conceptualization of the problem.

Furthermore, even if we assume for a moment that we reliably identify all of the appropriate components, we may not know how to appropriately weight them into some index of accessibility that we can apply to the GIS data layer. For example, are all of the variables of equal importance, or are some more important than others? If the latter is true, we will need to incorporate weighting values into the model. Is there a variable in the analysis that is mandatory? In other words, even if all of the variables are positive, thus allowing a woman to get to the clinic, would just one negative criterion, perhaps language, have the ability to counteract everything else in the model? Finally, even if we assume a valid index for the accessibility layer, we may not properly understand how that layer should interact with the other layers in the GIS. Accessibility may be one of several data layers incorporated into the final analysis.

So how can you possibly address all of these issues? It is unlikely that you can, or even that you need to. But to avoid the pitfalls that come with the abstraction process, what you should do is carefully consider each step in the process. Consider the input and opinions of other professionals in the discipline, others who use GIS, and the published literature. Working in a vacuum is a surefire way to overlook something that you should include in your model.

Test it

Test your models before assuming everything is working appropriately. When you arrive at the physical implementation phase, run a small sample dataset to pilot test the effectiveness of your analysis and to determine if you are getting results that make sense. One of the common pitfalls in using computers for analysis is that we can run very large datasets efficiently; however, these large datasets can mask errors that might be obvious in smaller datasets. This is especially true if you have the ability to get out on the ground to see, in person, if your results make sense. There is no better way to ensure that your model is doing a good job at representing reality than to ground truth it against reality. If there is a difference, you have a much better chance of identifying the cause of the error in a small dataset, for example, a single neighborhood as opposed to a large metropolitan area. Sometimes it is even useful to check your model with artificial data so that you can determine the correct result in advance and see if your GIS analysis provides what is expected.

Virtual reality is still *not* reality

Another pitfall in GIS-based analysis is being so engrossed by the technology that you blindly trust the output. Remember that what you do with the GIS is always a simplification of reality, and the results are based on that simplification, not on the reality behind it. The GIS is most effective as a tool to filter large amounts of data spatially to efficiently identify locations that meet a set of criteria, show a relationship, and so on. Computers do exactly what we tell them to do, but if we tell them the wrong thing, they will, precisely, do the wrong thing. We can analyze spatial data in GIS far faster and easier than we ever could with paper maps, rulers, and markers, but sometimes that just means a faster, more efficient way to make a bad map.

Collecting good data, thoughtfully conceptualizing your analysis, and at least occasionally checking results on the ground (or with a reliable surrogate) are essential before you can make a final decision based on a computer model. A very public example of blindly trusting the map was the erroneous US Central Intelligence Agency (CIA) bombing of the Chinese Embassy in Belgrade, Yugoslavia, in 1999. The CIA later attributed the error to a mistake on the map that was used in planning the attack intended to destroy a Yugoslavian arms agency. If you start with bad data, you will get a bad result, no matter how precise the GIS is in telling you the location of that result.

Finally, a more general caution, but one that all too often leads to unpleasant GIS experiences, is not to expect too much from your GIS. As mentioned early on in this text, preparing data for use in GIS analysis takes a great deal of time and effort. It is rarely, if ever, the case that all of the data you require for an analysis will exist in the appropriate format and be ready to go. Although GIS analysis can be done rapidly once everything is in place, it may take months or even years to get to that point. GIS can take a little getting used to at first, as can getting all of the appropriate data for your common analysis tasks. Once you have those things in place, GIS is a fantastic tool with immense potential. But just as installing a word processor on your computer won't have you writing Shakespeare overnight, installing GIS software is only the beginning of doing meaningful spatial analysis. With a little patience, time, and practice, it will happen.

A GIS when used with care and consideration can be a wonderful tool for social science research. Although as a tool, GIS was originally developed for natural resource management and has a longer history in those disciplines, the possibilities for spatial analysis of social science data with GIS are only recently being explored. There are untold possibilities for applying GIS to questions yet to be asked; no doubt, the answers such analyses will provide in the future will be important and meaningful.

Spatial statistics

In any research project, the type of statistics that a researcher chooses is determined by the questions that he or she wants to answer. Spatial statistics provide a unique set of analytical tools intended to analyze data with a locational component. Spatial statistics as a subfield of statistics are far more complex than can be illustrated in a single section of a book on GIS. The ArcGIS Spatial Statistics toolbox provides a robust set of operators for carrying out many of the most common analyses (figure 11.23).

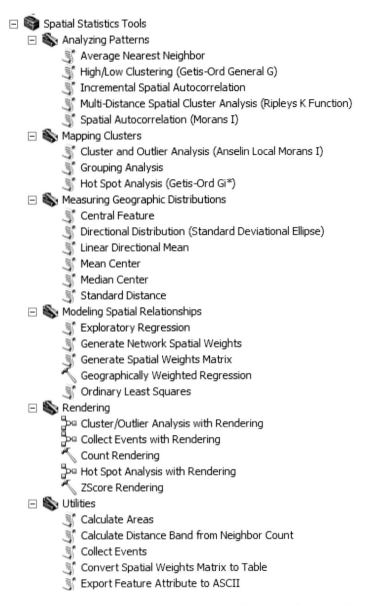

Figure 11.23 The ArcGIS Spatial Statistics toolbox provides multiple tools for the statistical analysis and testing of relationships in spatial data. Esri.

In exploring the names of the available tools, you might notice that many of these tools are designed to help determine characteristics of spatial patterns and clustering among others. Similar to traditional statistical tests, the role of spatial statistics is to determine if patterns observed in the data are nonrandom (i.e., there is some sort of cause-and-effect relationship).

Another statistical concept that is useful in GIS is a technique called **Monte Carlo simulation**, named for the popular resort town and gambling destination in the principality of Monaco. As in gambling, the approach is based on probabilities. A Monte Carlo simulation is not a specific tool available in ArcGIS but rather an approach in which multiple realizations of the output data are generated based on known or likely variations in the sample data. The results are then examined to assess the distribution of results obtained over the course of multiple runs.

To accomplish a Monte Carlo simulation in a practical sense, you would want to build your analysis using scripts or the ArcGIS Model Builder application to allow for the automation of your analysis.

For each of the measured values, or data layers, you incorporate into your analysis, you then randomly vary the numbers according to an appropriate distribution and within the specified range, for example, ±0.05. (The specific range of variation would be selected to match the known variation within the data, for example, based on the accuracy specification of tools or approaches used to collect and develop the underlying datasets.) Each time the model is run, values are adjusted randomly within the known range of variance. If you imagine running your analysis one hundred, one thousand, or even ten thousand times, you will find that certain realizations occur more often than others. Most likely plotting the full set of results will give you something like a normal distribution of results.

Fortunately, using the computer, it is feasible to manage such simulations efficiently. The benefit to some analyses is that you can get a sense of the likelihood of a particular result and the range of results instead of just a single answer. For example, consider the weather reports you see on the local news every evening. The forecaster will put a map on the screen and state, "There is a 60 percent chance of rain tomorrow." What does that mean to you? Will you take an umbrella or raincoat, or will you gamble that the 40 percent chance of no rain is in your favor? Mapping in probabilities puts the onus on the end user of the data, but it may actually be a more honest map. Of course, even if the report stated a 1 percent chance of rain, it could still rain. Probabilities are sometimes confusing or uncomfortable, so they need to be used with caution and consideration of the intended audience.

ArcGIS Spatial Analyst

A significant array of analysis tools with particular emphasis on raster, or gridded, data is included in the Spatial Analyst extension of ArcGIS (figure 11.24). In addition to a number of tools specifically designed to work with natural resource analysis (e.g., water, solar exposure), Spatial Analyst includes a variety of statistical tools (e.g., interpolation, multivariate analysis, and map algebra) that act on raster data. It also allows you to conduct analysis integrating cell-based raster data with vector data sources. From a more practical standpoint, ArcGIS enables a researcher to predict using statistical models. This ability becomes very important when the researcher is trying to perform analysis on land use types, predict where fires are likely to occur, conduct demographic analyses, or scientifically model what communities are going to be affected by sea level rise.

Figure 11.24　The ArcGIS Spatial Analyst extension provides a large number of tools classified into multiple toolboxes. Tools provided here are largely focused on the analysis of raster data types, whereas others provide methods for working with raster and vector data together as well as some additional statistical analysis operations beyond those available in the Spatial Statistics toolbox shown in figure 11.23. Esri.

Although a number of the tools included in the Spatial Analyst toolbox appear to be natural resource focused, keep in mind our prior discussion regarding creative use of such tools. For example, a researcher could use a tool designed to delineate a watershed or flow of a river to instead explore the flow of resources and finances in an economic setting or the level of fear in a social-psychological study of crime across a city.

Prior to using ArcGIS to actually conduct your spatial analysis, you want to consider what questions you or your team seek to answer, as this will ultimately drive your analysis. At the outset of your research project, you most likely established questions to be answered. If you do not have a statistician as a part of your research team, we advise consulting more specific documentation details provided with each tool in ArcGIS, along with one of the many statistics and spatial statistics textbooks on the market, to ensure that the analytical approach you choose is appropriate to the data that you plan to analyze.

Review questions

1. What is cartographic classification? How is it used?
2. What does the term *buffer* mean for GIS users? How might a researcher use a buffer? Please provide an example.
3. How might a GIS user use an overlay? Please think of a topic that you are interested in and come up with an example of how you might use an overlay in your analysis. Describe it here.
4. What is the difference between measurement error and location error?
5. What does spatial interpolation mean, and when might it be necessary to do this? Please provide an example.
6. What is the main idea behind a least cost path from a spatial perspective? What are you trying to accomplish?

Additional readings and references

Bivand, R. S., E. Pebesma, and V. Gomez-Rubio. 2013. *Applied Spatial Data Analysis with R (Use R!)*. New York: Springer.

Cressie, N., and C. K. Wikle. 2011. *Statistics for Spatio-Temporal Data*. New York: Wiley.

Haining, R. 2003. *Spatial Data Analysis: Theory and Practice*. Cambridge: Cambridge University Press.

Krivoruchko, K. 2011. *Spatial Statistical Data Analysis for GIS Users*. Redlands, CA: Esri Press.

Legendre, P. 1993. "Spatial Autocorrelation: Trouble or New Paradigm?" *Ecology* 74:1659–73. http://dx.doi.org/10.2307/1939924.

Lloyd, C. 2010. *Spatial Data Analysis: An Introduction for GIS Users*. Oxford: Oxford University Press.

Maguire, D., M. Batty, and M. Goodchild. 2005. *GIS, Spatial Analysis, and Modeling*. Redlands, CA: Esri Press.

Mitchell, A. 2005. *The Esri Guide to GIS Analysis: Volume 2. Spatial Measurement and Statistics*. Redlands, CA: Esri Press.

Ord, J., and A. Getis. 1995. "Local Spatial Autocorrelation Statistics: Distributional Issues and an Application." *Geographical Analysis* 27:4, 286–306. http://onlinelibrary.wiley.com/doi/10.1111/j.1538-4632.1995.tb00912.x/abstract.

Relevant websites

- **Esri, ArcGIS Spatial Analyst (http://www.esri.com/software/arcgis/extensions/spatialanalyst):** Overview of spatial modeling and analysis tools.
- **The Geographer's Craft (http://www.colorado.edu/geography/gcraft/notes/cartocom/cartocom_f.html):** Chapter on cartographic communication.
- **"PASSaGE: Pattern Analysis, Spatial Statistics, and Geographic Exegesis, Version 2,"** by M. S. Rosenberg and C. D. Anderson, *Methods in Ecology and Evolution* 2 (2011): 229–32, http://onlinelibrary.wiley.com/doi/10.1111/j.2041-210X.2010.00081.x/pdf.
- **"Spatial Statistics in ArcGIS,"** by L. M. Scott and M. V. Janikas, in *Handbook of Applied Spatial Analysis,* 24–41 (New York: Springer, 2010), http://link.springer.com/chapter/10.1007/978-3-642-03647-7_2.
- **"Network Analysis—Network versus Vector: A Comparison Study,"** by Jan Husdal (University of Leicester, 2000), http://www.husdal.com/mscgis/network.htm.

Chapter 12

Spatial analysis of qualitative data

In this chapter, you will learn about qualitative analysis techniques and their link to spatial data in ArcGIS. You will also learn about content analysis techniques and means for linking external or discipline-specific quantitative analyses with ArcGIS spatial outputs.

Learning objectives

- Learn to distinguish qualitative data
- Comprehend the differences between and values of inductive and deductive approaches to research
- Understand how qualitative data can be analyzed using spatial concepts
- Learn the various steps in the qualitative data analysis process
- Understand the process for theming and coding qualitative data

Key concepts

coding
content
data type
deductive approach
digital data
hard-copy data
inductive approach
latent spatial data collection
manifest spatial data collection
metadata
qualitative data
spatial qualitative analysis
variable definition table

Qualitative data and GIS

Researchers view both qualitative and quantitative data as useful to consider in the research process. Because GIS was first used for mapping and analysis of themes, such as natural resources, that had been surveyed or census of populations, traditional applications were built around quantitative data types—data with measured values and locations. However, ArcGIS can also be used to successfully map and assess qualitative data. Because computers are digital systems, the natural form of data is numeric, and of course, to map data, an x,y coordinate on the Cartesian grid is the numeric method for locating things. Mapping nonnumeric, or qualitative, data types requires you to follow a series of steps to prepare qualitative data for mapping and analysis in the ArcGIS environment. Most qualitative data require preparation and coding before they are ready to put into a GIS, which is our focus in this chapter.

What are qualitative data?

So what exactly are **qualitative data**? In their most simple form, qualitative data can be defined as nonnumeric data described by their qualities or characteristics but that are not readily measurable or quantified, including words, photos, pictures, videos, art, handwritten comments, or recorded interviews. The sources of such data are diverse and could include newspapers, open-ended interviews or survey questions, key-informant interviews, archival or museum collections, oral histories, public participation GIS sessions, case studies, or field observations.

However, different types of qualitative data require different considerations. Thematic maps often show categorical data that are nonnumeric; for example, vegetation types, such as grass, shrub, and tree, might be considered nonnumeric. This can be contrasted with qualitative data that come from a community survey in which responses are open-ended narrative or conceptual descriptions. The underlying difference between these might be formally definable, for example, the difference between a shrub and a tree typically includes a specifically defined height range and branching pattern (multiple or single stems), values that at their root can be quantified and measured with little or no ambiguity.

It should be noted that the process of analyzing qualitative data often takes a different direction than processes used to analyze quantitative data, referred to as the **inductive approach**. In the inductive approach, a researcher begins with observations about a topic of study and works his or her way backward from those individual observations to develop a theory and explain the phenomenon observed. Careful review of individual observations can yield the identification of thematic patterns. These in turn are assessed to generate higher-order ideas and theories. This is the direct opposite of the steps in the **deductive approach** (discussed in chapter 3). In the deductive approach, one first begins with theory and hypothesis and then collects and analyzes data to determine if the hypothesis is supported. When conducting research, there may be times when taking an inductive approach using qualitative data is preferable. This is especially true when you are conducting exploratory research and examining things for the first time.

From a practical standpoint, another way to think about the qualitative research approach is that if you have a general idea about a topic you are interested in exploring but lack a clearly defined hypothesis, you would choose to enter the situation with an open mind, collecting observations and reviewing existing data to begin exploring the question. Through persistent review, a picture or pattern of what is happening will begin to emerge, although this approach runs counter to the traditional scientific method using a deductive research approach, which begins with a hypothesis that is then tested by collecting data to assess it.

Spatial qualitative analysis

In the following pages, we focus on the steps in the process of conducting spatial qualitative analysis. **Spatial qualitative analysis** is the identification of spatial patterns associated with qualitative data. The researcher can take two approaches to conducting spatial qualitative data analysis: manifest spatial data collection or latent spatial data collection. The primary distinction between these two approaches is the researcher's intent, because both approaches involve a geographic location or place.

In **manifest spatial data collection**, the researcher actively plans to incorporate some geographic information into the case-by-case data collection process. For instance, you might have a survey that asks a series of questions about a particular topic and also establishes a geographic locator for the person or household that completes the survey. The level of geographic detail you collect will depend on the purpose of your project and issues of confidentiality. In some cases, you will seek more specific, micro-level location data (such as the specific street address or nearest cross street), and in other cases, macro-level data, such as ZIP Code, town, or county, may suffice. A more detailed location can always be generalized when designing maps or reporting results, but failure to collect adequate location information may limit your analysis options. This must be balanced against the likelihood that requesting detailed location information will result in lowered response rates. Of course, when you are collecting manifest spatial data, you have some level of control over what data you collect.

In the applied research world, it is not uncommon to have a dataset that you want to analyze that you did not personally collect. We call such data secondary data because somebody else collected them, perhaps with a different purpose in mind. In such cases, these data may or may not include a geographic or spatial component. We use the term **latent spatial data collection** when some form of geographic location may be attached or connected to the data but you have to find, tease out, or append that information to the data.

Often the people who think spatially and who want to conduct a spatial analysis of the data were not those who collected the data. For example, if you were to collect data from various newspapers around the country that discuss the topic of legalization of marijuana, each newspaper would be associated with a particular geographic area. One approach might be to geolocate each newspaper based on the region or town where it is published.

Earlier in the book, we explored different ways to collect some forms of qualitative data (e.g., chapters 6, 8, and 9). Here we explore the nuances of spatial qualitative data analysis.

How do you know if your data are qualitative?

One of the first things to address when discussing qualitative data is *how you know if your data are quantitative or qualitative*. If your data exist in a form that is not numeric, then they are qualitative data. In many areas of research, quantitative data are the norm. Some view qualitative data as less rigorous than quantitative data because they feel qualitative data analysis introduces too much of a researcher's bias and is therefore too subjective. The best way to combat this critique is to be meticulous in documenting the research methods you follow and decisions you make as part of your qualitative analysis. By keeping very good track of your methods, you are also tracking your coding decisions and data origins in an organized manner. These steps and processes are discussed in further detail later in the chapter.

Different data sources

Let's assume for a moment that you want to conduct a community assessment project. As part of your project, you begin looking at different types of qualitative data. Some sample data types you might choose to consider are described in what follows.

Historical data
Local libraries, museums, historical societies, and longtime residents of a community are excellent qualitative resources for data. A researcher can use information from such sources to better understand a community's past and to develop a detailed history and understanding of place. This research approach can shed light on issues such as, Who once lived in a place? What were the major sources of income and commerce? What was life once like? Who were the various groups who lived in a place? How did these groups use the local resources? How did they interact with one another? Knowing about the evolution of your community's history can provide context to what is going on today and, most important, can

help explain the reasons behind why some groups work or don't work together, or why some groups are left out of decision making based on history.

Figure 12.1 presents an example of what historical data might look like. Figure 12.1 presents a page from a historic travel diary kept by an early traveler to the center of Africa.

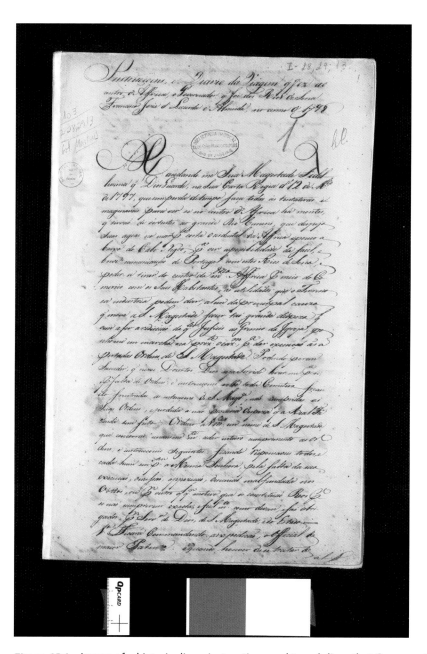

Figure 12.1   Image of a historic diary: Instructions and travel diary that Governor Francisco Jośe de Lacerda e Almeida wrote during his travels to the center of Africa, to the River of Sena in 1798.

By reading and assessing historic information, a researcher can get a sense for what the topography looked like and of the general geographic features and social and cultural aspects of a society through qualitative description.

Newspapers

Local newspapers can also be a source of qualitative data or information for your research project, especially if you examine the letters to the editor section of the paper. For instance, what is the media reporting about different groups that live in your community? What types of issues are being reported? As researchers, when we plan on going into a community, we spend time reviewing the newspapers for the last several years (or for a time frame relevant to your study question) to get a sense of the importance of local topics. How will this information help? While finding out what is going on in the local community, this research also aids in identifying important groups within the community of interest. Every community has groups that hold the power and make the decisions. This power is not always based on having numeric majority but may be based on other things such as a group's social connections or social status. There may be different groups of people who live in a place but have little influence over important decisions made in the community. Understanding a community's sociodemographic data helps a researcher to know what ethnic and socioeconomic groups are important to the community aside from a numbers standpoint.

Interviews with local leaders

By talking with local leaders, a researcher can get a sense of the community and different groups who live there, where they are working, and how best to interact with these groups (figure 12.2). This an important step to take in addition to reviewing local newspapers. Sometimes the news that appears in a local paper reflects the opinions, perspectives, and values of the newspaper owner, not necessarily the entire community. Talking to people from different parts of the community can provide an understanding of what is going on and what people's perceptions of each other are. Local leaders can be city council members, economic development professionals, prominent business owners, and directors of community organizations or nonprofits, among others.

Figure 12.2 A US Marine talks with a local resident in an Iraqi community. In this particular case, the marines were "posting questions to people they met on the street to assess the quality of life for the city's residents and to facilitate a positive relationship with their host country." Photo by Captain Paul L. Greenberg, US Marine Corps, and used courtesy of the US Marine Corps.

Technological equivalents

Because many historical documents, images, and other data are now stored digitally, one valuable opportunity is to review the technological equivalents to the preceding examples. Many online editions of newspapers include the opinions of the community not only in the form of letters to the editor but also as comments attached to individual stories. Variations on this theme can also be drawn from social media sources, which often include a location along with the information posted to the web or social media feed of text, photos, or other useful data. Historical libraries and museums place documents and images on their websites for access without requiring an in-person visit to their archives. In addition to the convenience of gathering these sources digitally, the fact that they are digital provides the added benefit of easy storage as part of your GIS database, for example, by linking a scanned document or photo to a map in ArcGIS to further enhance the information captured in your analysis.

Steps for spatial qualitative analysis

The next question to address is how one translates qualitative data into a form that can be readily incorporated into a spatial analysis. Simply linking images or documents to a map is not sufficient, as this relies on the viewer of the documents to assess and interpret their relevance. Rather, your objective is to convert the qualitative data into something that can be analyzed using the tools in ArcGIS. You can follow a series of steps as you prepare your data. These steps will help you to successfully conduct qualitative spatial analysis.

Step 1: Creating a list of data types and their origins

Data can come in various forms and formats, and as a researcher, you want to create a list of the different data types and their origins. By **data type**, we mean what format your data takes. Your data may exist in a hard-copy (paper) or a digital format. What is the difference between the two? If your data exist in a **hard copy** format, then they exist on paper, film, or some other physical medium. Maybe your data exist in the form of field data sheets, recorded interviews, or old newspapers in an archive. You will need to convert those hard-copy data into a format that ArcGIS can use. To do so means to engage in some sort of data coding and entry process (see step 3). Data in a **digital format** means that the data are already in some sort of computer file format.

Any research project could easily involve multiple types of data in both hard-copy and digital formats. The goal for a researcher is to describe and document these data in a systematic manner. A simple spreadsheet or database can serve to catalog your data sources and other key information about them. In this way, you are gathering essential information about your data, or what we referred to earlier in the book as **metadata**. Why do you need to bother collecting all of this? Researchers always want to know the background and origin of their data to be able to assess their reliability and validity. If the data are collected from a reputable source, you want to be able to reference this. For research purposes, it is central to understand the origin, time of collection, and methods used by whoever collected the original data, and if the data are not your own, the methods you applied in making the data ready for use in ArcGIS as well. If you lack this information, you are missing an important, central component to secondary data use—establishing the context:

1. Data file name
2. Data format: hard copy (paper, film, photograph, etc.) or digital

3. Data file format (SPSS, Excel, .mpg, .avi, etc.)
4. Data source (name of the organization that was the source of the data)
5. Methods used to collect original data
6. Contact information for the agency (a person at the organization or agency)
7. Website (if available)
8. Date collected

By compiling this information (and, as appropriate, additional relevant information about your data sources), you are documenting essential details that can be used to assess the empirical nature of your data. To say that your data are empirical means they were collected using replicable steps and can be considered more rigorous. Data that are not collected in a scientific or empirical manner may be viewed as suspect.

Where do you find such information? The answer to this question will most likely come from a variety of different sites. Information may be available in the files of the data or on the website of the organization or agency from which you downloaded the data. In some cases, you may have to track down data in an archive or locate someone familiar with the data source to ask questions about the data and how they were gathered. Unfortunately, not all data are adequately supplemented with properly prepared metadata, particularly historic data collected for another purpose. Sometimes the required information exists as knowledge stored in the head of one or two employees of the organization. There are obvious benefits to capturing and recording such metadata, not only for your project but for others who may wish to use the data in the future—especially because those who may have the knowledge may be difficult or impossible to track down as more time passes.

Step 2: Determining the purpose of your analysis

In this step, you want to clearly explain the primary reasons for conducting your research. Suppose you ask, What are the different health needs of the population of our town, and where can we locate a health center to best meet these needs? To answer this question, you would need several different types of data. First, you would want to map out the locations of existing health clinics. Next, in conjunction with city leaders, clinic staff, emergency responders, and others with relevant knowledge, you would want to determine who is being served by the existing health clinics and who has to travel the farthest to reach them. You could accomplish this through conducting focus groups or surveys of local people in the town who may be considered stakeholders and asking them how they travel around the town (using public transportation or private vehicles). If you find out that most people do not own cars, then you would want to situate the health clinic in a place that is easily accessible via public transit or along a safe walking or biking route. In other words, you would want to find a place that is in close proximity to where people already live, work, and recreate and along public transportation routes. Trying to answer this simple question would naturally require the consideration of various types of information and data, some traditional GIS data (e.g., infrastructure map layers for roads and clinics) and some qualitative data that need to be geolocated.

Step 3: Determining themes and creating variables

As we described earlier, analyzing qualitative information is typically an inductive and iterative process. In other words, it is a process that can take some time and that begins with carefully reading through the content-based information. Each time you read through the data, you will start to see emerging patterns. The process of qualitative analysis starts with identifying relevant themes in the data, defining these themes, and naming them. You will have a fairly good sense of the themes after the second read-through and should be able to identify your themes by the third or fourth time you read over the data.

Identifying relevant themes
As you review the qualitative data, the goal is to identify repeating ideas and themes. A good place to start is by creating a draft list of important themes and ideas that emerge throughout this process. You will work from this draft list, continually refining it during each pass you make of the data. Conducting effective qualitative analysis requires a number of careful rereadings of the data.

Defining themes
A standard rule for defining themes is to be parsimonious or limited in their creation. We often tell our students who are engaged in the qualitative data coding process to focus on developing five to seven themes in total. If a researcher makes the mistake of identifying too many themes, these can spread the data too thin and result in an analysis that doesn't reveal any patterns. Too few categories may not provide sufficient detail to be useful, either. Cartographically, restricting to five to seven categories is a good rule of thumb that will prevent the map from becoming cluttered.

Giving themes variable names
Once you have identified your themes, each theme needs to be assigned a variable name. We advise keeping variable names fairly short, perhaps one to three words. The variable name should encapsulate the meaning of the theme in a concise way.

Step 4: Coding themes and quantifying

The best way to organize the information around your qualitative variables is to create a **variable definition table**. In such a table, the variable names are listed along with a coded description of the themes. Table 12.1 presents a sample variable definition table. Notice the variable name is listed on the left, followed by a more detailed description of the theme on the right. We create these tables for empirical purposes. You want to provide a quick overview for the reader of the themes. Such a table is very useful in the data coding process as well.

Table 12.1 represents a compilation table that draws out the themes in answer to the question "Why do you choose to live in this community?" You can see that the table has themed the ideas into different variables, such as family, small town/good community, natural environment, and jobs/economy. These qualitative themes arise from a review of the content that was considered a part of the qualitative answer. You can refer to this table as you seek to identify within the content that you are reading where these themes emerge.

Table 12.1 Variable definition table: "Reason for living in the community"

Variable name (code)	Coded description of theme
Family (1)	Born and raised here, raise kids
Small town/good community (2)	Friendly people, social connections, safe, community involvement, rural, quiet
Natural environment (3)	Climate, weather, diverse geography
Jobs and economy (4)	Jobs, economy, moved here for job opportunity, business, own a business

Creating frequency tables

The next step in your analysis is to identify how many times these different themes occur. Before you can do this, you will need to develop a code book. A code book will list the question that you asked to get the information and variable names followed by the different definitions of your themes. When you engage in coding the qualitative data, you may choose to assign numeric codes to each of the different themes. Keep in mind that these numbers are being used as a means of identifying different themes occurring in the data—you could write the coded number. Referring back to the variable definition table (table 12.1), let's say that you read a qualitative answer to the earlier question posed: "Why do you choose to live in this community?" You may get the following answer in paragraph form to analyze:

> "I live in this town because it is where my great grandparents, grandparents, and parents grew up."

In such a case, you would code the answer as code 1, which is "family." When you are attempting to quantify existing qualitative data, you want to be able to code each comment into the themes. As you read through the qualitative comments, you'll code each one. After you have coded all the comments, you will tally them up to see how many comments you have for each category. It could also be that you have a few comments that don't neatly fit into any of the thematic areas in table 12.1. If this is the case, you could always come up with a category called "other" in which to place these comments.

Coding qualitative data can be more challenging than working with already predefined quantitative data because of the coding process. To engage in successful coding, you need to have a clearly defined coding scheme—one that you can follow to quantitatively assign codes to your content.

As you code your data, the goal is to develop a frequency count of how many pieces of content are coded as a given theme. The next step is to create a frequency table (table 12.2).

Table 12.2 Frequency table: "Reasons for living in the community"

Variable name	Frequency
Family	6
Small town/good community	7
Natural environment	4
Jobs and economy	3
Total	20

Table 12.2 lists the various themes and the number of people who answered the open-ended qualitative survey questions that produced these themes. Even though a comment may reflect multiple themes, to avoid confusion with your coding, a best practice is to code comments as belonging to one theme.

Step 5: Spatializing your data

The goal of this chapter is to discuss spatial analysis of qualitative data. At this point in the analysis, you are now at that point. To spatialize your data means that you have some geographic identifier that is associated with each different data point.

Thinking back to past projects, one comes to mind where the focus was food and food security. As a means of gathering data on what people's perceptions were about food and where they shopped for various foods, our team staged a public participation GIS session with members from different sectors of the community (Stubblefield et al. 2010). Our primary focus in this meeting was to facilitate the flow of bottom-up knowledge, meaning that we sought to gather the thoughts and ideas of the local people and their understandings of and access to food. This is an example of a manifest spatial data collection process because we collected data on space and place as a part of the original, primary data collection efforts.

About six groups of five to seven people worked on this task throughout the room. Each group was then asked to identify the various needs of their community related to food. As they talked, the groups were then asked to color code and mark the geographic locations of these needs on the map using differently colored markers (Stubblefield et al. 2010). As researchers, we took each of the individual maps and compiled the data into one master map. This process involved taking the color-coded data from each individual map and developing totals for the entire region.

Step 6: Identifying spatial patterns in your data

As part of the analysis process, we wanted to identify the frequency of the food system themes identified by geographic location. So, for instance, if someone in Millersberg said that there were not enough grocery stores that sold fresh vegetables, we would mark this. If many people say the same thing, then you can use multiple dots to indicate the frequency of occurrence and reflect the themes using different colors. Another data display tactic would be to use larger- or smaller-sized dots to represent that frequency of themes on the map. From an applied perspective, such a map is useful because it allows the map viewer to see the geographic spread of food needs for the county. Presenting such information by town location is very important because if you seek to meet these food needs, you'll know what the priorities are by town and can work to distribute your resources accordingly.

Hurricane Irene dumped about seven inches of rain and caused dangerous high winds and flooding in many places (Sekkes 2013). "Roads were closed, and thousands of residents lost electricity. The Chester County Department of Emergency Services response to Hurricane Irene is illustrated by this map which shows the concentration of storm-related 911 calls received [figure 12.3]. This information was used to help decide where to send damage assessment teams after the storm had passed" (Sekkes 2013, 41).

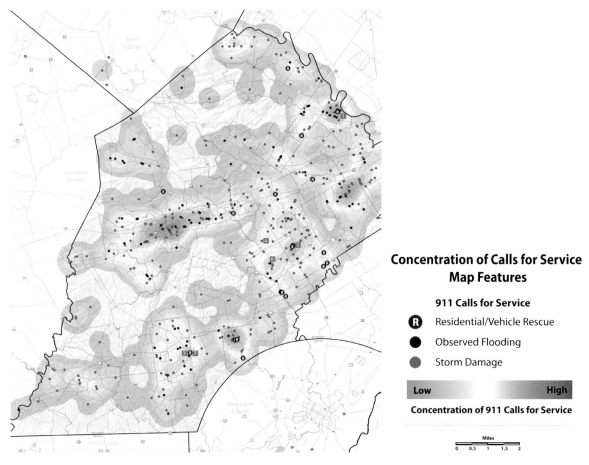

Figure 12.3   An applied example of how a GIS map created from qualitative data could be used to direct emergency responses. The reason that we say qualitative data is because this is a record of the "types" of 911 calls that occurred in Chester County, southeastern Pennsylvania, relative to the damage Hurricane Irene caused in August 2011, severely impacting much of the region. Map by David Sekkes, courtesy of Chester County Department of Emergency Services, West Chester, PA. From *Esri Map Book*, vol. 28 (Redlands, CA: Esri Press, 2013), 41. Data from Chester County Department of Emergency Services, National Aeronautics and Space Administration.

The content of responses came through the phone calls and was then categorized into one of three types of calls: "Residential/Vehicle Rescue," "Observed Flooding," and "Storm Damage." We say that the data were qualitative in this example because they were descriptive. The map used a dot density approach to focus on and illustrate types of incidents for planning and action purposes following the storm.

Here is an important point to consider: through the analysis process, you can choose to constrain responses and categories of input from your respondents, but often it is preferable to allow them to respond in whatever way is most natural for them. This lack of constraint and categorization may lead to responses or themes that you wouldn't have considered otherwise. Although this can be beneficial in drawing out the full range of responses, you will ultimately want to combine, categorize, and code the responses you receive into meaningful groupings that are relevant to your analysis in the GIS. These categorized responses, along with a frequency value (representing the number of individuals, focus group sessions, surveys, or other appropriate units of analysis for your study), can then be mapped in GIS with a location relating to where the responses originated (e.g., by city, ZIP Code, census tract) and the magnitude of the response represented by the frequency value.

What we have just described is a simple analysis. However, if you wanted to take your analysis further, you could implement a series of more intense analytical functions available in ArcGIS, such as determining proximity (nearness to) or distance between using the Near tool, along with analyzing various statistical features. For example, if you are examining access to the category for grocery stores, it might be relevant to use ArcGIS tools for proximity analysis to identify the service areas associated with each store. If the resulting areas are small (perhaps as defined by a reasonable travel distance or time via public transit, or walking distances), then there would appear to be good availability of grocery stores. Conversely, areas of the map with very large service area polygons might indicate an opportunity to improve access via a new store, improved transit options, or other solutions.

Another approach would be to use the Near tool to determine the distance from particular respondents to the nearest store or the distribution of distances relative to all of the responses in your analysis. In this case, respondents' locations would need to be fairly detailed to get good results, whereas mapping service areas is only dependent on knowing the locations of the services (grocery stores). Having multiple analysis options can be important, especially when secondary data lack the information required for a particular analytical tool. Analyzing a particular concept can almost always take multiple approaches, and you will sometimes need to be creative in considering what your options are and the best option to use for your study.

The collection of maps in figure 12.4 presents data collected using keywords indicating Internet activity at different points in time relating to a current event in the news. The creators of this collection identified the top one thousand most popular websites with activity related to the topic searched; in this example, "Burn Koran" was searched, and top websites were retrieved. Each website was mapped by converting the site's web address (URL) to a geolocation. Then, kernel density maps were created by weighting the popularity of the websites. This collection is especially interesting because of the method the creators used to geolocate where the ideas were being discussed. These maps are an excellent example of how qualitative data, or content, can be mapped and analyzed. Done over time, such a method can illustrate the spread of ideas over time and space. Each map presents different information about the data, helping to establish context for the information being discussed.

The important thing to remember when looking at the series of maps presented in figure 12.4 is that the maps do not represent an actual action but rather the discussion of an idea or action. The factors influencing how much an idea gets discussed in a particular place could directly stem from local or regional news coverage as well.

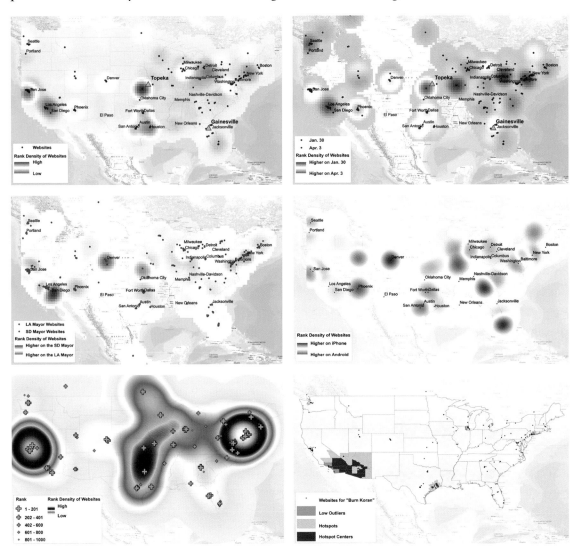

Figure 12.4 Mapping ideas from cyberspace to real space. These maps show the source locations of websites returned for a specific keyword at different points in time as kernel density maps. Maps courtesy of Ick-Hoi Kim, Sarah Wandersee, and Ming-Hsiang Tsou, Department of Geography, San Diego State University. Data from US Census and Ick-Hoi Kim. From *Esri Map Book*, vol. 27 (Redlands, CA: Esri Press, 2013), 16. Data sources: Project-generated data, decennial US Census.

Ultimately, when you work on the spatial analysis of data, you can either collect the spatial information at the moment or link it to a fitting spatial database at a later time. Sometimes when you work on qualitative data collection, depending on your topic, you may not know where geographically your ideas are going to appear until you begin that process of data collection. However, once you determine the answer to this question, you can always find geographic data at different levels of detail. If you take a more national approach to data in the United States, you may want to look for datasets from the federal government, versus if you choose to conduct a more state- or county-specific analysis, you would focus on data from one of the state or county databases. The linchpin in such spatial analysis is having a geographic identifier somehow connected to your data. As we have discussed here, it may be that you have to hunt a bit more to find geographically based elements in your data.

Step 7: Drawing conclusions

The final step in the process of spatial data analysis is to draw big-picture conclusions about your data. In the conclusion section of your analysis, you'll seek to answer the question, What does this analysis mean? Once you've identified spatial patterns in your data, it is easy to see where the next steps for policy, action, or further investigation should lead. A spatial qualitative analysis can make evident nuances in your data that, prior to your GIS investigation, were not clear. Why? As researchers, we don't always give credence to the more content-based data. We have a tendency to want to focus on data that are easy to analyze, assess, and map, which are usually quantitative data. But as we have pointed out throughout this chapter and book, at times, and with certain topics, you'll want to focus specifically on qualitative data—and this may mark the beginning of an exploratory research study. Similarly, your qualitative data analysis may be a more in-depth examination of a topic that emerged in earlier quantitative studies that you want to examine in more detail. In any case, the conclusions that you draw about your qualitative data can establish context and background for any type of research. Qualitative data can stand alone but are often combined with more quantitative data as well, to present a more well-rounded view of a research question. Sometimes, when you are engaged in qualitative research, you may uncover just the tip of the iceberg of a topic, which means that you'll want to focus future work on collecting more data.

Review questions

1. What are qualitative data? How are they different from quantitative data?
2. Why is it common to use an inductive approach for conducting qualitative research?
3. What is the difference between an inductive and a deductive approach to research?
4. In qualitative data analysis, why do you want to define your themes? What is a variable definition table?
5. What is the process for identifying the themes that emerge in qualitative data?
6. How does a researcher guard against introducing bias into his or her collection of qualitative data?

Additional readings and references

Caquard, S. 2013. "Cartography I: Mapping Narrative Cartography." *Progress in Human Geography* 37, no. 1: 135–44.

Cope, M., and S. Elwood. 2009. *Qualitative GIS: A Mixed Methods Approach*. Thousand Oaks, CA: Sage.

Sekkes, D. 2013. "Hurricane Irene Response: Chester County Department of Emergency Services." In *Esri Map Book*, vol. 28, 41. Redlands, CA: Esri Press.

Steinberg, S. L., and S. J. Steinberg. 2009. "A Sociospatial Approach to Globalization: Mapping Ecologies of Inequality." In *Understanding the Global Environment*, ed. S. Dasgupta, 99–117. Delhi: Pearson Longman.

Steinberg, S. J., and S. L. Steinberg. 2011. "Geospatial Analysis Technology and Social Science Research." In *Handbook of Emergent Technologies*, ed. S. Hesse-Biber, 563–91. Oxford: Oxford University Press.

Stubblefield, D., S. L. Steinberg, A. Ollar, A. Ybarra, and C. Stewart. 2010. *Humboldt County Community Food Assessment*. Arcata: Humboldt State University, California Center for Rural Policy.

Verd, J. M., and S. Porcel. 2012. "An Application of Qualitative Geographic Information Systems (GIS) in the Field of Urban Sociology Using ATLAS.ti: Uses and Reflections." *FQS* 13, no. 2 article 14 (2012). http://www.qualitative-research.net/index.php/fqs/article/view/1847/3373.

Relevant websites

- **Active Living Research (http://activelivingresearch.org/qualitative-gis-approach-mapping-urban-neighborhoods-children-promote-physical-activity):** This website discusses different types of active living research and specifically has an article written by P. Wridt, "A Qualitative GIS Approach to Mapping Urban Neighborhoods with Children to Promote Physical Activity" (2010), focusing on how to use GIS for qualitative mapping to map and promote physical activity in neighborhoods.
- **Esri, Understanding Statistical Data for Mapping Purposes (http://www.esri.com/esri-news/arcuser/winter-2013/understanding-statistical-data-for-mapping-purposes):** A piece written by Aileen Buckley from Esri news that discusses the differences between qualitative and quantitative data relative to spatial mapping.

Chapter 13

Communicating results and visualizing spatial information

In this chapter, you will be introduced to some of the key considerations in presenting your research findings. Effective and appropriate communication of your message, considering the audience and its needs, can make all the difference to a successful project. Of course, when using ArcGIS, one of your key means of communication will be a map. In this chapter, you will explore examples of excellent visualization using ArcGIS and strategies for effective communication using spatial technology. You will learn guidelines for putting together a final presentation of your data that effectively incorporates cartographic visualization tools in ArcGIS. You will also learn some of the other ways GIS can be used to share important research findings with a variety of audiences, including using outputs beyond the map.

Learning objectives

- Learn the keys to effectively communicating results
- Comprehend spatial visualization and how to achieve it
- Learn how to choose the appropriate method of communication
- Understand the guidelines for creating a final product

Key concepts

data visualization
geographic literacy
information
interactive map
map context
statistical output

Keys to effective communication

You have probably heard the saying that "a picture is worth a thousand words," and this is certainly true in the world of maps. Perhaps the most obvious output from a GIS is the map, a graphic form that has a long history in many cultures. Open any atlas and you will find maps that communicate information about demographics, economics, cultures, languages, history, politics, and a host of other information related to the cultural, physical, and natural geography of the world. The GIS provides a tool capable of most, if not all, of the software tools necessary for data collection, organization, storage, analysis, and output. However, before we get into the nuances of communicating your results, we'd like to present some things you should consider when developing an effective communication strategy.

When communicating the results of your work, it is essential that you answer two basic questions:

1. Who is your audience?
2. What is the purpose or message you want to get across?

Audience

Whereas some research projects fall into the category of "pure research" conducted solely to satisfy curiosity, more often, an underlying purpose drives the work. You may be reading this right now and thinking to yourself, "My goal is to empirically study something and present my findings to my academic peers." But consider the context for doing so. What is the venue for your presentation? Will you be presenting this at a meeting with maps projected on a screen for a few moments, or will your data be communicated in printed form as part of a paper or report? Is there an opportunity to develop your data into animated or interactive visualizations on the web? Does doing so provide something valuable to your audience, or will it simply add unnecessary complexity? We will come back to the basic question: What do you hope will be the final outcome of this presentation? Are you hoping to publish in a journal? Do you want to raise stimulating new points and generate novel discussion? Where do you want to go with this presentation?

Even if your audience comprises your peers, who are they really? Are they familiar with map-based data analysis and visualizations of results, or will this be foreign to them? In some disciplines, GIS analysis may be within the realm of experience of your audience—perhaps they had classes on GIS in school and even worked with GIS at some point. If so, you may have an audience with a basic understanding of your research methods and the types of outputs you are providing. However, in other cases, your audience may not have thought seriously about reading a map since they were in high school geography (or attempting to find their way on vacation). Although your peers may be well versed in the concepts of your discipline or profession, they may be unfamiliar with applying GIS to these questions.

Outside of the academic realm, your audience may be much more diverse. You may be conducting your research to improve an organization's services, to understand the root causes of a pressing social issue in a community, or to better allocate limited resources. In any of these situations, an individual or group will need to review and understand your findings and recommendations, whether this is the executive director or board of directors of an organization, the city council or school board, or the general public.

For example, you may be an employee of a company or agency that has collected spatial data on a topic and you want to be able to share this information with your colleagues in a manner that makes sense to them. Perhaps you might create an oral presentation with visuals. That is a good way to present new information to an audience, especially if it's new information that you plan to modify. Your presentation venue may be geared more toward your internal organization or agency. In such a case, depending on your stage in the data collection process, you may want to produce a short write up or report that includes your maps.

As you no doubt realize, each of these audiences will have a different level of understanding of GIS, maps, and the particulars of the issue or topic of your study. It will be important to keep this varying level of knowledge in mind when you are generating outputs and data visualizations. For example, an audience comprising policy makers will be interested in how your findings can be translated into something that can be implemented in a regulatory or legal framework. The director of a nonprofit may be more interested in the cost and effectiveness of delivering services to the organization's clients. If you are targeting the public, consider if a printed brochure or website is the best means to distribute the information. None of these audiences are likely to be experts with ArcGIS, so odds are you cannot simply hand them your data and project files and expect them to arrive at their own conclusions. Instead, you will need to present the information in a format and at a level of detail appropriate to your intended audience.

Purpose

The first aspect to consider when preparing to communicate your results is your goals. Are you planning to present your work at an academic conference? Are you trying to move people to action with the presentation of your results? Are you attempting to secure new funding for your organization? Are you simply sharing the findings from your project with the public because that is part of your funder's mandate? Is sharing your results simply meant to inform or educate the general public? As the researcher, you need to consider the purpose of your communication and develop products that are designed to communicate effectively before you go out and present, share, or distribute your results.

Most of the time, when conducting an analysis in GIS, you will draw in a fairly substantial dataset comprising numerous thematic layers, data tables, and variables. Not everything you explore, or even ultimately use, in your analysis will be equally important. Therefore, it is essential that you filter out the extraneous details and provide simple, clear, and direct messages. In the case of map output, this means limiting the information included in any single map. Similarly, whereas a lengthy and detailed report of your data sources, methods, analysis, and findings may be important to some audiences, a simpler executive summary or fact sheet boiling down key findings to a few pages, accompanied by charts, graphs, photos, and maps, is far more likely to be read and therefore to communicate your message effectively.

Consider the approach advertisers take. They use a central tenet of communication, which is to get people to understand and be interested in a product or service, something about the product must first resonate with them. In other words, information must be presented in a way that allows the viewer to relate to it. What better way to do this in the context of a spatial study than visually and with maps? Most people will innately recognize particular geographies and map shapes based on past experience and knowledge. Certainly most people recognize the outline of their nation and other local geographies such as their home state (and, quite likely, the outlines of many other states, countries, water bodies, and other geographic features). Most cultures use some sort of visualization (pictographs, patterns in artwork or on pottery or baskets, and maps) to convey stories, data, or locations. As we translate these concepts to communication of results of GIS-based research, it is important to keep in mind three basic tenets of effective communication: information, interest, and context.

Information

Anytime you present information, it is important to keep your story simple. If possible, you should limit each map to one primary message, and in designing the map, make that message clear to the reader by selecting appropriate colors, symbols, and labels. Because GIS offers you the ability to incorporate so much data into your work, sometimes there is a tendency to try to include it all. However, doing so can lead to information overload—a map with so many colors and symbols that it becomes a jumbled mess.

Minimally, you will need two components for an effective map visualization. First, you'll need a basemap, which provides context. Basemaps should incorporate a few key features, such as major roads, rivers, or boundaries, that will be familiar to and orient the reader. However, because these layers are not necessarily essential to the message, you may want to deemphasize them in the visual hierarchy by selecting muted colors that fall to the background. Second, you should have the map data that represent your core message. This information should be presented with brighter colors and clear labels or legends. Limiting the number of categories within a map is also important, for example, it may be preferable to produce a map communicating a limited number of data categories (figure 13.1) than one attempting to differentiate numerous categories that become difficult for the reader's eye to differentiate (figure 13.2).

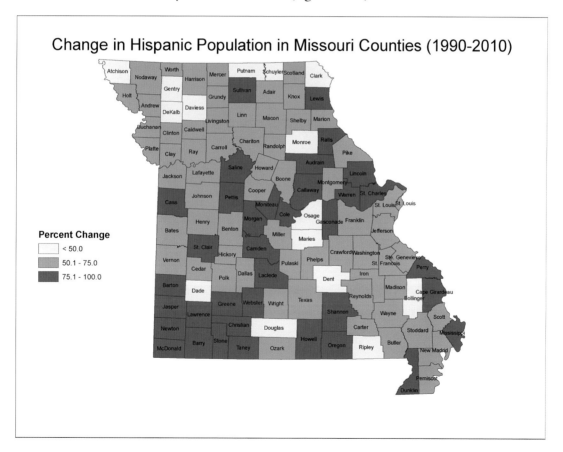

Figure 13.1 A map with only three categories is easier for a reader to decipher, making patterns in the data more apparent. In this map, larger percentage change in the Hispanic population during the study period is easy to identify across the state. Map by Steven Steinberg. Data from US Census.

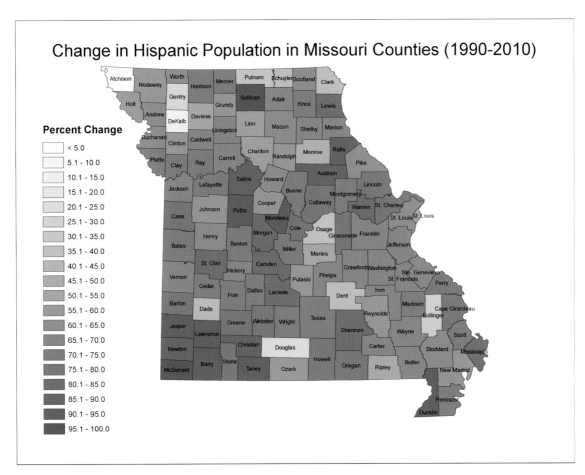

Figure 13.2 The same data presented in figure 13.1 are shown here using twenty categories. Although the map now provides more detailed information regarding the percentage change in the Hispanic population, it is difficult for a reader to decipher. Patterns in the data are less apparent than they are in figure 13.1. An identical color gradient was used for both maps. Map by Steven Steinberg. Data from US Census.

Figure 13.2 emphasizes that you should not include too much information in your visualization, but you do want to include enough so that the spatial visualization resonates and makes sense to the audience. In figure 13.3, we see a somewhat abstract representation of the continents overlaid with data that highlight the various populations that are at risk of earthquakes. Although this map may not match your initial expectations for an earthquake risk map, once you've taken a moment to understand what the patterns of color represent, the visualization is quite effective.

This map provides a focused message without extraneous information cluttering or complicating the visualization. The map shows large swaths of blue representing low risk, keeping in mind that risk occurs where there are both human populations and seismic activity. Therefore, for large portions of the globe that are ocean, few population centers are present apart from some islands, and risk is low. But these low-risk areas include significant regions of the continents where either few people live or there is limited seismic activity. Where these two datasets coincide, the map regions are colored in warmer tones of red. Of course, red is a good color choice given that to most people, it represents a warning or indicates danger. This striking portrayal creatively highlights the coastal regions of the Pacific ring of fire and other areas of the world at risk of a major earthquake. The interesting thing about this map is that the actual outlines of the continents are secondary to highlighting the risk. Continent information is there as a backdrop but is not overbearing.

Information is often used as another word for *data*. When a map effectively communicates, it clearly presents data to the audience. By creating clear and simple visualizations, without extraneous information, you should be able to convey your message. So when we talk about information, we are focused on sharing or conveying data, whether you accomplish that goal with a more traditional map, such as in figure 13.1; a stylized map or cartogram, such as in figure 13.3; or perhaps even an infographic, which represents your data visually but not in map form (figure 13.4). If you try to include too much information in one map or visual, you will create confusion. It's better to keep things focused and simple.

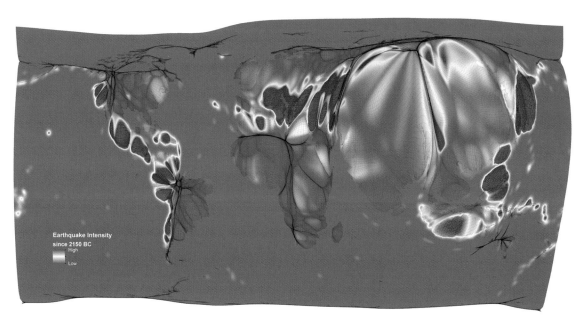

Figure 13.3 **People at risk: Humanity and global earthquake intensity.** Map courtesy of Benjamin D. Hennig, University of Sheffield, Sheffield, South Yorkshire, UK. From *Esri Map Book*, vol. 27 (Redlands, CA: Esri Press, 2013), 14. Data from National Oceanic and Atmospheric Administration National Geophysical Data Center, Socioeconomic Data and Applications Center (SEDAC), Columbia University.

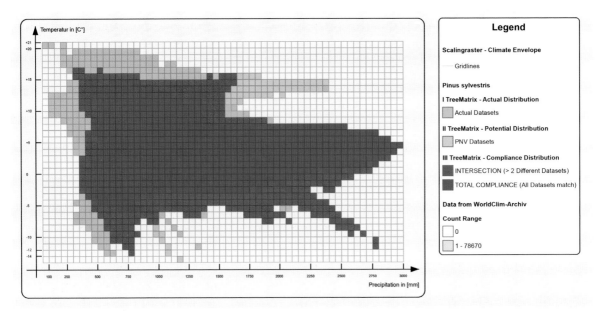

Figure 13.4 An infographic showing a graphical relationship between data in a nonmap format. In this case, the relationship between tree species and climate change factors (temperature and precipitation). While these concepts have a clear connection to geography, that is, we expect certain types of trees to grow in certain locations, as the climate changes the locations of particular habitat zones may also shift. Map courtesy of Holm Seifert, Technical University of Munich, Munich, Germany. From *Esri Map Book*, vol. 28 (Redlands, CA: Esri Press, 2014), 34. Data from Natural Earth Data, Germany; European Environmental Agency; AFE Data: Botanical Museum of Helsinki; ISPRA Data: Bavarian regional office for forest and forest management ICP Forest Level 1, Italy; Meusel Data: Bavarian regional office.

Interest in the topic

Your presentation of data should generate some level of interest in the topic. When conducting a spatial study at the request of an organization or group, there is likely to be a built-in level of interest. Nonetheless, your presentation of essential findings can be enhanced by an engaging map visualization. The patterns that jump off the map can help to tell the story embedded in the data and engage those who will use this information in decision making.

When your audience is less specifically defined, an engaging map can help to draw their interest in the topic. If you have an understanding of the audience, perhaps members of the public or recipients of a service your organization provides, you may be able to build off of an existing interest. However, you cannot assume that everyone has the same level of experience with or knowledge of the topic at hand. A hallmark of establishing interest means that you have created some hook or opening for the map viewer or consumer. This hook could be based on the way the information is presented or the interest a unique and creative map presentation can generate.

Map context

When we talk about **map context**, we are referring to background of the story that you want to tell from a spatial perspective. Maps have told stories for centuries. Maps by early explorers often depicted the many dangers lurking in the oceans (sea monsters, sirens, a plunging waterfall off the edge of the earth, etc.). Though the reality may not have been quite as fantastical, the dangers these visualizations imply were real, and the symbols the maps' creators used established the context of danger and uncertainty in certain unexplored regions. Modern maps also tell stories, and with the addition of interactive maps online, story maps have entered a new realm. An interactive map is a computer-generated spatial portrayal that enables the user to actively engage with the map. A number of excellent examples can be found at the Esri story maps website (http://storymaps.esri.com/home/), among others.

For any map or visualization that is created, there is some type of contextual backstory or meta-information. The degree of context that you need to provide should in part be based on the spatial geographic literacy of your audience. **Geographic literacy** refers to how familiar an audience is with the geography your visualized topic presents. It is also important to keep in mind that a person may be familiar with the geography but not with the spatial portrayal of data; it may take some people a bit of time to orient themselves to your visualization. This underscores the reason to have simple and clear visualizations. You can lose the message or the map consumer if you simultaneously overload a map with too much information.

Many people today have access to unlimited web-based information, including data presented as maps or spatially based data, infographics, and other formats. In our current media- and image-dominated culture, if you want to effectively communicate, you must be creative, thoughtful, and targeted in your approach.

In the next section, we present some additional considerations to help ensure that you effectively communicate your results and the important information contained in them.

Culture and map visualization

Every group or society is going to have aspects of communication and culture that are specific to the particular group. When you create a map, it is important to consider the culture of the group to which you are presenting. However, if you are creating a map to present to people across multiple cultures, you'll want to use symbols and visual artifacts to which people from different backgrounds can relate.

Figure 13.5 provides an illustration of how to present information on a map for people who are unfamiliar with the city of Luxor and who may be interested in visiting

Figure 13.5　A tourist map of Luxor, Egypt. Graphics such as a cruise ship orient tourists to one of the primary means of entering the city. Photographs highlight some of the key attractions, and well-understood symbols for food, restrooms, and other services provide additional visual information to the reader. Map courtesy of Dahlia Khalil, Esri Northeast Africa Cartography Department. Data from Esri Northeast Africa Data Center/Cartography Department. From *Esri Map Book*, vol. 26 (Redlands, CA: Esri Press, 2013), 11.

different sites. The map orients the viewer to the geography of the city by indicating some major geographic features, such as the bank of the Nile River and the location of major thoroughfares in the city. Streets are labeled in both English and Arabic to facilitate user comprehension. Additionally, the map creators thought carefully about their audience before they created this map. If you notice on the map, they've made good use of international symbols for food (the knife and fork) and for restrooms (another salient feature for travelers). The map is fairly simple and is not cluttered with too much information beyond the sorts of sites and services of interest to a tourist along with a road base to help orient the reader. If you know what is important to your audience, you can present that in a clear and straightforward way.

GIS output

Having a sense of the desired outputs from a GIS analysis can provide important direction to the data collection and analytical procedures incorporated early in the process. The following sections discuss two broad categories of **data visualization:** the graphic output of the GIS, most commonly in the form of a map, and **statistical output**, which may take the form of numbers, graphs, or mathematical models. Last we discuss exporting the results of your GIS analysis to other applications, both to facilitate analysis and to communicate final results.

Data visualization

Traditionally, printed maps were the most common type of graphic output from GIS—that is, maps designed to display the relationship between and distribution of a limited number of variables for a particular geographic extent (study area) at a particular point in time. By getting a bit more creative with cartographic design, a single map might have effectively shown change over time or a larger number of variables in combination, but doing so often resulted in a map that was difficult to interpret without significant effort on the part of the reader. Alternatively, sets of maps, such as shown in figure 13.6, could communicate multiple variables at multiple times. Though these individual maps are certainly easier to understand, mentally combining the messages of these multiple maps to fully understand the story they tell could be difficult for the reader.

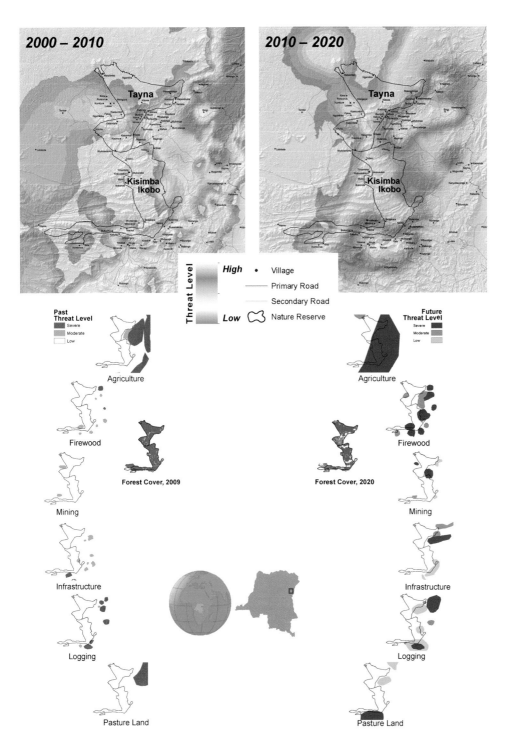

Figure 13.6 **Drivers and agents of deforestation in Tanya and Kisimba-Ikobo Reserves, Democratic Republic of Congo.** Map courtesy of Fabiano Godoy, Adam Dixon, Ryan Hawes, and Serge Omba, courtesy of Conservation International, Arlington, VA. From *Esri Map Book*, vol. 27 (Redlands, CA: Esri Press, 2013), 17. Data from World Resources Institute, Ministry of Environment, Conservation, and Tourism of the Democratic Republic of Congo.

The persistent challenge that a researcher or policy maker must consider is how to present the desired information without going overboard and giving too many details that cloud the main story or variables that are to be conveyed. Developing a series of maps may be preferable when you are attempting to communicate a good deal of information resulting from an analysis. Alternatively, if it is appropriate to your audience and purpose, you might consider using animated maps. Although animation requires your audience to have access to digital technology and perhaps connectivity to the Internet, animated or interactive maps allow for communication of more information by allowing the map to transform. This is particularly useful when communicating a change through time.

Figure 13.6 is an example of a change map with multiple variables. This map series provides a nice illustration of change over time. The large maps at the top illustrate the current and future overall threat to the reserves for two different decades. These maps include relevant base data, such as locations of villages and roads surrounding the reserve. However, because threats are derived from a variety of causes, the map authors provide additional detail showing these individual factors as simple symbolic illustrations beneath the primary maps, using the outline shapes of the reserve to orient the reader to detailed information related to resource use. This map series provides context to the reader about both past and future resource-related threats to the reserves and the relative degree of each threat. Therefore, in this example, we have the opportunity to visualize and understand two different portrayals of change. This provides an effective means of communicating both the central message and supporting details important to telling the data's whole story.

Of course, working in an interactive environment provides numerous opportunities to enhance the story your maps tell. For example, the viewer may be offered the option to select specific data layers he or she wishes to view and the geographic area on which to focus. Because the reader of the map is provided a level of control, additional detail may be made available as the viewer zooms in or turns different themes on or off. Supporting information in the form of photographs, charts, or text can be linked to the map without cluttering the visual interface. Nonetheless, interactivity does not relieve the author of a map from the responsibility to make the information understandable. As with a printed map, you do not want to put everything on the map at once. It is important to consider what the initial view will be and how various data will be symbolized to keep your maps readable. For example, do you want to limit certain details (such as local roads) until the viewer zooms in? How will you make sure the viewer of the map doesn't miss out on essential information that is not presented in the initial view or get confused by attempting to turn on all of the available information at once?

Though interactive maps can provide valuable capabilities, if not thought through, they can also overwhelm or confuse the audience. At their best, interactive maps will function much like static maps, presenting the essential information clearly and without clutter. At their worst, these may approach an online version of GIS, providing so much data and functionality that the user must be a GIS expert to successfully interact with the information.

Countless websites are now available for interactively exploring GIS data. The national atlases for a number of different countries are now provided in web-based format rather than in print. For example, the National Atlas of the United States (http://nationalatlas.gov) provides a variety of static and animated maps along with the underlying map data available to download for use in your own GIS software (figure 13.7). Other examples include the Atlas of Canada (http://atlas.gc.ca) and the Atlas of Sweden (http://www.sna.se), providing similar services for those nations.

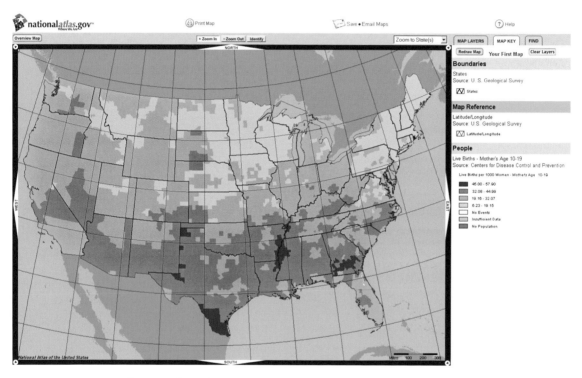

Figure 13.7   An example screen from the US National Atlas website (http://nationalatlas.gov). The site allows visitors to view data in a wide range of categories for the entire United States or any portion. The view shows the number of live births per one thousand women aged ten to nineteen years. Each dataset is linked to the source agency, in this case, the Centers for Disease Control and Prevention (http://www.cdc.gov). Interactive maps produced either on an individual computer with your own GIS software and data or via publicly available websites allow the user to explore variations and combinations of data much more effectively than would be achieved via static maps on paper or as scanned computer graphics. Map created by the authors using an open-source dataset. Data from the US National Atlas (http://nationalatlas.gov).

Animation is an effective tool if the results of your study will be communicated using video on a computer, video on a website, or even traditional video. The animated radar maps shown on your local news show's nightly weather report are a common example of this sort of video map. In the current version of ArcGIS, time-maps including animation can be enabled within the map properties. This can be useful for displaying change through time within the software. A time slider tool in the interface can be used to move through the data sequentially; of course, this assumes that the data with which you are working have a date field as one of the attributes for the layer. Additionally, time-enabled maps can be exported to a stand-alone .avi file format that can be viewed with freely available Windows media players, allowing you to share your animations with those who do not have access to ArcGIS software. You can achieve more sophisticated animation using third-party extensions to ArcGIS or other video editing tools.

Three-dimensional visualization is yet another powerful means for communicating data. The GIS can be used to create realistic perspective views from anywhere on the landscape. ArcGIS extensions such as 3D Analyst provide many of the tools for developing these perspective views. Often, when attempting to illustrate a proposed project or outcome, providing a picture to show how things will look on completion can make the difference at a community meeting (figure 13.8). For example, how will the construction of a new building on a vacant lot affect the appearance of the neighborhood? How will it alter the skyline of the city?

Figure 13.8 **A perspective view for a community development scenario. This image was derived from actual data of roads, buildings, and an available parcel in ArcGIS. A proposed community park was incorporated into the plan and visualized using the CommunityViz extension to ArcGIS.** Image courtesy of Michael Gough/Humboldt State University, with funding from the North Coast Environmental Center, Arcata, CA.

Even more dramatic 3D images can be animated to create a fly-through of the landscape, thus providing a variety of realistic views and perspectives in an animated movie format. A fly-through allows the user to interactively navigate through a 3D landscape, giving the user a feel for how things would appear as a person actually walked, drove, or flew through it. This is also sometimes referred to as virtual reality, and with the appropriate tools, data, and effort, such animations can be made to look as realistic as desired, even to the point of incorporating actual photographs of the local environment.

Statistical output

Perhaps less apparent to a newcomer to GIS are the nongraphic outputs, in other words, the statistics. It surprises many students taking an introductory GIS course to discover that some of the first GIS systems were developed before computers were capable of graphical output. In fact, the first GIS systems did not even produce maps as you would think of them today, and typically the computers used to carry out the analysis lacked a monitor capable of graphic display of the data. In the present day, this is fairly shocking, especially given the central role that maps have come to play in terms of communication and decision making. Having grown out of a nongraphical heritage, one of the most valuable outputs of a GIS analysis is its statistical or numeric output.

For example, if you were using a GIS to generate a list of addresses of individuals that need to be informed of a proposed construction project in the community (a common requirement in many jurisdictions), the analytical process might be to develop a buffer of a set distance around the project to identify owners of the parcels to be informed. The objective of this analysis may simply be to quantify the number of properties or people affected and to generate a set of mailing labels for the notice to be sent out. In such a situation, a map output is unnecessary. The desired outputs are instead some statistics and a list of addresses. As we discussed earlier in the book, one of the great values of using GIS is being able to quantify your data or information.

As an example, looking at figure 13.9, we see the visual presentation of a spatial-statistical analysis of data for emergency medical calls in Baltimore, Maryland. The Baltimore City Fire Department responds to more than 150,000 calls per year, and by visualizing which stations respond to calls at various times of day and types of calls, emergency response planners can better anticipate where, when, and what sorts of resources may be necessary in different areas of the city. The figure also indicates from how far away the response came. The value of this portrayal is that it highlights these data superimposed on the underlying geography of the city. This is an effective example of portraying statistics because it is clear and simple yet conveys a variety of important information without overwhelming the reader with extraneous details.

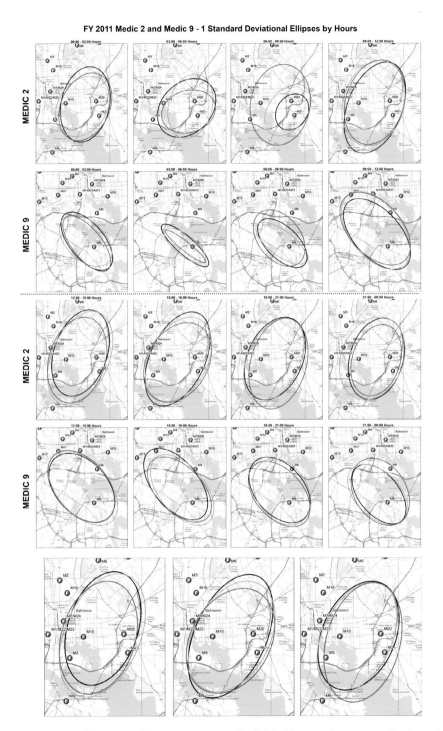

Figure 13.9 The maps show emergency calls divided into various types of calls such as "shootings, cuttings and cardiac arrests to show where the majority of the city's calls are occurring. The calls have also been grouped by hour of day to see how far units are responding to their station." Map courtesy of Baltimore City Fire Department, Peter Hanna. From *Esri Map Book*, vol. 28 (Redlands, CA: Esri Press, 2013), 40. Data from Baltimore City, 2011.

In answering many questions, the value of the map or visualization may be outweighed by the need for the numbers. This is especially true in decision-making applications where percentages, risks, and likelihoods, or dollars and cents are the more compelling data to the researcher. As discussed previously, ArcGIS includes a variety of tools for statistical analysis of your data. It also provides a number of capabilities to summarize data and results with common descriptive statistics, along with charts and graphs of the results, much like in a typical statistical analysis software package. Of course, the major strength of GIS in these analytical situations is the ability to carry out spatial operations on the data.

Regardless of whether you develop your analysis fully within ArcGIS or incorporate external tools for portions of the analysis, integrating these results into a comprehensive visualization makes for a compelling display of your results.

Exporting data to other applications

When the GIS software does not offer the more intensive statistical analysis tools you may require, the database of both spatial and nonspatial data contained within the GIS can be readily queried for the necessary variables and exported to a statistical software application of your choosing. In some cases, tools for the direct transport of data between the GIS and your preferred statistical software may be available to facilitate this process.

Just as data tables can be imported into the GIS, the GIS can be used to analyze spatial relationships to generate new tables of information to export back into other software packages. Imagine you have a GIS database of public health statistics for the entire United States but are doing a study of only a particular region. You could use the spatial analysis tools in ArcGIS to select those records associated with the region in question and create a new, smaller database table with the information you require for your study. In this way, you can use the GIS as a data extraction tool that allows you to select and extract data needed to run statistical tests or other analytical processes.

Of course, you might think, "I could do that in a database by querying the state or county names I require, so why use the GIS?" This is a valid point. If the database contained that information, you wouldn't need a GIS, but what if your study region was all households located within one mile of an industrial facility registered with the Environmental Protection Agency for release of registered toxins? Or if you were doing a study of the spread of West Nile virus, you might want to locate all residential areas within two kilometers of lakes and ponds (bodies of water where mosquitoes could breed). Spatial analysis such as this is not feasible with a standard database query but is readily accomplished in a GIS.

Selecting the mode of communication

Once you complete a research project or data collection and analysis effort, you are naturally going to want to share your results. Some people are tempted to adopt a one-size-fits-all approach for this. However, we caution against doing so. The best fit for your communication media should be governed by both your goals and your audience. Earlier in the chapter, we emphasized the importance of identifying your goals for communicating your results and the audience or audiences you intend to address with your end products. In many cases, you may find a multipronged approach is required.

Of course, any communication products you develop will take time and effort (not to mention dollars and cents), so it is important to consider the benefits of each. Do your results need to be available to a wide audience or just a small group or an individual? Will your presentation of the data be accompanied by other materials (a written report or executive summary, or perhaps an oral presentation), or must it stand entirely on its own? Is the project long-lived, or is it a one-time analysis that will not be returned to in the future?

Web-based information

A web-based presentation of your research can offer a number of benefits, including interactivity, animation, and a variety of supporting materials. For audiences that may be spread over a large area or are more general in nature, a website can be an effective means of communication. However, creating a website for your project can also take a significant amount of time and resources. For a long-lasting project or one with widespread interest, this may make sense. However, for an analysis intended for your boss to review and make a one-time decision, a website could be a bit too much.

Reports

If you are part of a more official line of data sharing then you might want to consider writing a report. Many organizations may desire a detailed description of the data, methods, and findings, which can be referred to and maintained for future reference. The spatial visualization of your data can be portrayed as figures or appendices within the report, and perhaps an executive summary can be included to summarize the essential findings or recommendations. Of course, reports are easy to make available online, especially if your organization already has a website. Posting a report or the associated static maps doesn't require extra software or expertise.

Pamphlet

A pamphlet is a good form of communication to use if you are communicating to a group of people who do not have access to technology or a population that has difficulty accessing technology. Pamphlets also provide an excellent means to advertise other communication formats you may have available. For example, a pamphlet of essential findings and maps might be provided in your office lobby as a means to catch the interest of the public and drive them to your website for more information or interactivity. Pamphlets can also serve as a complete product for populations who may not be inclined to read a lengthy report or who have limited or no access to computers. This could include some minority, elderly, or economically disadvantaged populations whose access may be limited to publicly available computers at a library or community center.

Poster

A poster is an effective way to communicate a lot of data, and it can become a permanent portrayal of this information. Posters are often used in workplaces to explain information that the employer or company wants to convey in a more self-paced manner, either internally or as a mechanism to inform visitors to the office. Posters present a nice opportunity for integrating spatial information/maps in a larger format that provides opportunities to include additional detail on your maps along with clearly written text to explain and highlight the various maps or visualizations. Posters are also popular in educational and professional settings such as schools or community and governmental facilities, and of course, at academic and professional conferences. Like a pamphlet, a poster is somewhat limited in the detail it can provide, but it can also serve as a means to advertise additional resources available online or in a published report. Posters also pair well with pamphlets; a poster can spark a viewer's interest and lead him or her to take a pamphlet for later reference.

Oral or slide presentations

In some cases, you will have an opportunity to present your research and results directly to your audience. In this case, you gain the benefit of sharing your enthusiasm for the work and the opportunity to answer questions and offer additional insight to your audience. Oral presentations accompanied by a slide presentation have the added potential to leverage the technology. With your laptop on hand, you can potentially include animated maps, a fly-through of the data, or simply the opportunity to zoom in on an area of interest to help the audience understand the data more fully.

A risk of presentations goes back to a point we made at the beginning of this chapter: be clear and avoid cluttering your message with too much information. Live presentations can run the risk of dumping too much information on the audience in a limited time or overwhelming them with too many animations and fly-throughs and, in doing so, lose the message. As always, keep it simple. Also keep in mind that your presentation is ephemeral, so it can be helpful to provide the audience with some things to take home with them, for example, printed copies of key maps and results or perhaps a pamphlet or executive summary. Don't expect your audience to hear, see, and absorb everything you present in the few moments you have that map on the screen.

Preparing the final product

As you prepare your final visualization product, you want to keep in mind the various bits of advice that we've provided throughout this chapter. To summarize, here is of list of things to consider:

- goal of the project
- geography
- audience
- resources
- temporality

Goal

Make sure that your goal is clear to you and your project team and stated in a way that is understandable to others. When preparing products to share with others, this means going through the results of your study to highlight key data or information you most want to portray in the spatial visualization.

Geography

What degree of importance does geography play in your map? In some of the spatial visualizations that we've portrayed here, geographic features were present in the background to help establish a context, but they were not overbearing. Normally you need some geographic component to root your data, but the way that you present that context is really up to you and the goals of your project.

Audience

Always consider your audience. Think about the types of information your audience is accustomed to viewing, the level and complexity of the material, the type of information that you are presenting, and the use of various approaches including visuals, charts, statistics, and text. Does your audience have some familiarity with the geography you present? Is your audience literate or illiterate? What is the primary language of your audience? What are some cultural practices of your audience? For instance, are they accustomed to viewing information visually and discussing it that way? How much time does the audience have to spend on the material? Is a one-page summary or a lengthy report more appropriate?

Resources

What sorts of resources do you have to present your final product? If you have minimal resources, we advise creating something that is digital instead of hard copy, when feasible. Consider if your audience has access to the technology needed to obtain and interact with digital products. It can cost a lot to print your maps in color, especially if you need lots of copies or larger format output. If your project is in process, you may opt for digital portrayals that can be changed and modified over time, whereas costs may be justified to print your final products.

Temporality

Where are you in the process of producing final results? Is your map at the beginning phases of the project (exploratory), or is it at the end (perhaps more of a final report)? Are you creating data that will need to be presented in a permanent manner, or can they be more fluid? Typically in the early stages of a project, maps and data are under development and will be more fluid than toward the end when people are expecting a more final portrayal of the data and information.

Once you have answered these points, you should be ready to produce your final visualization. Again, you have choices about how to present information. You may choose to highlight geography, or you may choose to downplay geography and instead highlight the data.

Conclusion

As stated earlier in the chapter, GIS acts as an integrator for a wide array of different data and information. The strength of the GIS is in its ability to analyze and explore this information in a spatial context. Data can be tied to basemaps at any geographic scale and, in many cases, to analyses at multiple scales. For example, data analyzed at the local level in a countywide study could later be aggregated at the county level for the purpose of a regional or statewide analysis. Spatial patterns and distributions in data may help to explain relationships that would not otherwise be apparent.

In addition to the ability to integrate existing information, the GIS offers us a platform to create and store new information collected through any of the traditional methods as well as to synthesize new data through combinations and queries of data already in the system. In combination, GIS provides a powerful set of analytical tools.

Last, GIS offers a variety of options for both visual and numeric output. We have presented a variety of examples in this chapter. These may include final products, such as maps, tables, and graphs, as well as exported data that are used for further analysis in statistical software or discipline-specific modeling. These external packages can subsequently output their results back into ArcGIS as new datasets for use in further analysis and development of the mapped outputs required to communicate results. In the end, remember that your goal is to communicate the essential information clearly and concisely, not to wow the audience with amazing technology and data-overloaded maps depicted in forty-seven shades of chartreuse labeled with twelve different fonts, just because you can.

Review questions

1. What are the two main questions to consider for effective communication?
2. What role does culture play in terms of map visualization?
3. What is map context, and why is it important to consider?
4. Why is it important to specify your audience? Can the same map be used for different audiences? Please explain.
5. How does a researcher choose the best method of communication? What are the factors to consider?
6. What are the guidelines to consider in preparing the final product?

Additional readings and references

Craighead, F. L., and C. Convis. 2013. *Conservation Planning: Shaping the Future*. Redlands, CA: Esri Press.

Dangermond, J. 2007. "GIS Is Providing a New Medium for Understanding: Geography and GIS—Communicating Our World." *ArcNews*, Winter. http://www.esri.com/news/arcnews/winter0607articles/gis-is-providing.html.

Sandstrom, P., T. G. Pahlen, L. Edenius, H. Tommervik, O. Hagner, L. Hemberg, H. Olsson, et al. 2003. "Conflict Resolution by Participatory Management: Remove Sensing and GIS as Tools for Communicating Land Use Needs for Reindeer Herding in Northern Sweden." *AMBIO: A Journal of the Human Environment* 32: 557–67. http://www.bioone.org/doi/abs/10.1579/0044-7447-32.8.557.

Taylor, A., D. Gadsden, J. J. Kerski, and H. Warren. 2012. *Tribal GIS: Supporting Native American Decision Making*. Redlands, CA: Esri Press.

Thomas, C., B. Parr, and B. Hinthorne. 2012. *Measuring Up: The Business Case for GIS*. Vol. 2. Redlands, CA: Esri Press.

Wuerzer, T., and P. Mallow. 2012. "Map versus Visualization: An Evaluation of Visual Communication Tools for Planning." Paper presented at ACSP, Cincinnati, OH.

Relevant websites

- **National Map of the United States (http://nationalmap.usgs.gov):** This site provides nearly 200 small-scale datasets (national coverage) for the United States. These data are particularly useful when examining questions across large geographic areas. Local detail is not included in these data. Links to over 200,000 traditional USGS topographic triangles are also provided here as well as the National Map Viewer, which provides a web-based viewer for data without requiring download by the user.
- **National Aeronautics and Space Administration (http://www.nasa.gov):** Provides a variety of imagery and monitoring data for the world or portions thereof.
- **US Census Bureau (http://www.census.gov):** The official site for the US Census, demographic data, TIGER files, and other related data.
- **US Geological Survey (http://www.usgs.gov):** Provides a variety of national datasets, including aerial images, geographic places, maps, and topographic information.
- **US Environmental Protection Agency (http://www.epa.gov):** Provides a variety of national databases related to environmental quality.

- **Atlas of Canada (http://atlas.gc.ca/)**: An example of an interactive GIS website.
- **Natural Resources Canada (http://www.nrcan-rncan.gc.ca/inter/index_e.html)**: Provides a variety of national datasets, including aerial images, geographic places, maps, and topographic information.
- **Atlas of Sweden (http://www.sna.se/)**: An example of an interactive GIS website.
- **Centers for Disease Control and Prevention (http://www.cdc.gov)**: Provides an array of national databases related to public health, disease, births, and deaths.

Chapter 14

Linking results to policy and action

In this chapter, you will learn how to translate your GIS analysis and findings into action. Throughout this book, we have explored various ways to use spatially based research methods. Carrying out an analysis in GIS is just the first step in answering spatial questions and informing decision making with your results. As a GIS analyst implementing a study, the effort you make to ask appropriate questions, identify relevant data, employ analytical methods, and present results effectively may provide essential information to those whose role it is. Conversely, if you find yourself in the role of decision maker, you will benefit from understanding the underlying methods used in the project. This includes understanding what occurred to carry out the GIS analysis intended to inform your work and knowledge about the people and place. In this chapter, you will learn a series of spatial steps to follow to achieve solid spatially based policy.

Learning objectives

- Understand the different forms of policy
- Learn how spatial analysis impacts policy
- Explore how stakeholder diversity affects policy
- Evaluate the challenges to creating good policy
- Learn about the importance of a decision support system
- Learn the relationship between politics and policy
- Learn about the function of place-based policy

Key concepts

community-based component
decision support system (DSS)
geographic boundaries
place-based policy
policy
policy creation
spatially based policy
stakeholder diversity

GIS and visualizing policy

One of the great benefits of GIS is providing a visual perspective of your data. In this chapter, we explore some of the politics of research findings and how spatially examining data may reveal information that may not have been apparent without a spatial perspective. When such findings enter a policy arena, they have the potential to create great forward momentum or, perhaps just as likely, result in significant pushback or denial of the findings. For these reasons, it is essential that you develop a strong, empirically based research protocol when embarking on any analysis. Following approaches presented throughout this book, from the development of your research questions to communicating your results, the GIS will help you to transform data to information that is an effective decision-making tool in the larger context of your organization, agency, or jurisdiction.

What is policy?

In simple terms, a **policy** is a plan for action. A policy can be directive, such as the restriction on parking from eight to eleven o'clock on Tuesday mornings for street sweeping (figure 14.1). This is a simple spatial-temporal policy according to which you are free to use the space and the street for parking at all times other than the three hours when street cleaning takes place each week. This simple example highlights how the same space can be used differently at various times.

Figure 14.1 **An example of communicating spatial-temporal policy via a no parking sign used by the New York City Department of Transportation.** Courtesy of the New York City Department of Transportation.

Typically, policy is developed after the decision maker has considered the available alternatives or options and then makes a decision that will address the condition in the desired manner. Often governmental bodies develop policies as a means to achieve particular goals or conditions in the area under their jurisdiction. In the case of most major cities, one such condition is to maintain clean streets, and this is accomplished by setting policies regarding when and where one may park so as not to interfere with the street sweeping schedule. At first glance, parking rules may not seem a policy decision that require GIS, but when you consider the complexity of the task of cleaning all of the streets in a major city on a consistent schedule that optimizes the use of costly staff and resources, the value of a GIS should be apparent.

Spatially based policy

A policy establishes a set of standards or expectations for response. Bringing a GIS analysis to bear on evaluating existing policies or developing new ones can provide excellent opportunities to evaluate options and solutions. We call this **spatially based policy**—when the plans and actions you develop result from analysis of geospatial data. The reason for creating policies is that they help a situation to run more smoothly by establishing standard protocols that follow a well-thought-out plan of action under a given set of circumstances. Without policies in place, each time a similar situation occurs or a problem arises, it would require a new assessment of the circumstances and a decision to be made. Policies improve efficiency, and the basic function of an organization or society, by removing the need for an independent assessment each time a decision must be made. Now that we've established what the term *policy* actually means, let's explore the connection between GIS, or spatial thinking, and policies.

To develop effective policy, policy makers must follow a well-considered decision-making process. For example, developing a policy may require considering a set of case studies of different potential scenarios and alternative responses or actions that might follow. Of course, the GIS provides an ideal tool for assessing a wide variety of situations and scenarios in a structured environment. This makes GIS a valuable tool for policy makers, not only in government, but across a wide spectrum of uses in the private and nonprofit sectors as well. GIS has and will continue to play a major role in the development of policy because of its unique ability to spatially integrate data from multiple sources.

How does someone create policy?

The first step to creating good policy is to have a strong grasp of the situation, or in context of conducting research, defining the problem or question you will address. You need to have an understanding of the various pieces and elements and how they fit together before you can do this. Should the questions and information fall outside your own area of expertise, you may need to draw on the experience of colleagues or other professionals. Does the GIS analyst for New York City understand the intricacies of street cleaning? More likely such an analysis required input from someone in the maintenance department who could provide insights on how many miles of street can be cleaned in a given time, and the uncertainties in covering that area to ensure the streets are cleaned while also minimizing the overall time a given block is unavailable for parking.

To begin addressing any policy question, one should always begin with relevant data before you take action. Sometimes the required data may be readily available, and in other cases, you may need to collect or acquire new data. In addition to providing the analytical framework, ArcGIS provides the tools to organize and document the data and its characteristics prior to developing scenarios to inform policy. In previous chapters we have discussed how to approach finding relevant secondary data—the local city or county website or the US Census is a great place to begin your search for relevant spatial data.

Challenges to creating good policy

There are many challenges that policy makers face. The number one challenge to making good policy is being able to make a complete decision that satisfies multiple users and perhaps that crosses multiple geographies. Sometimes a single decision may be appropriate for an entire jurisdiction such as a state or county. In other situations local differences may demand that decisions be customized to address these variations across the geography of a region for which a policy maker is responsible. As we have noted, ArcGIS can help with both of these issues because it allows for the incorporation of multiple types of data and for information that goes across multiple geographies.

Limited time to make decisions

Policy makers have many demands placed on their time and face time constraints when a decision is needed, particularly during times of emergency or crisis. When decisions must be made under pressure, and especially when a lack of prior planning was undertaken, there is a greater risk of making a bad or uninformed decision. Most decision makers should want to consider the various data or information on the topic before making their final decisions—but not all do. Often decision makers will be more influenced by who is proposing an idea versus the actual merits of the idea. As researchers, we can't do anything about the politics that surround decision making. However, we can offer suggestions for how to connect policy decisions to spatial data. As we have discussed throughout the book, spatial data resonate with a wider audience of people and make your results more accessible and easier to comprehend.

Working with decision makers

A linchpin of effective policy creation involves gaining a solid understanding of a situation, based on reliable data. ArcGIS can contribute to that goal as a means of analyzing, assessing, and collecting holistic data or information before a course of action is proposed. Leaders, managers, and community members who use ArcGIS to illustrate data and information often face similar challenges such as lack of time, limited knowledge of the subject, and competition from multiple sources. In this section, we explore some of these challenges and discuss strategies for how they can be successfully met.

Limited knowledge

There are times when the decision makers are limited by their lack of knowledge about the topic. Sometimes they must make a decision quickly, so having the ability to visualize the big picture can be an advantage. As always, the obstacle in this case is being able to make thoughtful decisions based on the best available information.

Politics of spatial data

In addition to facing time constraints, policy makers also encounter politics. Because spatial data portrayal brings such clear and present information in a visual format, it can sometimes elicit stronger reactions (or pushback) than you might expect. This can be both good and bad for the policy maker, because a GIS map can create momentum in support of a potential action or, conversely, generate strong feelings in opposition.

Anytime geospatial information is involved with the creation of policy, there is the opportunity for pushback on the results. For example, a number of years ago, we were working on a project in a small town. The research organization with which we were working was asked by some local city officials to provide a map of poverty levels for the region. Taking data from the 2000 Census, we developed maps showing levels of poverty in the community by census tract. The resulting maps were shared at a local city hall meeting where the policy question on the table was whether fluoride should be included in the local water system.

The topic of fluoridation had generated extensive, and sometimes heated, debate. Opposition to government intervention through fluoridation was strong and consistently challenged those who supported the case for improved dental health. At the meeting, as the data were being presented by the pro-fluoride constituency, a challenge came from the audience. The sentiment was strongly stated: "There are no poor people living in this town—they all live in the next town over." The perception put forward was essentially that everyone could afford adequate dental care and fluoride was not necessary. We realized at this point that the poverty maps that we had created were challenging commonly held, albeit erroneous, beliefs about the town (namely, that there were no poor people).

The counterargument in favor of fluoridation was that a significant proportion of the community was not able to afford to go the dentist to get fluoride treatments. In the end, we were able to stand behind our demographic poverty maps because we had a very well-documented methods section and could explain that the information used to derive the maps had been collected by the US Census—a reputable governmental organization, not simply opinions. The experience was interesting because it highlighted a common norm that people will often develop opinions based on their own experience and knowledge, even if it is incorrect. When those beliefs are held strongly enough, people may even find it difficult to accept actual data presented from another source that doesn't match those beliefs.

As researchers, we always have to make sure that we had a well-thought-out description of your data sources and research methods. Having the ability to stand up and clearly explain the information behind the creation of your GIS maps can go a long way to maintaining your credibility.

GIS in governmental decision making

As alluded to throughout this book, there are countless examples of GIS data and analysis being used in a variety of governmental efforts around the globe. Base layers ranging from infrastructure, to census of populations, tracking of disease, natural resources, and economic activity are commonplace in many countries and at all levels of governmental decision making. National and international standards for data and metadata help to ensure data are of known and documented quality, effectively serving a variety of uses within and between agencies.

In the United States, for example, there have been a number of efforts to enhance the quality and use of spatial information in support of policy. A 2009 report by the Congressional Research Service titled "Geospatial Information and Geographic Information Systems: Current Issues and Future Challenges" (Folger 2009) provided an accessible summary of the connection between GIS and the practice of policy making. The report also highlights examples of how efforts by the Federal Geographic Data Committee and National Geospatial Advisory Council have helped to provide support for effective use of geospatial data and analysis at the federal level and in support of a developing National Spatial Data Infrastructure.

In the report, Folger notes that

> the analytical power resulting from combining geospatial information with GIS more typically underscores its value to policy makers at all levels. GIS often provides for unique analyses of disparate types of information—linked by their spatial coordinates—to help resolve policy questions. For policy makers, this type of analysis can greatly assist in clarifying complex problems that may involve local, state and federal government, and may affect businesses, residential areas and federal installations. (Folger 2009, 7)

Folger's perspective is made in context of the keen ability of GIS to bring together different layers of information and to seamlessly integrate them in responding to wildfires, an issue affecting government agencies, citizens, and resources at multiple levels ranging from local to federal. However, the underlying sentiment of this statement can be just as easily applied to any number of policy issues. Bringing together data from multiple sources and perspectives in GIS provides the ability to draw connections between what is happening at a variety of levels. The report also draws on examples relating to other natural, infrastructure, and economic perspectives.

A fire example

In recent years, wildfires have become a significant concern in many western states. As development extends into areas adjacent to fire-prone vegetation and regions that are more difficult to access because of rough terrain and limited road networks, the risk of losses of property and life has increased. Coupled with hot, dry summers, these areas can be particularly vulnerable.

Sylmar fire

In the Congressional Research Office report, Folger (2009) mentions the 2008 Sylmar fire in California as an example of how GIS provided valuable decision-making capabilities. The Sylmar fire was a fast-moving fire that gained notoriety because of its close proximity to housing developments in southern California. More than six hundred structures were destroyed and thousands of people were evacuated. Providing information about the fire's mobility was key, and emergency personnel realized how important it was to share this information in a quick and easy manner. GIS factored into the scenario because it provided a platform of quick communication for various types of data. The report highlights the important role that GIS played in helping to clarify what action should happen when:

> To assist in real-time decision making, the fire's progress was posted on the Internet in near and real-time by several organizations. . . . The Sylmar fire example underscores the information power available when geospatial information is combined with tools for displaying the information, such as GIS and the Internet. In this instance, timeliness—the ability to post the geospatial information quickly, enhance the value to the data users, citizens trying to avoid the path of the fire (Folger 2009, 7).

The Sylmar fire is an excellent example of how, in an emergency situation, GIS aided effective and timely decision making by local leaders, law enforcement, and fire officials to route people in the safest way possible away from the fires. Additionally, the GIS-based data facilitated communication with the general public through using the Internet. This is an example of how the GIS could be used in conjunction with other forms of data to create a clear vision and path for immediate policy decisions. There are other times when a policy maker has more time to make a decision and to review a variety of different data.

Longitudinal fire data example

California is a state that has been plagued by many wildfires over the years. Table 14.1 presents data on the ten worst wildfires in US history based on financial loss. Of these, seven have occurred in California. Such wildfires are extremely costly and dangerous, not simply due to the loss of life and property but also the costs relating to fighting them while minimizing risks to the firefighters. The fire that tops the list in terms of loss is the Oakland Fire Storm from October 1991 (table 14.1), which resulted in a loss of $2.57 billion (in 2013 dollars).

Table 14.1 Ten worst wildfires in US history by financial loss

Fire name	Year	Month	Acres burned	Homes destroyed	Property loss (2013 dollars)
Oakland Fire Storm (CA)	1991	October	1,500	3,000	2.57 billion
Southern California Wildfires (CA)	2007	October	500,000	1,500	2.03 billion
The Cedar Fire (CA)	2003	October	280,278	2,000+	1.4 billion
Cerro Grande Fire (NM)	2000	May	48,000	400	1.36 billion
The Old Fire (CA)	2003	October	91,281	993	1.24 billion
Southern California Wildfires (CA)	2008	November	43,202	900	868 million
Laguna Beach Fire (CA)	1993	October	16,000	366	854 million
The Florida Wildfires (FL)	1998	May-June	500,000	150	566 million
Cloquet-Moose Lake Fire (MN)	1918	October	250,000	4,089	541 million
Painted Cave Fire (CA)	1990	June	5,000	550	424 million

Source: Reproduced from NFPA's website. © 2013 National Fire Protection Association. Largest Loss Wildland Fires (as of October, 2013). NFPA's Fire Incident Data Organization database and NFPA archive files.

When we visualize the amount of property loss and loss in dollars we can see the emergence of a pattern. For instance, table 14.1 illustrates a temporal pattern that shows that six of the ten fires occurred in the month of October, and the spatial pattern that shows that 40 percent, or four of the ten, of the fires occurred in southern California.

Policy makers might want to look at a number of issues based on the history of these various fires. For example, where and under what conditions should development of new homes be permitted? This might be limited by factors including the type of terrain and vegetation as well as building materials and landscaping. Too often homes are built somewhere because the land is cheap or because the developers may have some special connection or influence on local politicians. The issue of safety and the ability to defend against forest fires may not always be considered as part of this equation. In California, extensive use of GIS mapping of these and other factors are used to map fires, prescribed burns (used to reduce fuels), vegetation types, and a variety of infrastructure and resources to maintain and understand fire history and develop more effective policies to minimize risks to life and property. In addition to downloadable GIS data, an online interactive map entitled the Priority Landscape Mapper (figure 14.2) provides the public with a view of a variety of fire planning information.

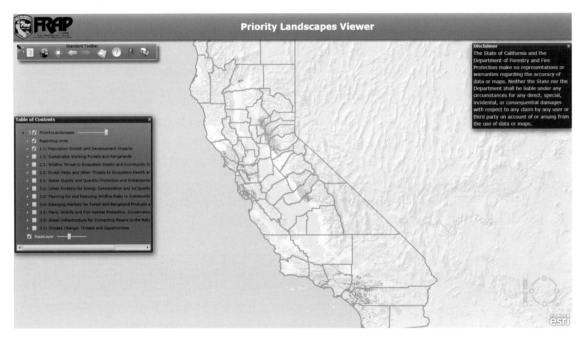

Figure 14.2   The Priority Landscape Mapper provides interactive map overlays for spatial analysis results from the California Forest and Range Assessment Program (FRAP). The site highlights priority areas overlain on map or satellite images. The site also provides access to input layers used in risk analyses, including assets and threats to them. http://frap.fire.ca.gov/assessment/assessment2010/mapper.html. Map composed by Steven J. Steinberg using publicly available California FRAP data. Basemap data from Esri, HERE, DeLorme, TomTom, Intermap, increment P Corp., GEBCO, USGS, FAO, NPS, NRCAN, GeoBase, IGN, Kadaster NL, Ordnance Survey, Esri Japan, METI, Esri China (Hong Kong), swisstopo, MapmyIndia, © OpenStreetMap contributors, GIS User Comm.

Although it is impossible for policy makers to know precisely where and when fires will occur, working with scientists and modelers, there are opportunities to develop an understanding of risk. Looking at the fire history for a region, along with associated data regarding topography, vegetation, climate, and other factors, emergency planners are able to develop risk maps such as that shown in figure 14.2. Of course, this does not mean cities at risk can simply pick up and relocate, but such maps can help to prioritize where resources are allocated.

For example, policy makers and emergency planners can use this sort of information when deciding how to allocate resources and to plan for prevention and response to fires. Priorities for educating the public living in high-risk areas about fire safety such as clearing of fuels around their homes, and planning for new development can be adjusted to minimize risks to the degree possible. In figure 14.3, we see a map of risk levels overlain by incorporated communities and several very large (in acres burned) historic fires for southern California. This information along with additional data is useful in planning for projects relating to fuels reduction and fire breaks, positioning of resources and fire stations, as well as planning for access to neighborhoods, ensuring safe evacuation routes for residents.

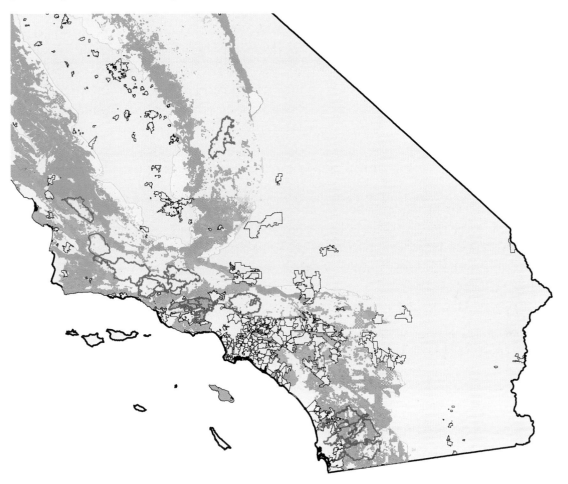

Figure 14.3   Map showing thirteen southern California fires of one hundred thousand acres or more (outlined in red) between 1922 and 2009. Incorporated cities are indicated by black polygons. Fire hazard ratings, derived from multiple GIS criteria, including vegetation type and amount, are indicated by the colored zones. Ratings included are moderate (yellow), high (pink), and very high (orange). In a map view, it becomes apparent that a number of very large fires have occurred near highly developed regions of San Diego, Los Angeles, and Ventura Counties, often in very high risk zones. Many incorporated cities are also clearly adjacent to or surrounded by very high risk zones.
Web map created by Steven Steinberg. Data from California Department of Forestry and Fire Protection FRAP website.

Coordinating data

There are times when decision making relies on data that come from different organizations or agencies. ArcGIS provides an effective platform for different groups to coordinate their efforts and information. Data can be shared between agencies and organizations in a variety of different ways, including data download or sharing of physical media, or using a service-based architecture providing data to the desktop using web services. Sharing of data with the public may also be done via web-based viewers, which typically provide the important information in a simpler, more understandable format designed for a general audience as opposed to GIS experts.

Of course, developing the ability to make an informed policy-based decision necessitates being able to locate and link various types of data. For instance, if you are going to present data on a fire, you would want to have data on major streets and roadways, fire station locations, terrain, property values, and escape routes, among other items. To facilitate this process, many agencies and even entire states or regions have begun producing data portals that make locating and accessing data much easier, and more centralized. Rather than contacting each agency independently to access current data, geoportals provide "one-stop shopping" for data. A wide variety of geoportals providing local, state, and national data around the world are being made available. Links to a variety of examples of live geoportal sites are maintained by Esri (http://www.esri.com/software/arcgis/geoportal/live-user-sites). When these data are provided as web services, the most up-to-date versions of the data are directly accessible to the end user simply by connecting to a data service in ArcGIS.

Decision support systems

A decision support system (DSS) is a tool that is used to assist decision makers by capturing key data and analysis capabilities into a user-friendly system. Typically, a DSS will capture the knowledge of domain experts in an analytical model that includes relevant base data as well as situational details relating to the specific decision at hand. A simple example of a DSS might be the Global Positioning System receiver in your car, which captures base information on roads and addresses along with situation-specific information such as your current location, and possibly even current traffic information relating to accidents, detours, or construction. The route recommendations take into account these inputs to give you options that are likely superior to what you could arrive at without the benefit of this additional historic and real-time data.

An example of a DSS for predicting wildfire behavior is shown in figure 14.4. This system provides planners a DSS relating to the probability of a fire's spread. Such a system can be useful when responding to rapidly changing conditions and better allocate fire suppression resources and risks to life and property. While such predictions can be done without a DSS, by capturing expert knowledge on fire behavior in a DSS, fire managers on the ground can quickly and easily generate probability maps with the most current information shared across multiple agencies and jurisdictions. The same DSS can also be useful for advance planning activities by exploring the likelihood of spread into particular areas should a fire ignite near a particular community or other area of interest.

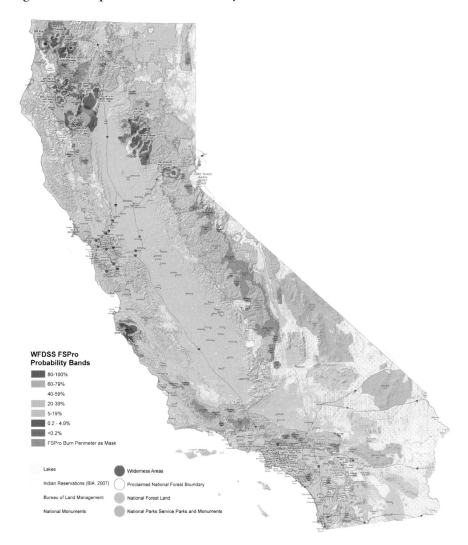

Figure 14.4 An example output from a fire-spread decision support system (DSS) developed by the US Forest Service. The DSS uses a probabilistic model named FSPro to predict the likelihood of a fire's spread from the initial point of ignition or current fire perimeter maps. Building a simulation model. Courtesy of Don Yasuda, US Department of Agriculture Forest Service Fire and Aviation Management, Sacramento (FAMSAC), from *Esri Map Book*, vol. 27 (Redlands, CA: Esri Press, 2013), 39. Wildland Fire Decision Support System: July 2008 Fire Spread Probability Analysis. Data from US Department of Agriculture, Forest Service, FAMSAC.

From maps to action

Policy creation is about deciding on an action based on information that you have. In this section, we focus on examples that explain how policies can emerge from spatially based analysis. It is important to remember that creative policy making also involves being able to harness what you learn through the entire data collection and analysis process.

Given our background and history using the spatial perspective, we were asked to conduct a research project to explore the connection between pesticide drift and various agricultural communities (Steinberg and Steinberg 2008a, 2008b). The project focused on two rural California counties (Monterey and Tulare) where agriculture is the primary industry. Through the project, we sought to examine the environmental health issue of pesticide drift (figure 14.5). We undertook a mixed methods approach to the research, including key-informant interviews with community leaders, farm workers, and others familiar with the issue; fieldwork; mapping via public participation GIS (PPGIS); and analysis of existing, secondary data. New data were developed into spatial data layers and integrated with existing geospatial data using ArcGIS 9.2.

Figure 14.5 The agricultural counties under study are surrounded (and in some cases infiltrated) by agricultural fields, which closely abut neighborhoods. In this photo, you can see how the farm workers wear clothing that fully covers their bodies, with many also wearing bandanas over their faces. Photo by Steven Steinberg.

Existing spatial data included county land parcels, school locations, city boundaries and zoning, US Census data, and pesticide use data from the State Department of Pesticide Regulation. In the project we strove to incorporate empirical data along with community-based knowledge. Fieldwork conducted as a component of our research included PPGIS with various groups of farm workers and community leaders to verify some of the information contained in geospatial data we obtained from government sources.

For the PPGIS sessions, we developed map layouts overlaying streets and parcels on National Aerial Photography Program color images of the region. Through the PPGIS process, we asked community members to indicate local landmarks and gathering places such as parks and schools on the maps. Participants were also asked to indicate locations of pesticide spills, exposure, accidents, and general information regarding the types of crops grown.

The project was unique because it included both quantitative and qualitative primary data in combination with a variety of secondary data sources (Steinberg and Steinberg 2008a, 2008b). We used this multiple methods approach to gather a broad view of the environmental, health, and social contexts faced by these agricultural communities. The final products included a pesticide atlas of the different types of pesticides that had been sprayed in the two different counties along with a final report (see references at the end of this chapter for links). Additional media exposure came from a Spanish radio interview with Radio Bilingue, which covered the state of California and various newspaper interviews and stories.

One of the main issues for the local people in these agricultural communities was concern about the health and safety of their children. Figure 14.6 highlights the danger associated with the pesticide spraying as evidenced by the universal danger sign of skull and crossbones and the warning posted in both Spanish and English. Anecdotal reports from community members indicated that some children were getting sick after going to the school or playing in their family's yard or at a local park. Many neighborhoods and schools in these communities are located immediately adjacent to agricultural fields that were aerially sprayed with pesticides, in many instances, separated by only a chain-link fence as portrayed in figure 14.7.

As researchers, we realized that in addition to our review of existing pesticide data records, we would need to also incorporate a community-based component to our work. A **community-based component**, an actively local people in a place, should be involved in the research process. Through community engagement, we learned that in several of the communities, pesticide spraying occurred at different times and sometimes in the middle of the night. We wondered why this might be the local practice and realized that some may have thought that nighttime spraying was advantageous because there were fewer people outside at that time, and spraying would be less apparent to the public

Figure 14.6   When fields are sprayed with pesticides, regulations require they be posted with signs such as this. However, these signs were often placed only at intervals along the field boundary and do not account for pesticide drift into adjacent areas. In our fieldwork, we found such signs were placed as much as a half-mile or more apart along a recently sprayed field boundary, potentially going unseen by an individual entering a field at a location between signs. Photo by Steven Steinberg.

Figure 14.7   This photo shows an example of an elementary school playground immediately adjacent to fields of grapes. No barrier to drift is present, potentially allowing pesticides to drift onto surfaces of playground equipment and the schoolyard. This particular school was surrounded on all four sides by fields, with only a chain-link fence separating it from the fields. Photo by Steven Steinberg.

and perhaps to visitors to these regions. However, we went back and reviewed weather patterns and wind data for the typical growing season and found a pattern that suggested nighttime as the time when winds are the strongest, potentially creating maximum pesticide drift.

Using a community engagement process we learned that various concerned parents had tried to raise the concern about pesticide use to the local school board but never properly got on the agenda. When farm worker parents tried to raise the issue or concern about children's health in other arenas, they were discounted as being uninformed, uneducated, or troublemakers.

A proposal emerged from our discussions with the community members: they wanted to create some sort of a limited "no-spray" buffer zone around the schools (Steinberg and Steinberg 2008a, 2008b). A buffer zone is a region of some specified distance around school grounds with limitations on how and when particular pesticides may be applied. A buffer zone was proposed as a possible policy solution that could protect the health of the children while minimizing the impact to the agricultural community, though there was a concern about the amount of acreage that would be affected. Of course, using the buffer tools in ArcGIS made answering this question trivial, and as it turned out, also trivial was the amount of farmland acreage that would be affected by the proposed buffer zone.

Therefore, we quantified the amount of land/acres that would need to be set aside under various buffer zone options. If you are unable to quantify something, the idea could be shot down as being too expensive or intrusive. Because of this, we created some maps that documented the number of acres that would be required to establish pesticide buffer zones of differing sizes and types, including buffers around schools only as well as another analysis that also considered buffers around all residential areas.

Ultimately, the concept of buffer zones gained traction in some communities but not in others. A positive policy impact of our research project is that it provided actual data and numbers as a starting point for some of the conversations regarding solutions. Community members were able to communicate their concerns and some potential solutions to local government officials and eventually establishing an advance notification system to alert the community before aerial spraying was to occur near schools. Additionally, a five hundred foot pesticide buffer zone around the schools was established in some communities. This case is an example of policy that evolved through the effective use of GIS to capture the data and stories of a community in a way that could be successfully brought to bear in the policy realm and lead to real change. To generate positive impacts the idea had to (1) use scientific data to illustrate the problem; (2) use spatial data to illustrate the geography of the people in the context of the agricultural fields; and finally, (3) use social data from the community to highlight their issues and concerns. The ability to successfully communicate data in a spatial format to the right people is what gained the community traction and led to the creation of workable policy solutions that could be accepted by everyone.

How to create good place-based policy

There are some basic things to consider when trying to develop policy that actively considers the factors of space and place. In this section we present some overarching suggestions and ideas for effective place-based policy creation. **Place-based policy** means decisions or actions that are created in sync with a particular geography or place—it is specific to a particular geography.

Use multiple methods

The underlying central component to creating successful policy begins with conducting the research project in a way that facilitates a view from multiple angles. When you do that it provides a more comprehensive picture of the situation and potential alternatives. This will help to develop better options than simply viewing the issue based on one type of data. We always try to use multiple methods whenever we conduct a research project, no matter what the topic. Employing multiple methods creates opportunities to identify key elements or influencing factors central to the issue, and provides a more holistic perspective of the issue at hand. A more diverse methods approach could also uncover elements that would otherwise be missed, providing new information about the topic.

Actively consider data

It is imperative that whenever a policy-making group gets together to make important decisions, they actually consider data. It is surprising how often decisions get made based on mere impression and anecdotal evidence. A group should not get together merely to rubber stamp a predetermined choice often based on perceptions rather than real data. The maps and data incorporated into a GIS analysis provide an opportunity to truly engage decision makers and stakeholders in a meaningful discussion. As a GIS analyst, you can help them to remain open and objective enough to actually consider the information and evidence pertaining to the decision at hand. The biggest challenge is to make sure that your policy makers first understand the data they are viewing and, second view the data before they have made up their minds about a decision. In most cases, policy makers really do want to make the best decision possible and don't want to make the mistake of glossing over the facts, views, and perspectives of multiple stakeholders and multiple data sources. But policy makers are human and sometimes take the expedient path, making decisions based on what their friends and colleagues tell them versus what the data say about the reality of a particular situation or problem. As a policy maker, if you adopt the view of people whom you already know, you can contribute to producing staid and ineffective policies. To be effective, the best policy should be reflective of reality.

Interdisciplinary approaches

Often, ideas come from unusual and perhaps unexpected sources. For instance, when you bring together people to interpret and understand the data from different backgrounds, they may naturally generate synergistic thoughts or suggestions. Views may emerge that had previously not been considered and are, in fact, innovative. The best solutions to situations often arise when you bring together people with different experiences and views to the topics and questions under discussion. In essence, when you integrate the people from different career, cultural, and disciplinary backgrounds, you generate an interdisciplinary environment.

Stakeholder diversity

At the beginning of any policy-making process, it is best to develop an understanding of your stakeholders, their cultures, orientations, responses, practices, and reactions—we refer to this as **stakeholder diversity**. While this may involve a little more leg work at the front end, it is definitely worth it. As you identify and work with various stakeholder groups, it will become clear what matters to them, what they really value, what technologies and solutions will potentially work for them, and which of those technologies and solutions can be rolled into proposed policy developments. For instance, for a group that values communal ties and social interaction over individual achievement, you would need to approach the creation of policy in a more consensus-based versus individualistic, manner. By the same token, you may have a group that values natural spaces and wants to set some of those aside as part of the development process instead of using every inch of land for houses or buildings. A good starting point for any policy maker is to have a good grasp of the various cultures that will be affected by the decisions that you make. Identifying these different groups and getting to know what they value will help to design a more fitting policy.

Geographic boundaries

In creating spatially based policy, the issue of boundaries and where they exist emerges. Therefore, it is crucial to identify, and delineate, the **geographic boundaries** or different geographies that will be affected by your policies. In any community or location there are going to be people who already have established thought patterns regarding ownership of resources and property that do not necessarily match the formal or designated boundaries on a map. In such cases it is important to consider the origin of the existing map and who made it to understand why the boundaries were drawn that way. So as a researcher/policy maker, you want to be able to gauge *all* data and information in light of diverging views on space and place. We raise this topic because if as a researcher you fail to consider the perceived boundaries of place, it can pose a threat to your policy. For example, consider the creation of many of the countries in the Middle East following World War I. Such boundaries were determined by European entities, groups not from the region, who did not understand perhaps the tribal and nomadic nature of the groups who lived in that space. As a result, more Westernized views of space and place were imposed on those groups and societies who did not view the world in terms of national boundaries. A similar situation happened in the United States when the American government sought to establish reservations for Native Americans, often mixing tribes of the people who did not get along and who used different types of resources to sustain their lives. The result was bad policy because it ended up creating problems through the wrong mixes of cultures and people instead of producing effective solutions.

Natural resources, place, and use

For some stakeholder groups, the notion and importance of natural resources will vary based on the particular value system of that particular group. Are there natural resources of value there? For instance, if your group uses oil as a natural resource, is there a lot of oil there? The structure of a society is going to drive the pursuit of resources and the level of conflict and competition over these resources. For instance, as Americans, we will go to great lengths to pursue the discovery and extraction of oil in our own country as well as in other countries. We can translate this to the observation of animal species and the notion of the home range.

In nature, wildlife species will develop their own home ranges based on access to food and shelter resources they need to survive—humans also establish home ranges. These home ranges may include the workplace (to earn money), grocery store (to gather food and resources), specialty coffee house (to get energy), gym (to get in shape to be more competitive for resources), and mall (to clothe themselves). Major controversies about place are often rooted in people's perceived access and right to resources, just like for animals.

For some natural resources use determines ownership, but this is not that common. For example, consider the use of water resources in the American West: those are based on patterns of historic water use. Water rights are rooted in historic use—the first to use the water has first rights, but failure to maintain use over time can lead to a loss of access. In some societies and cultures, it was (and still is) not unusual to engage in common property use of resources, where the resources are shared. However, this is a foreign concept to other cultures and economic systems where private property rights are more typical and policy makers seek to clearly define boundaries around resources and access rights to such resources.

Temporality and control

As you design your policy, one major consideration is to determine the time period that your policy will cover. Is it a short-term, temporary policy or a longer-term one? Temporality is an important consideration. Additionally, you want to consider who or what governing body will be in charge of your policy. Does that group already exist, or does it need to be created, and how will it maintain the resources necessary to implement and enforce the policy? Is the policy open ended? Does it need to be reviewed in the future, or is this a one-point-in-time policy?

Foster transparency

When a group develops a policy, they want to be both truthful and transparent in their actions. Failure to do so will result in conflict and suspicion about hidden agendas. The notion of being open and visible through the policy creation process is key. A policy-making body wants to demonstrate that they are following accepted protocols and procedures in their own process rather than being arbitrary, blown by winds of politics, the strongest lobbyist, or people's opinions. Policy makers will quickly face controversy when they are perceived as being biased or in the pocket of some special interest group.

Final thoughts

In this chapter, we have presented several examples that illustrate effective spatially based policy making. While there is no one-size-fits-all recipe for success that will work every time, our goal here is to provide you with a variety of ideas about presenting your research data which can help in crafting policy with successful results. Policy makers who consider space and place will be more likely to generate policy that is effective in the long run because it has a better opportunity to consider the concerns, needs, and perspectives that tie to data, people, and place.

Review questions

1. What is policy?
2. What are some challenges to creating good policy? Please explain and provide examples.
3. What is spatially based policy? How do you create it?
4. What is the value of using multiple research methods to gather data?
5. Why is stakeholder diversity important? As a researcher or decision maker, how do you ensure stakeholder diversity?
6. What is a decision support system (DSS)? Under what conditions would someone want to design a DSS?
7. What is place-based policy? Why would someone want to develop place-based policy?

Additional readings and references

Folger, P. 2009. *Geospatial Information and Geographic Information Systems (GIS): Current Issues and Future Challenges*. Congressional Research Service Report R40625. http://www.crs.gov.

Fortmann, L., ed. 2008. *Participatory Research in Conservation and Rural Livelihoods*. Hoboken, NJ: Wiley Blackwell.

Hacker, K. 2013. *Community Based Participatory Research*. Thousand Oaks, CA: Sage.

Steinberg, S. J., and S. L. Steinberg. 2008a. "People, Place and Health: A Pesticide Atlas of Monterey County and Tulare County, California." http://hdl.handle.net/2148/429.

———. 2008b. "People, Place, and Health: A Sociospatial Perspective of Agricultural Workers and Their Environment." http://hdl.handle.net/2148/428.

Steinberg, S. L., S. J. Steinberg, J. L. Kauffman, and J. R. Eckert. 2008. "Public Participation GIS Research and Agricultural Farmworkers in California." Paper presented at the twenty-eighth annual Esri International User Conference, San Diego, CA.

Relevant websites

- **National Spatial Data Infrastructure (NSDI) (http://www.fgdc.gov/nsdi/nsdi.html):** This website presents information about the Federal Geographic Data Committee, such as the strategic plan and Executive Order 12906 that established the NSDI.

- **National Geospatial Advisory Committee (NGAC) (http://www.fgdc.gov/ngac):** This website presents information on the National Geospatial Committee, including its charter and bylaws. The website hosts key documents such as the NGAC Geospatial Strategy Paper.
- **Federal Geographic Data Committee (FGDC) (http://www.fgdc.gov):** This website presents information specific to the Federal Geographic Data Committee, including the National Geospatial Data Asset Management Plan and the FGDC Annual Report.
- **US Census Bureau (http://www.census.gov/geo/maps-data/):** This website presents geographic information, including maps and data that can be used by the general public. Specifically, it presents the Census Data Mapper, a simple application to highlight demographic information, reference maps, and thematic maps. The website also links to statistical data tools.

Chapter 15

Future directions for geospatial use

In this chapter, you will learn directions, trends, and emerging applications of spatial technology. Although it can be difficult to accurately predict the future, particularly as it may evolve around computing technology, a number of trends are already evolving that incorporate spatial information in new ways. Not so many years ago, Global Positioning System (GPS)–enabled smartphones were uncommon; now, they are prevalent. We are now seeing the emergence of wearable computers in the form of glasses and "smart clothing," which not only have location intelligence but also may include additional sensors to monitor environmental or individual conditions such as the heart rate of the wearer. The emergence of the Internet of Things has also begun, a world in which almost everything is in some way "connected."

Learning objectives

- Explore future trends regarding GIS use
- Understand the difference between data and images
- Explore how GIS can integrate data from social media
- Illustrate that GIS is an art form
- Illustrate various spatial GIS topics for students

Key concepts

bad data
change technology
geofencing
geospatial agility
geospatial crowdsourcing
infographics
Internet of Things
visualization

Imagine the future

The world of GIS is continually evolving, having moved over the past several decades from backroom workstations to the desktop to the Internet, cloud, and mobile device. Analytical capabilities that just a few decades ago required thousands of dollars of hardware and specially trained GIS analysts working with command line interfaces are now readily available to staff in the field in real time with the swipe of a finger or a spoken command. These, and no doubt other emerging and still unimaginable technologies, will continue to be integrated to provide new opportunities for the use and application of spatial analysis, thinking, and understanding.

Although such technologies can and will present a host of ethical and privacy issues, it is not our intention to contemplate or resolve them here. Rather, we present some possibilities for how these and other emerging technologies might provide new and innovative opportunities for spatialized data collection and analysis, adding to the toolbox of any researcher, scientist, or practitioner. Every year, GIS continues to expand as a research tool, being used in new and innovative applications in more disciplines. In recent years, as digital mapping has become commonplace both online and via mobile devices such as GPS-enabled smartphones, tablets, cameras, and vehicles, many of the concepts discussed throughout this book are available at some level, to almost everybody—most often as an application built around finding destinations such as a store, train station, or other location of interest.

These same applications are already being linked to a plethora of additional data, both quantitative and qualitative in nature. For example, at this moment, we could ask a smartphone to show "Italian restaurants near me," and within a few seconds have a map of options that include ratings, user reviews, a phone number to call for a reservation, directions to get there (as well as traffic information), business hours, and links to the restaurant web pages, which include menus, prices, and images. Although this information is valuable to the hungry author or traveler in an unfamiliar place, it also provides a rich, readily accessible, up-to-date, digital resource to a researcher seeking spatial data. Although not all of these data are directly linked to our traditional desktop copy of ArcGIS, the data are not difficult to capture and integrate into a geodatabase for inclusion in an analysis. The point here is that you can find geographic information on pretty much any business using the web, but the ability to analyze such data can only come through using GIS. Combined with other data sources, such as those discussed throughout this text, there is significant opportunity to leverage these data for project planning, secondary data collection (e.g., a virtual site visit), or primary data collection by providing study participants a mobile application that collects the desired data at specified times or locations throughout the day.

Because so many people openly post spatial information on social networks, another emerging opportunity is to capture some of these data as they relate to particular behaviors or events. What do we mean by that? You can find such data in social media. A number of researchers are also leveraging spatial data from social media sites, including Twitter, Foursquare, Facebook, Flickr, and others, to map trends in society in space and time. Several examples of integrating live streaming social feeds with ArcGIS maps online have already been developed for mapping events such as earthquakes and severe weather (https://dev.twitter.com/case-studies/esri-enriches-maps-tweets-and-streaming-api). To tell a story or to create a whole picture with maps requires the ability to bring in different types of data.

With evolving sensor networks and the emergence of the **Internet of Things** (a world in which almost everything from your home appliances to door locks and even your clothing is connected to the Internet), some other interesting applications are emerging. For example, working with spatial data coming from a GPS-enabled vehicle or mobile device, combined with the ArcGIS Tracking Analyst extension, operations can be triggered by locations. For example, a vehicle crossing into a particular zone or within a distance of a delivery location could alert the recipient of the impending arrival, or a driver of a customer needing a package pickup, applications of a concept called geofencing. **Geofencing** defines particular boundaries (fences) that filter analytical operations or alerts based on specified attributes or spatial positions. For example, imagine your smart house ensuring the doors are locked, thermostat is adjusted, and appliances are turned off once you are a specified distance from home. There is no more wondering if you left the stove on as you pull into the airport parking lot for a well-deserved vacation. And conversely, dinner might be set to warm up based on your distance from home and current traffic conditions at the end of the workday.

As these devices and capabilities become more commonplace, researchers will begin to consider how to leverage them to facilitate spatial research capabilities in the field, including data collection, editing, and analysis. Tools such as ArcGIS Online already provide many GIS capabilities that just a few years ago were limited to the desktop. With these new tools, creative individuals are exploring and applying GIS in fresh and innovative ways.

As we have discussed throughout this text, a primary advantage to integrating a geospatial perspective is that it provides a comprehensive perspective to many research topics. GIS can draw together data from a variety of sources, providing a holistic view. This new view adds both breadth and depth to your topic of interest. As new technologies and new data sources become accessible to researchers, opportunities arise to further enhance your exploration of almost any topic and, perhaps more important, provide new opportunities never before possible. Just a few years ago, who would have thought it possible to map the thoughts of thousands of individuals based on their tweets? After all, prior to 2006, a tweet was the sound a bird made, and it was primarily wildlife biologists who were interested in mapping them. Today an online search for "Twitter Map" returns thousands of examples of data drawn from tweets to develop maps addressing just as many topics.

As the global society continues to hone its spatial awareness, we will quite likely witness continued and increasing demand for spatial analytical capabilities. It is now possible to create many different types of maps, and many more people are asking spatial questions and making maps. Alas, they may not always refer to these as maps, and many of those in the popular press trend toward infographics. **Infographics** can be defined as visualizations of data, which in many cases present the data using a cartographic perspective. One of the most famous infographics, and one familiar to students of both geography and European history, is Charles Minard's information graphic of Napoleon's invasion of Russia (figure 15.1).

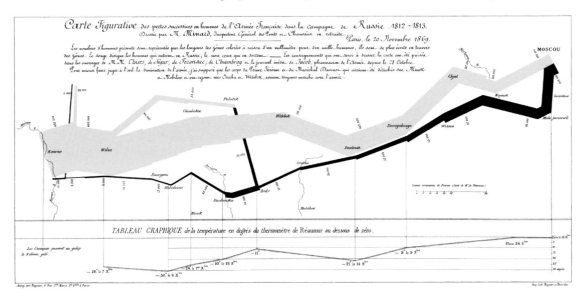

Figure 15.1 Figurative map of the successive losses in men of the French Army in the Russian campaign, 1812–13. Drawn up by M. Minard, Inspector General of Bridges and Roads in retirement. Paris, November 20, 1869.
Map courtesy of Daniel Hatch, Chris Clarke, Kenneth Field, and James O'Brien, Kingston University London, Kingston upon Thames, Surrey, UK. From *Esri Map Book*, vol. 27 (Redlands, CA: Esri Press, 2013), 60. Open-source data.

A somewhat more modern and maplike example of an infographic is shown in figure 15.2. In this example, the general appearance, geographic positions, and connectivity of the United States are retained, while color and size are used to communicate the candidate receiving electoral votes.

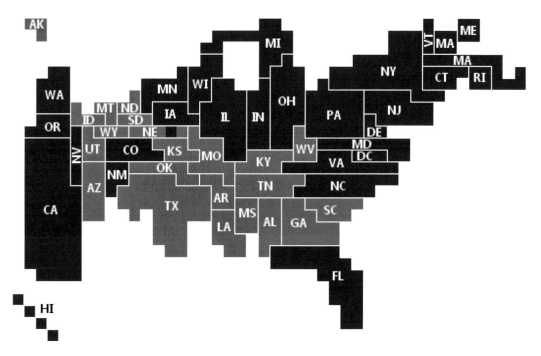

Figure 15.2 Final 2008 electoral college cartogram. Blue squares represent electoral votes for Obama–Biden. Red squares represent electoral votes for McCain–Palin. Note that this cartogram is topologically correct in that any two states that show as adjacent on the cartogram do not touch the ground in reality. Cartogram previously published in J. R. Gott and W. N. Colley, "Median Statistics in Polling," Mathematical and Computer Modeling 48: 1396–1408, 2008. It is licensed under the Creative Commons Attribution-Share Alike 3.0 Unported. http://en.wikipedia.org/wiki/File:Final_2008_electoral_cartogram.png.

Geospatial agility

Many researchers and scientists are drawn to GIS because of the added analytical and visualization possibilities it can provide. In essence, it offers the researcher geospatial agility. If you are agile, you are able to move quickly and easily. **Geospatial agility** means being able to grasp the spatial concepts and patterns within your data and to move around your data quickly and easily. You gain an advantage in being both broad and focused at the same time. In the animal kingdom, being able to quickly gather information about your environment and the mobility of your prey is central to survival. For instance, for a hawk to have a successful hunt, it would need to see the big picture of the meadow while simultaneously being able to hone in on the single field mouse it targets as its prey. The hunter goes through a series of analytical steps before choosing and successfully striking for the kill, and perhaps for the hawk, tweeting its success.

When you conceptualize, gather, analyze, and finally communicate your research results, a map-based **visualization** of data is often the best way to tell the story without confusing statistical results and complexity. In particular, when your research is targeted to decision makers, a simple and effective visualization of your data can help to translate research to real action and make a difference. This is part of geospatial agility. This is true from people who make decisions or policies in any field, ranging from corporate partners to nonprofit organizations to school boards and politicians. As scientists who come from both natural and social science backgrounds, we believe in the power of research and investigation to improve society. As a result, we have always looked for tools that will provide us with the best analytical capabilities and understanding.

For a researcher, an important advantage of a spatial perspective is that it facilitates both the macro and micro views in the same place, and simultaneously. Using GIS enables the user to zoom out to view data, conduct analysis at multiple scales, and examine patterns and trends across large regions or landscapes. You also can get into the details, zooming in for a micro view of the data. A GIS provides researchers the capacity to engage with data from both perspectives interchangeably, integrating the most appropriate data sources at each stage of the process. For this reason, adopting a spatial perspective is elementary to research success when examining topics from multiple scales and perspectives.

Image versus data

A centrally important point for researchers to consider is the distinction between being able to see images spatially and actually having access to real "data." Why does this matter? Why would we care? In many everyday uses of GIS, a person simply wants to find out where something is located and directions to get there. For the general public, this may be the most common use of spatial mapping. In these consumer uses of spatial analysis, there is no need to access the underlying data and analyze it; the tools required are built into the device or accessed via wireless communication to the cloud. Simply being able to access the desired information through an online search via phone or computer or the GPS in one's car is enough. We want to highlight that having the ability to visualize information or data is actually very different from having access to the spatial data in a form that is relevant to a researcher. We also want to point out that sometimes the data on the Internet are incorrect. For instance, when we first moved into our new home in a new housing subdivision, the online provider of maps had an earlier configuration of our street that did not match reality. As a result, no one (from the newspaper delivery person to the pizza delivery truck) could find us for the first six months we lived here. It was as if our "real" location had been supplanted by false data. Because no one could find us virtually, it was as if we did not exist.

We call that the power of bad data. **Bad data** are data that are inaccurate but are used as if they were correct.

As a researcher, student, practitioner, or policy maker, you will have many questions requiring analytical capabilities. You will want to be able to do more than just see data that somebody else provided for you in the limited ways they provided them. In those instances, you will want to be able to create, analyze, and manipulate data using your own GIS software. Fortunately, many of the common datasets you may require (streets, boundaries, census data, crime data, and many other kinds of data) are readily available in GIS-compatible formats such as shapefiles, geodatabases, raster grids, or images. Throughout this book, we have covered many ways that a person can find and analyze spatial data for research and/or applied purposes. However, it is important always to keep in mind potentially new and different ways to identify and acquire data that may be of value in your work. Stay abreast of emerging data sources such as those coming from social networks, new consumer devices, and technologies, as well as more traditional and newly developing sensor networks maintained by government agencies (weather, traffic, pollution, and others).

Many of the most innovative and enlightening spatial analyses come from someone being creative. Although many tried-and-true analyses continue to be valuable, particularly in production environments, research is a very different animal, one that may wish to tweet occasionally. Consider how you can take advantage of those sources, or how you can collect new, primary data in innovative ways. Currently, in the United States, more than 140 million (45 percent of the population) people carry a smartphone. This provides a huge potential for the collection of data, including locations, photographs, and much more. As mentioned earlier, many of these individuals volunteer useful data through their social network postings, and of course, some of these data become available via other channels (mobile companies certainly obtain valuable information about their customers, law enforcement can subpoena data, etc.).

However, a savvy researcher may simply seek out and enroll participants in a study. You might consider providing study participants with a mobile app, asking them to put a small GPS receiver on their dog's collar for a month, or asking them to deploy environmental sensors on their windowsills to build a high-density spatial network of data loggers. With cellular coverage prevalent, as well as public Wi-Fi and home networks, the opportunities to collect spatial data are far greater than in the past and may require little to no time for you (or your staff) to deploy to the field. As costs for the hardware and software continue to decrease, the opportunities grow exponentially. Communities of citizen scientists are emerging and developing new tools for collecting data on a regular basis, and they may have developed exactly the tool you need to collect the data you require. These options may not be available for every study, but consider your options and be creative.

A rebirth of spatial awareness

Over the last decade or so, enough adventurous students and faculty began to explore GIS as a means of expanding their disciplinary toolboxes that applications in these disciplines began to appear. Though by no means commonplace, GIS is now being increasingly incorporated into curricula across some campuses including K–12 education. Maps are in front of the public like no other time in history. Spatial applications are used by many on a daily basis, be it the GPS in the car or on a smartphone, or simply the spatial queries made to find a place for lunch. This extension of spatial thinking and essentially of GIS into new areas and applications will no doubt continue as creative individuals find new ways to apply the technology to address their questions. What this means in terms of society is that we are generating a new breed of spatial thinkers who will grow up and continue to employ geospatial agility and the sociospatial perspective into their daily work and lives.

Nonetheless, if you are a social scientist and are beginning to use GIS, you will most likely find that you are still in a group of fairly early adopters. The good news is that as you incorporate GIS into your own work, you will be one of the people to whom others come for suggestions and creative ideas. This is true even if you have minimal exposure to the use of GIS because it is still a new frontier. As professors, we share this adage with our college classes—to let them know that even though the students are not "GIS experts," they will be in demand for their skills and their ability to think spatially. Inevitably, this has proven true.

Although GIS has found its way into many more local and regional governments, nonprofit organizations, and schools, you may still find yourself being the first one to bring GIS into your particular organization. While this presents an exciting opportunity, it can also make it difficult to build a community of colleagues. If you work in a public agency, we suggest finding out who else is using GIS in your workplace. It may be colleagues down the hall in a department focused on infrastructure, parks, or other natural resources. If you are in academia, you may find people who use GIS in one of several different departments on campus. Of course, check what the geography department is doing with GIS. There are sure to be individuals who have a strong understanding of the types of analysis in which you are interested. Also look to traditional natural resource departments, where GIS has a long history. Computer science is another place you might find people with insights into GIS. In the private sector, many energy and telecommunication companies, logistics and shipping companies, and a variety of engineering and environmental consulting firms and nongovernmental organizations use GIS. In many areas there may be a local GIS user's group, regional or state meetings with a GIS emphasis, and of course, national and international meetings such as the annual Esri International Users Conference and a variety of discipline-specific meetings (http://www.esri.com/events), among others.

Of course, being at the leading edge as a user of any new technology also means you may need to be creative in approaching these tools. It may not be immediately obvious how you can translate a GIS study of wildlife habitats to a social science analysis situation, but try to be imaginative. Of course, the reasons that an animal in the wild chooses to live where it does may be far more similar to how people select and develop their communities than is first apparent. Both animals and humans have a set of needs that they are attempting to meet, such as being close to resources like food and water and having a safe place to live away from noise and congestion. The point here is that the same tool can be used across the varied disciplines.

GIS is an art form

Using GIS as an analytical tool is as much an art as it is a science. The art is in conceptualizing your analysis and implementing it logically in the GIS. Of course, the actual processing, map analysis, and statistical results are the science behind the GIS. As we pointed out early in this book, GIS is an integrative technology that allows you to bring together data from a variety of sources and disciplines. Being creative in how you apply these tools is one of the most important components of successful GIS analysis, particularly when applying it in new situations, as is often the case when conducting research. Of course, once you work out a reliable process that can be operationalized and potentially even automated, you'll want to stick with it for future work.

Although GIS has existed for more than five decades, it is beginning to gain more traction as a tool within the social sciences. With the exception of some large government agencies, most social science researchers did not start to rediscover their spatial roots until recently. Whatever your specific discipline within the social sciences, odds are that if you go back to the roots of your discipline, you will find that maps were integral to the analytical process at some point in its development. Over time, many social science disciplines got away from these roots, looking to new and different approaches. Even the academic geographers largely ignored GIS in its early years of development. Nonetheless, GIS continued to develop within other disciplines, primarily in the natural resource sciences during those intervening years, remaining largely unnoticed by social scientists.

However, as GIS software has become more accessible and easier to use (in large part with the introduction of the graphical interface provided in the early 1990s with ArcView and continuing with the introduction of ArcGIS for Desktop, ArcGIS Online, and ArcGIS for Mobile products), GIS is now more familiar, accessible, and easier to use than at any point in its development. Social scientists are once again realizing the value of a spatial perspective and are exploring applications of GIS in their disciplines. What was old is new again. Countless agencies, private companies, community organizations, and nongovernmental organizations could be assisted in better achieving their goals by incorporating GIS approaches into their work. As we have discussed throughout this book, GIS provides a more holistic and broader view of topics than other approaches. It also helps to make scientific data more accessible and easier to comprehend because of its visual nature. Of course, GIS is a technology that is not without its faults. Most important, one cannot rely on developing a technology without also developing an equal consideration for the people who will use the technology (the human side of GIS).

GIS as change technology

Many scientists deal with the concepts of change over time. To assess change requires data from two or more points in time. You would begin with what is known as the first set of data also known as baseline data and then create or discover data for other points in time. We refer to GIS as **change technology** because taking an action or developing a policy is often in direct response to a certain state of affairs. Noting what these state of affairs are can be visualized and charted using the ArcGIS software.

The social scientist is often interested in changes that influence our society and might ask questions such as the following: What social changes are impacting society today? What type of a society does one live in? What political or environmental changes are influencing society today? Is there inequality that is occurring? Social scientists try to give different types of inequality different names, such as ageism, sexism, or racism.

Essentially any of these questions might benefit by considering the spatial context. For example, where are different types of inequality occurring? Who is being most affected by these types of inequality? Because environmental and social resources that support certain societies and social structures throughout the world are dwindling, questions of geographic location and inequality are gaining increasing importance. One academic area of research that covers such topics is the field of environmental justice. This field gives equal weight to the intertwined role that the social and physical environment have with one another. It is not a coincidence that poorer people and people of color often find themselves living in substandard physical environments, from both health and safety standpoints. Another area within the social science field that can readily be enhanced through the use of geospatial technology is the area of social inequality. Increasingly, as information sharing continues, the question of where geographically these different types of inequality occur will become increasingly relevant. Being able to target inequality hotspots could lead to predicting potential areas for political unrest and the need for certain types of resources.

Identifying social inequality

Social inequality can exist at any level, be it an organization, community, neighborhood, state, or nation. In the world in which we live today, the gap continues to widen between the haves and the have-nots, the rich and the poor. GIS is an important tool that can be used to help document and analyze where, to what degree, and what types of social inequality occur in various places throughout the world. More important, to truly assess social inequality, you need to be able to have baseline data that document a situation as it occurs today and at different points in time. Doing so can enable the researcher to create maps that visually map out the changes (both positive and negative) as they occur geographically. For instance, let's say a group of people lived in a polluted environment that was slowly being cleaned up. One could chart the progress of this environmental cleanup on the map. They could also chart the mobility of people out of the polluted environment and into other, presumably, healthier and safer environments. All of this could be mapped, tracked, and analyzed in ArcGIS. Figure 15.3 illustrates a polluted environment that exists in Lachine Canal in Montreal, Canada.

Figure 15.3 **Pollution in the Lachine Canal in Montreal, Canada.** Photo by Aarchiba.

Additionally, by exploring the social structure of the locations where inequality occurs, we might gain a better understanding of the causes and potential solutions. This is valuable information because it can provide a faster response time for dealing with a variety of social, economic, medical, environmental, and political issues that arise around the world. Solutions based on GIS analysis allow for a geographic perspective to be considered. The solution might be as simple as determining access to clean drinking water by using GIS to explore the distance to water supplies in a community suffering from high levels of waterborne disease. Much as the famous drinking water analysis by Dr. John Snow in the mid-1800s, which identified the contaminated pump causing cholera, used a map-based approach, a modern study in ArcGIS today could assess opportunities to resolve and improve on a similar situation in a developing country. Perhaps drilling a new drinking water well in a location convenient to the community would solve the problem. Both the analysis of where people get their water and where a new well should be located would be well served by a GIS. This process of determining factors or variables causing the problem and pointing to an appropriate solution would be an example of applied social change in a concrete and real sense.

Finding remedies to different types of social inequality involves finding ways to improve society; improve the world; solve existing problems; and right existing wrongs and inequalities in a society, community, or organization. Each of these goals can be assisted through the use of GIS, specifically ArcGIS software. The applied or practical view of social change encourages people to work out the steps that must be taken to right the wrongs and concretely determine the steps that are necessary to fix a specific social problem. An example of such an application is establishing greater equality for groups who have previously experienced unfair or unequal treatment.

The role of geospatial crowdsourcing

As people around the world gain increased access to cell phones—their ability to gather and share information spatially in a real-time way has grown exponentially. In essence, everyone becomes a potential source of data—based on information that they know and now have the ability to share via social media sites and/or the web—**geospatial crowdsourcing**. This is because of the increased ease of access to cell phones in many countries. In fact in many rural areas in developing countries, it is common to see people who have cell phones in communities that never had traditional landlines. The ability to spatialize information and share this type of data has become huge. Increased information access has produced an increased reliance on and ability to use spatial technology through phones as individuals react to emergency situations or socially mobilize. Cell phones are equipped with GPS that can turn local people into on-the-ground reporters and conveyors of important information. This has been true after many disasters, both natural, as in the case of a hurricane, tornado, flood, or earthquake, and man-made, such as major oil spills. Figure 15.4 illustrates one such emergency situation that followed an earthquake in the country of Haiti.

Figure 15.4 Life in the aftermath of the Haiti earthquake. A man exits a restaurant after looking for his belongings. Photograph by Marco Dormino, United Nations Development Programme; originally posted to Flickr as "Haiti Earthquake"; licensed under the Creative Commons Attribution 2.0 Generic license.

New directions for GIS-based research

As GIS continues to penetrate our societal consciousness, it is inevitable that, increasingly, space will continue to be central to all disciplines and sciences. The general public now demands that a spatial component be included in the way information is conveyed and shared because spatial thinking has become a norm in many societies. In fact, many people expect it.

Spatial data and smartphones

GIS has made its way into many people's lives via people's phones and the numerous Internet websites that incorporate maps and mapping technology. Furthermore, a spatial perspective is regularly presented in the news, both in print form via newspapers and magazines, as well as in the form of compelling 3D maps or fly-through visualizations often incorporated into television news broadcasts. GIS software and digital mapping technologies such as GPS are already being presented in many middle and elementary schools around the world.

We think that it's important again to underscore the value of spatializing your data for the long term. Many tools will allow you to quickly produce a map, but these tools do not often enable you to conduct analysis or to have long-term access and the ability to analyze data that lie behind that map. As a researcher or someone concerned with creating policy, you want to make sure that you have access to the data and the long-term creation of any maps that you produce. This can be accomplished using the ArcGIS technology. Such technology has wide-ranging applicability (see figure 15.5, which presents an example of a ArcGIS mobile map).

Figure 15.5   An example of an ArcGIS mobile map, deployable to the field on a smartphone. Esri.

With so much exposure to these tools and their numerous applications in understanding the world around us, it would be naïve to assume anything other than a growing demand for more spatial information and analysis from the public.

Visualization and GIS

The purpose of GIS is to create, analyze, manipulate, and portray spatial data. GIS enables the user to visualize social and physical elements of a certain space over time. Therefore, this could allow for the historical examination of changing physical environments, shifting human settlement patterns, the spread of disease, and access to resources. From a research standpoint, the value of GIS is in the added capabilities it provides to facilitate the integration of different types of data, such as community stories, sociodemographic profiles, and community needs assessments. From a natural science or medical science standpoint, the possibilities are virtually limitless. As we noted, the health and medical sciences have become very reliant on spatial technology to do their daily work and research. We can also expect that the natural sciences, the earliest field to adopt GIS, will continue to be a major player in the future development and use of GIS technology. Increasingly, too, we will continue to see GIS be used frequently by businesses and corporations as they continue to employ the spatial technology to manage existing businesses and help chart the locations for new ones.

GIS allows for integrating multiple sources of information, which ultimately produces a picture of issues or problems experienced by a specific geographic area or multiple geographic areas. We hope that by now you understand its importance for natural science, medical science, social science, arts or humanities, or any other given field. GIS gives the user the flexibility and power to understand both recent and historic trends and changes that have occurred over time, both environmental and social.

A number of tools exist that allow for the researcher, city planner, or community developer to run different scenarios regarding future development issues based on population growth and available resources. One such example that supports both dynamic modeling and 3D visualization of results is CommunityViz (http://placeways.com/communityviz/), which works hand in hand with ArcGIS to provide powerful dynamic models. By inputting different parameters into the program, the individual is able to model and visualize how things would appear in the future, given a set of specified conditions (see figure 15.6).

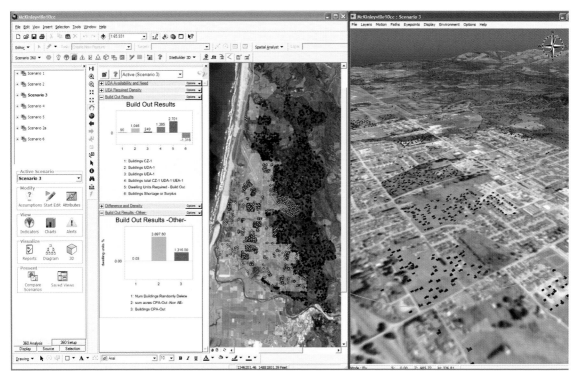

Figure 15.6 **An example of visualizations developed for a city planning process that explores alternative scenarios for infrastructure capacity or expansion, housing density, and protected areas, using the CommunityViz extension to ArcGIS.** Image courtesy Michael Gough/Humboldt State University, with funding from the North Coast Environmental Center, Arcata, CA.

For example, let's assume that a community just received an economic development grant to revitalize the downtown. This revitalization will include the construction of some new buildings (to add to the downtown's existing Victorian-style building structure) and will involve the installation of a new bridge. The construction of the bridge will take at least a year. During this time, various roads will be closed and the traffic demand patterns will affect some new buildings in the downtown area. How and where a community decides to locate these buildings, what height the buildings will be, how much parking will be needed, and so on, could all be visualized using GIS.

As computers continue to become more powerful and less expensive, photo-realistic 3D visualization and simulation models of various options will become more readily available and easier to use. It is likely that in the next few years, we will have ready access to augmented realities in which scenarios such as this might even be presented in real time and space via overlaid data on a smartphone image of one's surroundings or images projected via wearable devices such as virtual reality goggles.

Faster response time

Using a GIS to assemble information regarding communities and integrating it immediately into a database can improve analysis and save time. In the past, much of this information may have existed in diffuse form. That is, each agency may have particular information regarding its particular issues or needs but may be unaware of the data other agencies and organizations possess. The same information might have existed on a variety of maps or files located in filing cabinets. Such data would rarely, if ever, be brought together in analyzing social issues or planning processes. When you use a GIS to bring this multitude of different types of data together into one form, it sets the stage for data sharing and a consolidation of effort that can then result in better data overall. Having access to better data overall regarding communities can be invaluable in the planning process. But again, simply because data are available digitally does not mean that people or agencies are going to freely share this information. If you look at the response to Hurricane Katrina that hit New Orleans and much of the southeastern United States in August 2005, the major failure in agency preparation and response following the hurricane was due to a lack of communication and data sharing between different types of agencies who needed to be in communication with one another before, during, and after the crisis. On-the-fly decision making in a crisis situation still requires good data; otherwise, bad or ineffective decisions can result. Sometime inaction is the result of lack of data.

That major data sharing snafu highlighted the need for different agencies with various governmental ties (federal, state, regional, local) to be able to communicate with one another easily and efficiently, especially in a time of crisis. The cost of not being able to rapidly communicate and share data is just too high.

Parting thoughts

This book has led the reader through the salient topics that users of GIS, and more specifically users of ArcGIS, currently face. Additionally, we have provided a background on GIS, how to identify topics that would be appropriate for using ArcGIS technology, research design and data collection, measurement, preparation, and analysis. It is our hope through this book that you develop an appreciation for some of the opportunities that a spatial perspective can bring to your work and ways that this can be facilitated using ArcGIS tools.

If we have motivated you to venture forward to try to use GIS approaches to meet your own needs, we have succeeded in realizing our primary goal. Often, the greatest barrier to using a new technology is gathering the courage to take the first step and give it a try. As with anything new you might undertake, the more you work at it, the easier it will become.

Start off with something simple and work up to the complex. As the saying goes, you must learn to walk before you learn to run.

Geographic information systems are a powerful tool that can be harnessed equally by researchers, scientists, and practitioners from a variety of fields. As these technologies continue to develop and become more prevalent in our society, so, too, will the opportunities to apply GIS in new and creative ways. With increased applications of spatial technologies, the need for competent analysts with a strong understanding of GIS will be even stronger. Gaining a good grasp of ArcGIS will put you in a great position to examine and understand a variety of problems or issues using a geospatial approach; in other words, you can increase your geospatial agility.

Almost every datum, including physical, natural, social, and even conceptual, has a geographic aspect that can contribute important information to an analysis and ultimately to our understanding of a range of topics and subsequent policy development. Our goal in writing this book is to assist individuals like you to take the first steps toward embracing a new and perhaps foreign technology. An equally important goal is to help you to understand how this technology could benefit your own data collection, analysis, and decision making in the social sciences. GIS can truly help researchers, policy makers, practitioners, and scientists move closer toward meeting the needs of groups of people throughout the world. It enables the researcher or policy maker to incorporate important contextual variables into policy-making decisions. We hope that perhaps you will assist your fellow scientists or researchers and practitioners in moving forward in this direction, one that is simultaneously the history and the future of geospatial science.

As we suggested at the beginning of this text, spatial analysis techniques can be applied and provide valuable insight into a wide array of issues, including many that have not yet been considered. The applicability of these tools in a geospatial context, and the valuable information and analysis results they can provide, are endless, limited only by the creativity of the practitioners who use them in their work. It is our hope that some of the ideas presented in this text will prompt you to begin to think creatively about GIS in your own work and to share your experiences with others.

Suggestions for student research projects

Any set of variables that has a geographic location is ripe for being included in a GIS-based study. As a starting point, we provide some suggestions that you might use directly, or in modified form, in developing a GIS-based social science research project in an area of your own interest.

The following are some example hypotheses that could be investigated using a GIS. (For each subject, discuss in a small group how one could capture relevant information related to the topic.)

1. Environmental justice: Poor and minority communities are more likely to be located nearer to toxic waste facilities. (A geographic proximity question)
2. Health: Cities that are closer to international airline connections are more likely to have higher rates of contagious diseases. (A network proximity question)
3. Crime: Crime rates are higher in areas that have a higher density of teenagers. (A spatial correspondence–containment question)
4. Anthropology: Rural indigenous communities that are farther from cities have a greater preservation of their culture than those communities located closer to cities. (A proximity question)
5. Political science: Communities that are located closer to international borders tend to be more supportive of international trade agreements such as the North American Free Trade Agreement (NAFTA). (A geographic proximity question)
6. Sociology: Communities that have many local service organizations have greater social networks. (A spatial correspondence–containment question)
7. Wildlife science: Home ranges of various wildlife species are being affected more by encroaching development in suburban versus rural areas. (A geographic proximity question, including buffer zones)

Review questions

1. How would you describe the Internet of Things? What role does GIS play in this?
2. What is geofencing? What role does geofencing play today? Be creative and think about what roles geofencing might play in the future.
3. Why can we say that GIS is an art form?
4. What is change technology? Why do we refer to GIS as change technology?
5. Why is map-based visualization often the most effective way to present your data?
6. How can GIS be used to highlight issues of inequality? What are some of the potential challenges to doing this? Please provide an example.
7. What is the most effective way that you think GIS will be used in the future? Is there a social, environmental, health, or economic problem that you think could be improved through using GIS? Please describe.
8. What is crowdsourcing? What role do you predict crowdsourcing will play in the future of GIS?

Additional readings and references

Crossley, R. 2012. "Future of GIS." *Geocommunity*. http://spatialnews.geocomm.com/features/futuregis.html.

Goodchild, M. 2012. "Looking Forward: Five Thoughts on the Future of GIS." *ArcWatch*. http://www.esri.com/news/arcwatch/0211/future-of-gis.html.

Radke, S. L., R. Johnson, and J. Barayni. 2013. *Enable Comprehensive Situational Awareness*. Redlands, CA: Esri Press.

Yang, C., R. Raskin, M. Goodchild, and M. Gahegan. 2010. "Computers, Environment and Urban Systems." *Geospatial Cyberinfrastructure* 34: 264–77.

Relevant websites

- **Esri events (http://www.esri.com/events):** Lists upcoming and previous Esri meetings, including discipline-specific, regional, and international meetings. Such events provide excellent opportunities to build a network of GIS colleagues and learn new and innovative ways to apply GIS in addressing your own questions.
- **Esri Enriches Maps with Tweets and the Streaming API (https://dev.twitter.com/case-studies/esri-enriches-maps-tweets-and-streaming-api):** This website presents a case study of how real-time data can be incorporated into tweets that include mapping. A conversation can be enhanced because it provides realistic perspectives on the situation along with a visual map.
- **The New Age of Real-Time GIS, Matt Artz, April 1, 2013 (http://blogs.esri.com/esri/esri-insider/2013/04/01/the-new-age-of-real-time-gis/):** An article on the use of real-time analysis in ArcGIS with the operations dashboard to implement geofencing.
- **US Census 2010 GIS data: General Information (http://www.census.gov/2010census):** This is the main website for the US Census 2010 and contains a sample interactive map along with Census briefs and reports.
- **Geography Page, Including Downloads of 2010 Data in Shapefile and Geodatabase Formats (http://www.census.gov/geo/maps-data/data/tiger-data.html):** This website contains TIGER/Line shapefiles that are available with data in a geodatabase. This is where you can download the American Community Survey and 2010 Census various shapefiles for geospatial analysis.
- **Geospatial Platform (http://www.geoplatform.gov):** A source for a wide variety of federal geospatial datasets.

- **Data.gov (https://catalog.data.gov/dataset):** A source for a wide variety of federal geospatial datasets.
- **Placeways, CommunityViz (http://placeways.com/communityviz/):** An example of a modeling and visualization tool useful in creating informative analytical outputs to visualize multiple planning scenarios.
- **US National Institute of Health (http://www.nih.gov):** Provides spatial data and tools with an emphasis on heath and epidemiology.
- **First Gov (http://www.firstgov.gov):** A web portal to all US federal and state government websites.
- **Open Geospatial Consortium, Inc. (OGC) (http://www.opengeospatial.org):** A nonprofit, international, voluntary consensus standards organization that is leading the development of standards for geospatial and location-based services.

Index

abstraction, phases of, 40–47
abstraction process, 300
accessibility, 38
accuracy issues, 100–101, 176
Active Living Research, 327
Addams, Jane, 13
Add Data function, 187
address matching, 174–75
aerial imagery, 178–79, 224–25
Amazon Georeferenced Socio-Environmental Information Network (RAISG), 197
American Community Survey, 206
American Evaluation Association, 264
analysis questions, 51–53
analysis tools: buffers, 271–74; cartographic classification, 268–71; least cost analysis, 285–89; modeling, 295–98; nearest neighbor analysis, 278–79; overlay, 268, 274–77; pitfalls of GIS, 299–302; proximity polygons, 277–78; simulations, 295; social networks, 279–85; Spatial Analyst, 304–6; spatial interpolation, 290–95; spatial statistics, 302–4
analytical approach, 40–47
animated maps, 341–43
annotations, 201
anonymity, 106
ArcCatalog, 111, 233
ArcGIS, 33, 42, 58, 111, 113, 156, 181, 387, 395–96; 3D Analyst, 343–44; Add Data function, 187; alignment of datasets in, 100; buffers, 268–71; cartographic classification using, 268–71; data analysis using, 265–307; extensions and add-ons, 43–45; formatting considerations, 182–83; Near tool, 278–79, 323; numbers in, 185; overlays in, 274–77; Proximity tools, 277–79; qualitative data analysis and, 310; Schematics application, 283–84; Spatial Analyst toolbox, 304–6; Spatial Statistics toolbox, 180, 302–4; Tracking Analyst, 381
ArcGIS Online, 174, 199, 212–13, 381
archival data, 158
ArcView, 387
art, of GIS, 387–88
ASCII delimited text, 186–87
Atlas of Canada, 162, 189, 342, 353
Atlas of Sweden, 189, 342, 353
ATLAS.ti, 79
attributes, 48, 125–26, 146, 181
audience, 330–31, 350
available subjects sampling, 132

background research, 119–20
bad data, 385
baseline data, 60–61
basemaps, 33–34, 37–38; access to, 223; creating, 177–79; linking data to existing, 173–74, 182, 184; for public participation GIS, 200–202
Battle of the Greasy Grass, 12
Battle of the Little Bighorn, 12
benchmarks, 243, 255
Booth, Charles, 20–21
boundaries: conceptual, 52–53, 248; conceptualization of, 139; defining, for PPGIS research project, 204–5; edge effects and, 134–36; functional, 248; geographic, 373–74; modifiable area unit problem and, 136–37; naturally defined, 248; physical, 52, 120; policy making and, 373–74; selection of, 138–39; for study, 52–53, 55, 120, 134–39, 248
brainstorming, 73
British Museum, 235
buffers, 271–74
buffer zones, 371
built environment, 119, 249–52

Cartesian coordinates, 310
cartograms, 28
cartographic classification, 268–71
Cartographic Communication website, 24
Cartographic Excellence in Creating a Community Basemap, 202
case studies, 229–30
categorization of data, 29–32
cells, 181
Center for Spatially Integrated Social Sciences (CSISS), 24, 70
Centers for Disease Control and Prevention, 162, 189, 264, 353
Central Intelligence Agency (CIA), 302
change technology, 388–90
character fields, 185
Chicago slums, 13–15
civilized society, railroads and, 15–16
close-ended questions, 167, 258
cloud-based technologies, 199
code data, 35
coding schemes, 174, 183–86, 320
color coding, 214
Columbia CNMTL-QMSS e-Lessons, 140
comma delimited files, 186–87
commercial databases, 33, 144–45
communication: cultural issues and, 337–39; of data, 16–17; keys to effective, 330–39; mode of,

347–49; public, 227; of results, 80, 210–11, 227, 329–53; using visualizations, 16–17, 339–44, 384
community-based component, 369–71
community of interest, 205–8
CommunityViz, 393–94, 399
comparative study, 150
compliment, 275–76
CompStat system, 2–4
computer animation, 29, 341–43
computers: in the field, 198–99, 222; for recording data, 173, 198–99
concept questions, 48–49
concepts, 123, 139, 150
conceptual boundaries, 52–53, 248
conceptual geography, 6
conceptualization, 123
conceptual maps, 36
conceptual model, 37–40; defined, 73; development of, 73–74
conceptual networks, 283–84
conceptual space, 12, 27–28
conclusion, of evaluation report, 260–62
confidentiality, 92–93, 106, 109
content analysis, 232
content-based field notes, 203
contextualization, 17
control, 375
coordinates, 26, 145, 300, 310
Cornell Office for Research on Evaluation, 264
cost surface, 289
Create Thiessen Polygons tool, 277–78
crime mapping, 2–4
crime rates, 65
cross-examination, 75–76
crowdsourcing, geospatial, 390–91
CSISS. *See* Center for Spatially Integrated Social Sciences (CSISS)
cultural implications, of research, 94
cultural issues, map visualization and, 337–39
cultural perceptions of technology, 226

data, 5; accuracy issues, 100–101, 176; archival, 158; attributes, 146; availability of, 39; backups, 113, 199; bad, 385; baseline, 60–61; case studies, 229–30; categorization of, 29–32; classification, 268–71; code, 35; coding schemes, 183–86, 320; commercial, 144–45; communication of, 16–17; community, 205–8; contextualization of, 17; conversion of, 156–57; coordinating, 366; creation, 77; deleting, 113; demographic, 207–8; developing own, 166–71; digital, 33, 141, 155, 158, 315, 316; discrete, 124; economic, 206; evaluation of, 259; exporting to other applications, 346; geocoding, 84; geographic, 26–32; GPS, 175–77; ground truthing, 78–79, 86, 224–25; hard copy, 316; historic, 158–60, 206–7, 312–14; holistic understanding of, 18; images vs., 384–85; integration of, 18; Internet, 142, 144–46, 384; intersection of, 275, 276; from interviews, 35, 169–72, 198–99; level of analysis, 180; linking to basemap, 173–74, 182, 184; linking to geographic location, 173–75; logical, 46; masking, 109; metadata, 99, 110–11, 122, 316–17; nominal, 126–28; nondiscrete, 124; nonindependence considerations, 291–92; ordinal, 128–29; personal, 92–93, 105–7; preliminary, 120; preparation of, 77–78; primary, 110–12, 121–23, 142, 153, 154–55, 166, 192; processing, 100–101; qualitative, 79, 198, 203, 221, 309–27; quantitative, 312; questions about, 49–50, 147–52; realizations, 294; recategorization of, 29–30; reliability, 152, 154–58, 384–85; reputable, 193; sampling, 120–23, 130–34; secondary, 77–78, 98–100, 110–12, 121–23, 141–63, 166, 192–93, 259; sharing, 112; spatial, 207–8, 359–60, 392–93; spatializing, 29, 320–21; storage, 35, 112–13, 199; suitability of, 143; trades, 49; trends in, 29; types of, 126–30; US Census, 33, 36, 97, 110, 159, 167, 205–8, 220, 398; validity, 152–54; value-added, 144; volunteered, 200
data aggregation, 27, 106–8, 110–12, 174, 180
data analysis, 79, 86–87. *See also* spatial analysis; accessibility example, 300–301; approach to, 266–67; buffers for, 268–71; cartographic classification for, 268–71; errors in, 98–101; least cost analysis, 285–89; modeling, 295–98; nearest neighbor analysis, 278–79; network analysis, 281–85; organization of, 267; overlay for, 268, 274–77; pitfalls of GIS for, 299–302; proximity polygons for, 277–78; social networks and, 279–81; Spatial Analyst toolbox for, 304–6; spatial interpolation, 290–95; spatial statistics and, 302–4; techniques, 267–68; topographic tools for, 289–90; using ArcGIS, 265–307; when to use GIS for, 298–99
databases, 5, 33, 180–81; attributes, 181; cells, 181; character fields (strings), 185; commercial, 33, 144–45; creating GIS-friendly, 182–87; data format and coding considerations, 183–86; development of, 181–87; entities in, 181; field names, 181; formatting considerations, 182–83; number fields, 185; output formats, 186–87;

records, 181; software for, 180–81, 185–87; space and case issues, 182–83; spatial identifiers, 182
data cataloging, 233
data collection: basemaps and, 173–74; considerations for, 173–79; errors, 102; ethics and, 105–12; in field, 119–21, 173; forms of, 77–78; ground theory approach and, 84; interview-based, 169–72; latent spatial, 311–12; low-tech approach to, 198; manifest spatial, 311; oral history interviews, 230–32; participant observation, 232–34; PPGIS, 203–11; primary data, 192; privacy issues and, 93; questions about, 49–52; spatializing, 172; survey-based, 166–69; technology for, 222; using computer, 173, 198–99; using GPS system, 175–77; volunteered geographic information, 200; without computer, 172–73
data dictionary, 41, 54, 56, 184–85
Data.gov, 399
data model, 36, 145; logical, 41–43; physical, 46–47
data portals, 366
datasets, 49–50, 86, 99, 122, 142
data sources, 49–50, 77, 112; choosing, 33; emerging, 385; evaluation of, 142; existing, 98–100, 141, 151; Internet, 142, 144–46; local, 225–26; news as, 160–61; offline, 158–60; for qualitative data, 312–15; for secondary data, 142–43
data types, 316
data visualization, 16–17, 32, 65, 332–36, 339–44. See also visualization
datums, 26, 145, 300
dBASE, 186–87
DBF file format, 186
decennial census, 98, 148, 150, 167
decision makers, 359

decision making: data coordination and, 366; governmental, and GIS, 360–61; time constraints on, 358–59
decision support systems (DSS), 366–67
deductive research, 204, 310–11; vs. inductive research, 65–66
delimited text formats, 186–87
demographic data, 207–8
dependent variables, 72
DERC. See Digital Ethnography Research Centre (DERC)
descriptive research, 60–61, 148
digital annotations, 201
digital data, 33, 141, 155, 158, 315, 316
Digital Ethnography Research Centre (DERC), 235
digital maps, 212–13
digital orthophoto quadrangles (DOQs), 179
discrete data, 124
distance, role of, 151–52
DOQs. See digital orthophoto quadrangles (DOQs)
doughnut buffers, 274
DSS. See decision support systems (DSS)
dummy coding, 127–28
dynamic modeling, 393–94

ecological fallacy, 65, 103–5, 180
edge effects, 134–36
Electronic Statistics Textbook, 140
empowerment, 194
entities, 181
Environmental Protection Agency (EPA), 56, 58, 144, 163, 189, 352; Toxic Release Inventory (TRI), 55, 72
errors, 302; accuracy issues, 100–101; caused by analysis, 98–101; in human inquiry, 102–5; in interpolation, 293–95; location,

294, 295; measurement, 293–94
Esri, 23, 33, 34, 44–45, 58, 90, 162, 205, 217, 327, 337, 366, 398
Esri International Users Conference, 386
ethics. See research ethics
ethnographic research: case studies, 229–30; oral histories, 230–32; participant observation, 232–33; qualitative spatial, 219–35
ethnography, 227–29
Ethnography Matters, 235
evaluation interviews, 257
evaluation research, 237–64; benchmarks for, 255; challenges and benefits of, 263; defined, 238; determination of subject of, 254; focus groups, 256–57; group vs. individual focus for, 255–56; interviews, 257; presentation of evaluation, 259–62; project design, 254–59; questions, 239–46; reasons to use, 238–39; secondary data use in, 259; sociospatial, 246–62
evaluation survey, 258
exact interpolators, 294
exclusive or, 275–76
existing data sources, 98–100, 141, 151
explanatory studies, 65
exploratory research, 62–64
export, of data, 346

false zero, 130
feasibility, 38–39, 67
Federal Election Commission, 96
Federal Geographic Data Committee (FGDC), 115, 122, 361, 377
field computers, 173
field journals, 233
field names, 181
field notes, 203
field research, 119–21; basemaps for, 223; boundaries for, 120, 134–39; cultural perceptions of technology

and, 226; ground truthing map data, 224–25; integrating GIS into, 222–27; local sources of data for, 225–26; qualitative spatial ethnographic, 219–35; reliability in, 154–58; results from, 226; safety issues in, 120; sampling methods for, 121–23, 130–34; using computer, 198–99, 222; using GIS in, 172–73, 198–99; validity in, 152–54; without computer, 172–73

file compatibility, 186–87

file formats: dBASE, 186; delimited text, 186–87

fire regimes map, 61

fires, 242, 361–67

First Gov, 399

fixed (circular) buffers, 273

flowcharts, 46–47

fly-throughs, 344

focus groups, 256–57

formal spatial analysis, 13

freeways, 249–50

frequency tables, 319–20

functional boundaries, 248

fuzzy GIS systems, 36

garbage in, garbage out (GIGO), 143

geocoding, 84

geofencing, 381

geographic area of interest, 204–5

geographic boundaries, 373–74

geographic context, 26–32

geographic information, volunteered, 191, 200

geographic information system (GIS): about, 4–6; analysis tools used in, 9, 267–89; as art form, 387–88; buffering in, 268–71; case study research and, 229–30; as change technology, 388–90; data analysis using. *See* data analysis; databases and, 180–87; for field research, 172–73, 198–99; future directions for, 379–95; fuzzy, 36; geographic component of, 26–32; grounded theory using, 82–88; information component of, 33–35; integration of, into field research, 222–27; modeling using, 295–98; new directions for, 391–95; oral histories and, 231–32; output, 339–46; participant observation and, 233; pitfalls of, for analysis, 299–302; policy and, 356–77; as problem-solving tool, 298–99; professionals, 37; public participation. *See* public participation GIS (PPGIS); qualitative data and, 203, 310; reliability and, 155–58; research ethics and, 92–97; response times, 395; software programs, 42–46; spatial advantage and, 10–11; system component of, 35–37; as tool for voice and empowerment, 194; understanding, 6; usefulness of, 22; use of, in research, 16–22; validity and, 153–54; variables, 146–47; visualization and, 393–94; in the workplace, 386

geographic literacy, 337

geographic location: choosing, 83–84, 211–12, 248–49; comparison of, 151; GPS systems and, 175–77; health and, 220; human-constructed features of, 249–52; identifying, for research, 53–55; impact on target population, 248–49; linking data to, 173–75; natural features of, 248–49; questions about, 51, 150

geographic scale, 138

geographic target area, 248

geoportals, 366

geospatial agility, 383–84

geospatial crowdsourcing, 390–91

"Geospatial Information and Geographic Information Systems" (Folger), 361

Geospatial Platform, 398

geospatial technology. *See also* geographic information system (GIS): future directions for, 379–95

GIS. *See* geographic information system (GIS)

GIS data. *See also* data: from Internet, 144–46; reliability, 155–58; validity, 153–54

GIS Lounge, 24

Global Positioning System (GPS), 8, 9, 155, 173, 222, 295; as decision support system, 366; prevalence of, 379, 380; using, 175–77

goals: benchmarks for, 255; communication of results and, 332, 349; defining, for study, 148–50; setting, 47–48; understanding, and data analysis, 300

government agencies, 206; data from, 143, 144; decision making by, 360–61

GPS. *See* Global Positioning System (GPS)

graphical user interfaces (GUIs), 5–6

grounded theory, 80–88, 230

ground truthing, 78–79, 86, 224–25

group focus, for evaluation, 255–56

GUIs. *See* graphical user interfaces (GUIs)

hard-copy data, 316

hard-copy maps, 10, 199, 200, 212, 339

Harvard University Graduate School of Design, 163

hierarchy, 12

historical data, 158–60, 206–7, 312–14

historical studies, 13–16

historic poverty, 20–21

holistic understanding, 18

Homeless Organization Evaluation Benchmarks Rubric, 244–45

home range, 11

Hull House, 13–14

human-constructed features, 119, 249–52
human inquiry, errors in, 102–5
Hurricane Katrina, 395
HyperRESEARCH, 79
hypothesis, 66, 204, 310–11; development of, 70–73

identity, 275, 276
IDW. *See* inverse distance weighting (IDW)
illogical reasoning, 103–4
images: aerial, 178–79, 224–25; vs. data, 384–85
implementation, of policy, 18–19
inaccurate observations, 102
independence, of observations, 179
independent variables, 72
indicators, 241
indicator variables, 224
individual focus, for evaluation, 255–56
inductive research, 65–66, 80–88, 192, 204–11, 310–11
inexact interpolators, 294
infographics, 335–36, 382–83
informal spatial analysis, 13
information, 33–35; local, 194; personal, 92–93, 105–7; presentation of, 332–36; web-based, 347
information systems, 18. *See also* geographic information system (GIS)
institutional review boards, 105
integration, of data, 18
interactive maps, 341–43
interdisciplinary approaches, 373
interest, establishing in topic, 336–37
internal variables, 129
International Standards Organization (ISO), 115
Internet: data from, 144–46, 384; datasets on, 142; for public participation GIS, 199
Internet of Things, 379, 381

interpolation, 290–95
intersection, of data, 275, 276
interventions, evaluation of, 239–46
interviews: data from, 35, 169–72, 198–99; evaluation, 257; key-informant, 170; with local leaders, 314; oral history, 230–32; spatializing, 172; when to use, 170
interview sampling, 171
introduction, of evaluation report, 260
inverse distance weighting (IDW), 293

Jefferson, Mark, 15–16
Joins and Relates function, 182

Kelly, Florence, 13–15
key-informant interviews, 170
key informants, 35
knowledge: limited, and policy making, 359; local, 194, 196

latent effects, 241–42
latent population, 241–42
latent spatial data collection, 311–12
latitude, 32
least cost analysis, 285–89
level of analysis, 180
Library of Congress, 189
Likert scale, 128–29, 186
linear regression, 294
lines, 36, 276–77
literature review, 39–40, 68–70
local community: earning trust of, 209–11; meeting with, 208–9, 210; sharing findings with, 210–11, 227; as source of data, 225–26
local groups, meetings with, 208–9
local knowledge, 194, 196
local leaders, 208, 314
local newspapers, 207
location errors, 294, 295
location questions, 51
logical data model, 41–43, 46
longitude, 32

manifest population, 241
manifest spatial data collection, 311
map algebra, 277, 296–98
map context, 337
maps: animated, 341–43; color coding, 214; for communication, 339–44, 384; conceptual, 36; cultural issues and, 337–39; digital, 212–13; ground truthing, 224–25; hard-copy, 199, 200, 212; hard-copy maps, 10, 339; interactive, 341–43; online, 212–13; paper, 10, 199, 200, 212, 339; portrayal of qualitative data using, 221; presentation of, 332–36; probability, 295; raster, 145, 292–93; of research area, 200–202, 212–13; scale of, 145; stories told by, 337; thematic, 310–11; time-maps, 343
masking, 109
mathematical models, 298
MAUP. *See* modifiable area unit problem (MAUP)
measurement, 76–77
measurement errors, 293–94
Megan's Law, 96, 97
metadata, 99, 110–11, 122, 316–17
methods section, of evaluation report, 260
metrics, 255
Microsoft Access, 185
Microsoft Excel, 42, 181, 187
modeling, 295–98; dynamic, 393–94; mathematical, 298; raster, 296–98; testing, 301
modes of communication, 347–49
modifiable area unit problem (MAUP), 134, 136–37, 179
Monte Carlo simulation method, 136, 303–4
multigeographies, 248–49
multiple research methods, 19, 192, 194–97, 254, 372
multivariate regression techniques, 290–91, 298

National Aeronautics and Space Administration (NASA), 144, 163, 188, 352
National Agriculture Imagery Program, 195
National Atlas of the United States, 188, 342
National Geospatial Advisory Council (NGAC), 361, 377
National Highway System (NHS), 10
National Institutes of Health (NIH), 399
National Institutes of Health TOXMAP, 58
National Map of the United States, 188, 352
National Spatial Data Infrastructure (NSDI), 361, 376
natural environments, 119, 248–49, 253–54
naturally defined boundaries, 248
natural resources, 36, 374–75
Natural Resources Canada, 163, 189, 353
natural sciences, 36
Natural Sciences and Engineering Research Council of Canada, 264
nearest neighbor analysis, 278–79
Near tool, 278–79, 323
negative occurrence, 274
network analysis, 281–85
news: analysis, 161; as source of data, 160–61
newspapers, 207, 314, 315
nodes, in networks, 281–82
nominal variables, 126–28
noncorrespondence, 274
nondiscrete data, 124
non-Euclidian buffers, 274
nonindependence considerations, 291–92
nonlinear regression, 294–95
nonprobability sampling, 131–33
nonprofit organizations, data from, 144

number fields, 185
NVivo, 79

observations, spatial, 258–59
offline data sources, 158–60
online maps, 212–13
open-ended questions, 167, 170, 258
Open Geospatial Consortium, 45, 399
open-source GIS software, 46
OpenStreetMap, 200
operationalization, 76–77, 125–26
opinion polls, 131
oral history, 35, 230–32
oral presentations, 348–49
ordinal variables, 128–29
overgeneralization, 102–3
overlay, 268, 274–77

pamphlets, 348
paper maps, 10, 199, 200, 212, 339
participant observation, 232–33
participatory GIS (PGIS), 192. *See also* public participation GIS (PPGIS)
personal experience, 118
personal information, 92–93, 105–7
pesticide study, 368–71
phases of abstraction, 40–47
PhoneBooth, 24
physical boundaries, 52, 120
physical data model, 46–47
physical entities, 125
physical space, 12
place, 12, 138
place-based policy, 372–75
place-names, 12
point density analysis, 261–62
points, 36, 276–77
policy: challenges to good, 358–65; creation of, 358–65, 368–75; defined, 18, 356–57; evaluation of, 239–46; fires and, 361–65; GIS and, 356–77; implementation of, 18–19; linking results to, 355–77; place-based, 372–75; politics and, 359–60;

spatially based, 357; time constraints and, 358–59
political implications, 95–97
politics, of spatial data, 359–60
polygons, 36
population: latent, 241–42; manifest, 241; target, 240–41, 243, 248–52
posters, 348
poverty studies, 20–21
PPGIS. *See* public participation GIS (PPGIS)
PPGIS.net, 217
precoding answers, 167
predictive study, 150
preliminary data, 120
primary data, 110–12, 121–23, 142, 153, 154–55, 166, 192
privacy, 27, 29, 56–57, 93, 94, 105–7, 174
probability maps, 295
probability sampling, 130–31
problem definition, 68
problem-solving tool, GIS as, 298–99
programs, evaluation of, 239–46
project background, in evaluation reports, 260
project goals, determining, 47–48
projection, 300
project resources, 152
proximity, 151–52, 254
proximity polygons, 277–78
public opinion, 166
public participation GIS (PPGIS), 94, 191–217; approaching local groups about, 208–9; choosing location for, 211–12; data collected via, 192; data collection, 203–11; defining community of interest for, 205–8; directions for participants in, 213–14; empowerment and, 194; group monitors for, 214; incentives or prizes in, 215; inductive spatial research, 204–11; Internet-based methods, 199; local community and, 205–11;

low-tech approach and, 198; maps for, 200–202, 212–13; as part of mixed methods, 194–97; policy creation and, 368–69; preparation for, 211–15; time for, 212; type of, 204; use of computers in, 198–99
punctuation, in data tables, 183
purposive sampling, 132

quadrangle maps, 144
qualitative analysis, 79, 86–87
qualitative data, 79, 198, 203, 221; coding, 320; defined, 310–11; drawing conclusions about, 325; GIS and, 310; identification of, 312; identifying spatial patterns in, 321–25; sources of, 312–15; spatial analysis of, 309–27; spatializing, 320–21
qualitative research, 198, 310–11; spatial qualitative analysis, 311–25
qualitative spatial ethnographic research, 219–35
quantiles, 31
quantitative analysis: dummy coding in, 127–28; zero in, 130
quantitative data, 312
quantitative survey measurements, 120
questions: about data, 49–50, 147–52; about location, 50; for analysis, 51–53; analysis, 51–53; close-ended, 167, 258; concept, 48–49; for data analysis, 266–67; evaluation research, 239–46; open-ended, 167, 170, 258; sociospatial evaluation research, 247–54
quota sampling, 133

railroads, 15–16
RAISG. See Amazon Georeferenced Socio-Environmental Information Network (RAISG)
random sampling, 131, 134, 168, 171
range, 294
raster-based least cost analysis, 286–89

raster-based spatial interpolation, 290–95
raster data layers, 128
raster maps, 145, 292–93
raster modeling, 296–98
ratio value, 130
reality, 40–41, 301–2
realizations, of data, 294
recategorization, 29–30
recommendations, in evaluation reports, 260–62
records, 181
recreational areas, 251–52
regional economic data, 206
reliability, 152, 154–58, 384–85
reports, 80, 347
reputable data, 193
research: background, 119–20; case study, 229–30; choosing topic for, 67, 82; cultural implications of, 94; deductive, 65–80, 204, 310–11; defining problem for, 68; descriptive, 60–61, 148; design, 59–90; evaluation research. See evaluation research; explanatory, 65; exploratory, 62–64; feasibility of, 67; field. See field research; geographic location for, 83–84; goals, 47–48; guiding questions for, 48–53; hypothesis, 66, 70–73; inductive, 65–66, 80–88, 192, 204–11, 310–11; political implications of, 95–97; purpose of, 55, 57, 60–65, 148–50; qualitative, 198, 310–11; sharing results of, 80, 210–11, 226–27; social implications of, 93–94; sociospatial, 66–80, 82–88, 246–62; spatial thinking in, 16–19
research area: boundaries of, 52–53, 55, 120, 134–39, 248; defining, for study, 204–5; maps of, 200–202, 212–13
research ethics, 91–115; data collection and, 105–12; data sharing and, 112;

data storage and, 112–13; defined, 92; errors caused by analysis and, 98–101; errors in human inquiry and, 102–5; GIS and, 92–97; personal information and, 92–93; privacy and, 105–7
research findings: communication of, 80, 210–11, 227, 329–53; in evaluation reports, 260; keys to effective communication of, 330–39; linking to policy and action, 355–77; mode of communication for, 347–49; preparation of final product, 349–50; purpose of, 332; visualizations for, 332–44
research methods, 6–9, 12. See also specific methods; case studies, 229–30; choosing, 75; multiple, 19, 192, 194–97, 254, 372; oral history interviews, 230–32; participant observation, 232–33; sociospatial, 20–21, 66–80
research process: in deductive research, 66–80; in inductive research, 80–88; steps in, 53–57, 66–80
research reports, in evaluation research, 259–62
research subjects: anonymity of, 106; privacy of, 105–7
resources: for presentation, 350; for project, 152
respondents, 167
results. See research findings
rubrics, for evaluation, 243–45

safety, 120
sample unit boundaries, 134–39
sampling, 120–21; available subjects, 132; frame, 168; GIS and, 130–34; interview, 171; methods, 121–23; nonprobability, 131–33; probability, 130–31; purposive, 132; quota, 133; random, 131, 134, 168, 171; snowball, 133; survey, 168–69

scale, 26, 145, 300
Schematics application, 283–84
scientific method, 66–80
SDelete, 113
secondary data, 77–78, 98–100, 110–12, 121–23, 141–63, 166, 192–93; evaluation of, 142, 143, 259; from Internet, 142, 144–46; from news, 160–61; from offline sources, 158–60; questions to consider about, 147–52; reliability of, 152, 154–58; reputable, 193; sources of, 142–43; validity of, 152–54
selective observation, 103
sensor networks, 385
simulations, 295
slide shows, 261, 348–49
slope, 289
smartphones, 222, 379, 380, 385, 390–91, 392–93
snap distance, 157–58
Snow, John, 70, 390
snowball sampling, 133
social implications, 93–94
social inequality, 13–15, 389–90
social media, 315, 381, 382
social networks, 279–81, 381, 385; mapping, 27–28, 32
social organization, 227
social sciences, 26–27, 36, 387–88
social status, 12
sociodemographic variables, 52
socioeconomic demographics, 207–8
sociospatial documentation, 220–21
sociospatial evaluation, 246
sociospatial evaluation research, 246–62; challenges and benefits of, 263; focus groups, 256–57; interviews, 257; presentation of evaluation, 259–62; project design, 254–59; questions, 247–54; secondary data use in, 259; spatial observations, 258–59; surveys, 258
sociospatial research, 246; grounded theory, 82–88; stages of, 66–80
sociospatial theming, 203
sociospatial thinking, 20–21
software programs. See also ArcGIS: choosing, 43–46; database, 180–81, 185–87; open-source, 46; output formats, 186–87; spreadsheet, 181, 187; statistical, 79; variety of, 42–43
space, 6, 12
spatial advantage, 10–11
spatial analysis, 6–9, 13. See also data analysis; historical examples of, 13–16; of qualitative data, 309–27; questions for, 51–53; sampling methods for, 121–23; tools for, 267–89; unit of analysis and, 179–80
Spatial Analyst toolbox, 304–6
spatial autocorrelation, 179–80
spatial awareness, 386–87
spatial conceptualization, 37–40
spatial correspondence, 274–77
spatial data, 207–8, 359–60, 392–93
spatial ethnography, 228–29
spatial evaluation interviews, 257
spatial identifiers, 182
spatial information, 5
spatial interpolation, 290–95
spatial knowledge, use of, 2–4
spatially based ethnographic research, 219–35
spatially based policy, 357
spatial observations, 258–59
spatial patterns, identifying, 321–25
spatial qualitative analysis, 311–25; data types for, 316–17; drawing conclusions, 325; pattern identification, 321–25; purpose of, 317–18; steps for, 316–25; themes in, 318–20
spatial research, 204–11. See also research
spatial statistics, 302–4
Spatial Statistics toolbox, 180, 302–4
spatial thinking, 7, 9, 11–12; in Chicago slums, 13–15; rebirth of, 386–87; in research, 16–19
spreadsheet software, 181, 187
spurious correlation, 104–5
stakeholder diversity, 373
statistical interpolation methods, 160
statistical output, 339, 344–46
statistical software, 79
statistics, spatial, 302–4
storage, of data, 112–13, 199
A Street Corner Society (Whyte), 227
strings, 185
student research projects, 396–97
studies. See research
study feasibility, 38–39
suitability, of data, 143
survey-based data collection, 166–69
surveys: designing, 167–68; evaluation, 258; spatializing, 172; types of, 167; uses of, 166
survey sampling, 168–69
Sylmar fire, 362
symbology, 268, 270
Symmetrical Difference tool, 275–76
systems, 35–37

tables: data. See databases; frequency, 319–20; variable definition, 319
tablets, 222
target population, 240–41, 243; impact of human-constructed features on, 249–52; impact of natural features on, 249
technology. See also specific types: change, 388–90; cloud-based, 199; cultural perceptions of, 226; in the field, 222; web-based, 199, 200
Temple University Libraries, 90
temporality, 350, 375
temporal scale, 29
terminology, 39–40
thematic maps, 310
themes, 203, 318–20